Impact of Teleworking and Remote Work on Business:

Productivity, Retention, Advancement, and Bottom Line

Harish Chandra Chandan
Independent Researcher, USA

A volume in the Advances in Marketing, Customer Relationship Management, and E-Services (AMCRMES) Book Series

IGI Global
PUBLISHER of TIMELY KNOWLEDGE

Published in the United States of America by
IGI Global
Business Science Reference (an imprint of IGI Global)
701 E. Chocolate Avenue
Hershey PA, USA 17033
Tel: 717-533-8845
Fax: 717-533-8661
E-mail: cust@igi-global.com
Web site: http://www.igi-global.com

Library of Congress Cataloging-in-Publication Data

Names: Chandan, Harish C., 1947- editor.
Title: Impact of teleworking and remote work on business : productivity, retention, advancement, and bottom line / edited by Harish Chandan.
Description: Hershey, PA : Business Science Reference, [2024] | Includes bibliographical references and index. | Summary: "This book delves into the multifaceted impact of teleworking on businesses, exploring how different organizations grapple with these challenges, drawing on the experiences of industry giants like Google and IBM"-- Provided by publisher.
Identifiers: LCCN 2023059457 (print) | LCCN 2023059458 (ebook) | ISBN 9798369313145 (hardcover) | ISBN 9798369313152 (ebook)
Subjects: LCSH: Telecommuting. | Organizational effectiveness. | Organizational behavior. | Personnel management.
Classification: LCC HD2336.3 .I47 2024 (print) | LCC HD2336.3 (ebook) | DDC 658.3/123--dc23/eng/20231227
LC record available at https://lccn.loc.gov/2023059457
LC ebook record available at https://lccn.loc.gov/2023059458

This book is published in the IGI Global book series Advances in Marketing, Customer Relationship Management, and E-Services (AMCRMES) (ISSN: 2327-5502; eISSN: 2327-5529)

British Cataloguing in Publication Data
A Cataloguing in Publication record for this book is available from the British Library.

Advances in Marketing, Customer Relationship Management, and E-Services (AMCRMES) Book Series

Eldon Y. Li
National Chengchi University, Taiwan &
California Polytechnic State University, USA

ISSN:2327-5502
EISSN:2327-5529

MISSION

Business processes, services, and communications are important factors in the management of good customer relationship, which is the foundation of any well organized business. Technology continues to play a vital role in the organization and automation of business processes for marketing, sales, and customer service. These features aid in the attraction of new clients and maintaining existing relationships.

The Advances in Marketing, Customer Relationship Management, and E-Services (AMCRMES) Book Series addresses success factors for customer relationship management, marketing, and electronic services and its performance outcomes. This collection of reference source covers aspects of consumer behavior and marketing business strategies aiming towards researchers, scholars, and practitioners in the fields of marketing management.

COVERAGE

- B2B marketing
- Mobile CRM
- Customer Retention
- Electronic Services
- Cases on CRM Implementation
- Social Networking and Marketing
- Mobile Services
- CRM strategies
- Relationship Marketing
- Data mining and marketing

IGI Global is currently accepting manuscripts for publication within this series. To submit a proposal for a volume in this series, please contact our Acquisition Editors at Acquisitions@igi-global.com or visit: http://www.igi-global.com/publish/.

The Advances in Marketing, Customer Relationship Management, and E-Services (AMCRMES) Book Series (ISSN 2327-5502) is published by IGI Global, 701 E. Chocolate Avenue, Hershey, PA 17033-1240, USA, www.igi-global.com. This series is composed of titles available for purchase individually; each title is edited to be contextually exclusive from any other title within the series. For pricing and ordering information please visit http://www.igi-global.com/book-series/advances-marketing-customer-relationship-management/37150. Postmaster: Send all address changes to above address.

Titles in this Series

For a list of additional titles in this series, please visit:
http://www.igi-global.com/book-series/advances-marketing-customer-relationship-management/37150

Marketing Innovation Strategies and Consumer Behavior
Belem Barbosa (University of Porto, Porugal)
Business Science Reference • © 2024 • 469pp • H/C (ISBN: 9798369341957) • US $325.00

Corporate Governance and CSR Strategies for Sustainability
Maja Pucelj (Faculty of Organisation Studies, University of Novo Mesto, Slovenia) and Rado Bohinc (Scientific Research Centre Koper, Sloenia)
Information Science Reference • © 2024 • 368pp • H/C (ISBN: 9798369358634) • US $290.00

Balancing Automation and Human Interaction in Modern Marketing
Arjun J. Nair (St. Lawrence College, Canada) Sridhar Manohar (Chitkara Business School, Chitkara University, India) Amit Mittal (Chitkara Business School, Chitkara University, India) and Wasim Ahmed (University of Hull, UK)
Business Science Reference • © 2024 • 326pp • H/C (ISBN: 9798369322765) • US $290.00

Building Strong Brands and Engaging Customers With Sound
Minna-Maarit Jaskari (School of Marketing and Communication, University of Vaasa, Finland)
Business Science Reference • © 2024 • 267pp • H/C (ISBN: 9798369307786) • US $265.00

Globalized Consumer Insights in the Digital Era
Fatih Sahin (Bandirma Onyedi Eylul University, Turkey) and Cevat Soylemez (Kutahya Dumlupinar University, Turkey)
Business Science Reference • © 2024 • 296pp • H/C (ISBN: 9798369338117) • US $325.00

AI Impacts in Digital Consumer Behavior
Thomas Heinrich Musiolik (Woxsen University, India & University of Europe for Applied Sciences, Germany) Raul Villamarin Rodriguez (Woxsen University, India) and Hemachandran Kannan (Woxsen University, India)
Business Science Reference • © 2024 • 372pp • H/C (ISBN: 9798369319185) • US $325.00

Contemporary Trends in Innovative Marketing Strategies
Belem Barbosa (University of Porto, Portugal)
Business Science Reference • © 2024 • 394pp • H/C (ISBN: 9798369312315) • US $275.00

IGI Global
PUBLISHER of TIMELY KNOWLEDGE

701 East Chocolate Avenue, Hershey, PA 17033, USA
Tel: 717-533-8845 x100 • Fax: 717-533-8661
E-Mail: cust@igi-global.com • www.igi-global.com

Table of Contents

Preface xii

Chapter 1
Enhancing Work-Life Balance in Remote Working via Good Health to Enhance Organizational Performance 1
Bhupinder Singh, Sharda University, India
Christian Kaunert, Dublin City University, Ireland
Komal Vig, Sharda University, India

Chapter 2
The Influence of Teleworking on Work-Life Balance and Employee Psychological Health: Evidence From Mauritius, a Small Island Economy 32
Zuberia Aminah Hosanoo, Edinburgh Napier University, UK
Soujata Rughoobur-Seetah, Curtin Mauritius, Mauritius

Chapter 3
Remote Work and Job Satisfaction: Navigating the Complex Landscape of Modern Employment 57
Likitha Venkatesh, PES University, India
Zidan Kachhi, PES University, India

Chapter 4
Understanding Rational Decision-Making in Virtual Collaborations 81
Yan Zou, Nexi Group, Germany
Frederico V. R. Castro, Éstudo, Belgium
Petros Chamakiotis, ESCP Business School, Spain

Chapter 5
Strategies to Overcome Isolation in Remote Work 99
Nisha Thundiyil, St. Joseph's College, India

Chapter 6
Relationship Between Remote Working and Work Outcomes of Constantly Connected BYOD Knowledge Workers 117
Melina Doargajudhur, Edinburgh Napier University, UK
Zuberia Hosanoo, Edinburgh Napier University, UK
Geshwaree Huzooree, Edinburgh Napier University, UK

Chapter 7
Arduousness in Technological Aspects of Remote Work 151
Kritika, Independent Researcher, India

Chapter 8
Workplace Transformation: Navigating the Transition From Remote to Hybrid Work 178
Ridhima Goel, Institute of Management Studies and Research, Maharshi Dayanand University, Rohtak, India
Jagdeep Singla, HPKV Business School, India

Chapter 9
Fostering Manager-Employee Relationships in the Era of Remote Work: A Literature Review 208
Nitish Kumar Minz, K.R. Mangalam University, India

Chapter 10
Telework Perspectives of Employers, Managers, and Employees 229
Harish C. Chandan, Independent Researcher, USA

Compilation of References 245

Related References 278

About the Contributors 303

Index 309

Detailed Table of Contents

Preface...xii

Chapter 1
Enhancing Work-Life Balance in Remote Working via Good Health to Enhance Organizational Performance...1
Bhupinder Singh, Sharda University, India
Christian Kaunert, Dublin City University, Ireland
Komal Vig, Sharda University, India

Technology is advancing so quickly that it has completely changed the way we work, with remote work growing in popularity. While there are many advantages to working remotely, like flexibility and work-life balance, there are drawbacks as well, such a potential for blurring work and personal life boundaries. Improving worker health and well-being is essential to raising output and performance levels. Employees are more likely to have a pleasant work attitude and be more productive when they are well. Remote working can exacerbate existing health issues, such as stress, burnout, and musculoskeletal problems, due to prolonged sitting and lack of physical activity. Implementing strategies such as flexible work hours, clear communication, remote work policies, and employee support programs can play a crucial role in achieving a more productive and balanced work environment. This chapter comprehensively explores the various dimensions impact of teleworking and remote work on business and how to uplift the productivity, retention, and business advancement in this technological era.

Chapter 2
The Influence of Teleworking on Work-Life Balance and Employee Psychological Health: Evidence From Mauritius, a Small Island Economy.......32
Zuberia Aminah Hosanoo, Edinburgh Napier University, UK
Soujata Rughoobur-Seetah, Curtin Mauritius, Mauritius

This chapter investigates the impact of teleworking on the psychological health of employees in the new era of work shaped by the pandemic. Adopting a positivist approach in the post-pandemic world, this study aims to assess the employees' perception of their psychological health resulting from the drastic change in the work sphere in the small island economy of Mauritius. Teleworking has been observed as a core behaviour emerging from the global pandemic has become enduring and gradually institutionalised. Widespread continuation and advocacy of teleworking may fundamentally change job characteristics and outcomes. This research assesses the impact of teleworking on the psychological health of employees in the new era of work shaped by the pandemic. This chapter adopted a positivist approach and conducted a survey in the post-pandemic world to assess the employees' perception of their psychological health resulting from the drastic change in the work sphere in the small island economy of Mauritius.

Chapter 3
Remote Work and Job Satisfaction: Navigating the Complex Landscape of Modern Employment 57
Likitha Venkatesh, PES University, India
Zidan Kachhi, PES University, India

This chapter delves into the impact of technology and COVID-19 on the modern workplace, with a focus on remote work. It analyzes various factors that affect job satisfaction, such as relationships, environment, structure, and growth opportunities. The chapter explains remote work, highlighting its benefits and challenges, and provides strategies to enhance job satisfaction in remote work settings. Furthermore, it discusses the future of work and the shift towards decentralized structures. Ultimately, the chapter reveals the complex relationship between remote work and job satisfaction, shedding light on how they affect each other.

Chapter 4
Understanding Rational Decision-Making in Virtual Collaborations 81
Yan Zou, Nexi Group, Germany
Frederico V. R. Castro, Éstudo, Belgium
Petros Chamakiotis, ESCP Business School, Spain

Despite the wealth of studies on virtual teams, there is a knowledge gap in relation to how virtual collaboration might influence rational decision-making in this context. With the study presented in this chapter, the authors address this gap by presenting the research findings drawing on eight interviews with practitioners who work in virtual teams. They identify three factors that influence rational decision-making: (a) efficient information integration via the usage of ICT tools, (b) reduced emotional impact on the decision-maker, and (c) better control and enhanced confidence of the decision-maker. The chapter explains how the findings contribute to existing literature and what this means for practitioners on the ground.

Chapter 5
Strategies to Overcome Isolation in Remote Work...99
Nisha Thundiyil, St. Joseph's College, India

The chapter provides insight on the role of employers and also self-help by the employees to ward off isolation due to remote work. Hybrid work, remote work, and telework have become part of the new normal era. Remote work has added its own measure of woes to the trail of negative effects in terms of bonding within family left as residue by the overuse of mobile phones. Isolation and in turn depression are two undisputable negative impacts of remote work. This chapter lists an array of strategies that can be inculcated by the corporate sector to minimize and help their employees overcome the ill effects of isolation. The chapter starts with the meaning and conceptual background of remote work. The strategies to overcome isolation due to remote work form the focal point of this chapter, and these strategies have been listed point wise and explained.

Chapter 6
Relationship Between Remote Working and Work Outcomes of Constantly Connected BYOD Knowledge Workers...117
Melina Doargajudhur, Edinburgh Napier University, UK
Zuberia Hosanoo, Edinburgh Napier University, UK
Geshwaree Huzooree, Edinburgh Napier University, UK

Remote working is growing in popularity globally, and the COVID-19 pandemic has contributed to this trend. While the idea of remote work is not new, the reality is that it is now more common than ever in all industries and organisations where it can be implemented, and it is becoming a preferred method of working globally. The issue of remote work is current, and given how quickly it has spread throughout the globe, it is critical to look at how this new working paradigm impacts work outcomes of constantly connected BYOD (bring your own device) workers. The chapter examines the effects of working remotely for this type of employee. It also presents the remote working trend across the globe, the theoretical lens, and the reason for choosing the job demands-resources theory. A research model is proposed by carefully analysing the key constructs related to the impacts of remote working for BYOD users. The chapter explores strategies and recommendations tailored to enhance the remote work experience for these employees, while also providing insights into future directions.

Chapter 7
Arduousness in Technological Aspects of Remote Work 151
Kritika, Independent Researcher, India

The evolution of internet and Web 3.0 poses an ever-growing trend of remote working allowing people to work at their ease and raising concern over cyber security issues for secured transaction of work in remote environment. The necessity of implementing a thorough cybersecurity framework that includes encrypted data transmission, secure communication channels, and reliable access control is examined. The chapter also explores the importance of continuous monitoring systems and endpoint security solutions in identifying and addressing possible threats quickly. It highlights the challenges and safeguard measures to be taken to prevent the cyber threats in the near future and eventualities of remote working in the years ahead.

Chapter 8
Workplace Transformation: Navigating the Transition From Remote to Hybrid Work 178
Ridhima Goel, Institute of Management Studies and Research, Maharshi Dayanand University, Rohtak, India
Jagdeep Singla, HPKV Business School, India

Recent years have witnessed a dramatic transformation in the workplace when it comes to flexibility, mostly as a result of advances in technology and changes in societal perceptions regarding the nature of work. This chapter covers the primary influences that lead to this change, the significant effects it has on businesses and the employees they hire, as well as the prospects and obstacles that may follow. In addition, this chapter discusses implications for the path of work, taking into consideration the profound and complex impacts of this fundamental shift on businesses, workers, and society at large. Based on these findings, the chapter provides recommendations to organizations striving to make their way amid this evolving environment so they may establish a more adaptable, long-lasting, favorable work environment.

Chapter 9
Fostering Manager-Employee Relationships in the Era of Remote Work: A Literature Review 208
Nitish Kumar Minz, K.R. Mangalam University, India

Amid the global COVID-19 pandemic, remote work has emerged as a transformative influence, fundamentally altering organizational dynamics. This research explores the critical need to cultivate robust manager-employee relationships in the realm of remote work. Tackling challenges of communication, teamwork, and sustaining employee engagement, the study examines diverse methodologies and best practices transcending physical distances. Drawing insights from relevant literature, real-world case studies, and practical experiences, it contributes to a holistic understanding of the evolving landscape of remote work. Emphasizing the importance of nurturing connections, assessing the evolving managerial role, and recognizing the central role of technology, the research aims to empower organizations with insights and strategies to fortify manager-employee relationships, ultimately enhancing productivity, job satisfaction, and overall team performance in the era of remote work.

Chapter 10
Telework Perspectives of Employers, Managers, and Employees 229
Harish C. Chandan, Independent Researcher, USA

The three stakeholders of work—employers, managers, and employees—have varying viewpoints about telework. During the COVID-19 pandemic, telework grew using the internet and video communication technologies. However, the nature of work is still evolving. This chapter examines the perspectives of employers, managers, and employees on telework and summarizes the most recent work trend of hybrid work. With the end of the pandemic, the hybrid work is becoming a popular choice as it combines the benefits of telework for the employees like flexibility and lack of commute and in-office work that promotes collaboration, innovation, and a sense of loyalty for the company.

Compilation of References 245

Related References 278

About the Contributors 303

Index 309

Preface

INTRODUCTION

In recent years, the landscape of work has undergone a profound transformation driven by globalization and advancements in Information and communication technology (ICT). The recent COVID-19 pandemic gave a boost to remote work. Remote work has revolutionized traditional notions of workplace dynamics. Work is no longer a place you go to. Work is what you do where ever you are. Remote work offers employees the opportunity of doing their jobs from a remote location or home. The remote location could be some other shared space or a private office. Now, even though the pandemic is gone, the remote work has emerged as a partial solution to work. In-office and remote work together is emerging as the new hybrid work.

This book, *Impact of Teleworking and Remote Work on Business: Productivity, Retention, Advancement, and Bottom Line*, explores the intricate interplay between teleworking and its effects on employers, managers, employees and society at large. This book deals with the current trends of work from home, shared space or remote work. Remote/hybrid work has created new opportunities and challenges and benefits for each of the three stakeholders- employers, managers and employees.

The three stakeholders of work – employers, managers and employees have different perspectives on the organizational culture, employee productivity, organizational productivity, employee advancement in the firm, employee retention and improvement in the firm's bottom line. The current trend of Hybrid work which refers to a combination of remote work and work at the corporate office is growing. More and more firms are adopting hybrid work. The widespread use of remote/hybrid work has raised new questions about organizational culture, employee productivity, career advancement, job satisfaction, and firm's bottom line.

The rise of remote work represents a paradigm shift in how organizations operate, offering flexibility and opportunities for both employers and employees. As more individuals embrace remote or hybrid work models, it is imperative to understand the myriad implications of this trend. From increased productivity and cost savings to challenges such as isolation and cybersecurity concerns, remote work presents a complex mix of advantages and disadvantages. The remote/hybrid work has brought new efficiencies and opportunities. It has also raised questions about employee productivity, job satisfaction, and organizational performance. It is essential for businesses to navigate these challenges effectively while leveraging the benefits of remote work to drive innovation and growth.

In this edited reference book, we delve into the latest research surrounding teleworking, examining its management challenges, employee dynamics, social implications, and technological considerations. By synthesizing insights from leading scholars and industry experts, we aim to provide a comprehensive understanding of the remote work phenomenon and its impact on various stakeholders of work and society at large.

The objectives of this book are ambitious and has practical considerations. We discuss current research challenges and benefits of teleworking on the employers, management and employee. We explore the social and technological dimensions of remote work, and shed light on the emerging trend of hybrid remote working. By addressing these objectives, we hope to equip small, medium, and large-sized companies, government agencies, and universities with the knowledge and tools necessary to thrive in an increasingly remote-centric world.

As the editor of this book, I recognize the importance of research in guiding strategic decision-making and fostering a deeper understanding of remote work dynamics. Through rigorous inquiry and analysis, we endeavor to contribute to the ongoing dialogue surrounding teleworking and its implications for businesses and society. Ultimately, our goal is to empower organizations to embrace the remote work revolution while navigating its complexities with confidence and foresight.

In the following chapters, readers will embark on a journey through the multifaceted landscape of remote work, exploring its nuances, challenges, and opportunities. From examining the social benefits of teleworking to dissecting the technological challenges of remote work, this book offers a comprehensive framework for understanding and harnessing the transformative power of teleworking.

Join us as we unravel the impact of teleworking on business dynamics and pave the way for a more connected, resilient, and innovative future of work. The emergence of teleworking as a prominent feature of the modern work landscape has prompted a reevaluation of traditional business practices and organizational structures. Enabled by advancements in technology and accelerated by the COVID-19 pandemic, remote work has become a defining characteristic of the contemporary workplace. As businesses strive to adapt to this new reality, it is essential to understand the implications of remote work on productivity, retention, advancement, and the bottom line.

This edited book seeks to explore the importance of research in comprehensively examining the impact of teleworking on businesses and organizations. By synthesizing insights from diverse sources and disciplines, it aims to provide a holistic understanding of remote work's transformative potential and its implications for stakeholders at all levels.

The Importance of Evidence-Based Research

Research serves as the cornerstone of informed decision-making and strategy formulation in the realm of remote work. By systematically investigating the various facets of teleworking, researchers are uncovering valuable insights into its effects on employer choices, management challenges, employee behavior, organizational dynamics, and societal benefits. At its core, research enables businesses to identify emerging trends and anticipate future challenges in the remote work landscape. Through empirical studies and data analysis, researchers are discovering patterns, correlations, and causation shedding light on the complex interplay between teleworking, job satisfaction, and organizational performance.

Research also plays a crucial role in highlighting the socioeconomic implications of remote work. By examining issues such as spatial access, skill polarization, and disparities in job distribution, researchers can advocate for inclusive policies and initiatives that promote equitable opportunities for all individuals, regardless of their geographic location or background.

Opportunities for Employers

The remote work policy offers employers a huge increase in the talent pool to choose from. The physical location of the employee is no longer a concern. It offers an opportunity to hire people from different countries and cultures. This increases the creativity and innovation potential for the employee teams. The employers save on acquiring and maintaining the office space. They can hire the best talent for the job regardless of the location. However, some jobs like the retail, transportation, warehousing, hotel and food services jobs are not well-suited to remote work.

Navigating Management Challenges Through Research

For managers, there are additional new challenges in managing remote workers, especially if the employees come from different cultures and time zones. Managing remote employees requires some additional training and skills for the managers. They have new challenges for managing employee productivity remotely.

Organizations worldwide have shifted to working from home, requiring managers to engage in distance management using information and communication technologies (ICT). Studies show that managers experienced higher job demands and inadequate guidance during COVID-19. This has an impact on managers' well-being and organizational productivity.

Managers and employees need to understand that remote work is more suitable for conscientious and introverted people. Extraverted people need in-person human connections more. Personality traits of a remote worker affects their productivity. New employees need to be in office for some time before they can start remote work. Employees with expertise in information and technology, finance, accounting, business services and education may be more suited for remote work. One of the primary areas of focus in remote work research is the identification and mitigation of management challenges. As businesses transition to remote or hybrid work models, they must contend with a myriad of logistical, operational, and cultural hurdles

Research helps businesses understand the intricacies of managing remote teams using the three C's- Communication, Control and collaboration. Fostering trust, communication and collaboration helps to maintain productivity and morale. By studying best practices and case studies, researchers can offer practical recommendations for overcoming common challenges such as isolation, loss of work-life balance, and cybersecurity concerns.

Finally, research enables businesses to adapt their organizational structures and policies to accommodate the needs of remote employees effectively. Through flexible work arrangements, digital infrastructure investments, remote-friendly culture initiatives, management and employee raining, research provides valuable insights into optimizing remote work strategies for long-term success of the organization.

EMPOWERING EMPLOYEES

Remote work offers the employees to save on commute time, expense and have flexible work schedule and create a good work-life balance. However, remote work can result in isolation, less connections and visibility to management and peers and less connection and access to leadership. Based on surveys, both employees and employers feel that remote work makes them more productive, improves morale, reduce absenteeism and save on operational costs. However, there are differences in perspectives of employers, managers and employees. In this remote/hybrid setting, the organizations have to make adjustments to allow business processes to operate productively at a distance. The organizations have to make sure that remote/hybrid work engages all the employees with different personalities effectively in different countries and cultures.

Studies have shown that remote work can have both positive and negative effects on employee morale and productivity. While remote workers often cite benefits such as flexibility, autonomy, and reduced commute times, they may also experience feelings of isolation, burnout, and difficulty disconnecting from work. The advantages of tele-working include better work-life balance, higher productivity, timeliness, reduced absenteeism and decreased turnover, cost savings, flexibility, incentive to improve workplace technology and reduced environmental impact due to lack of employee commute. The distractions include distractions at home, isolation, loss of work-life balance, increased need for meetings, cybersecurity concerns, difficulty maintaining confidentiality, unstable/inconsistent internet access. The new employees need to spend some time in-office to understand the organizational culture and establish personal connections with the management.

Based on the data from fully remote on-line platform, it was observed that the work is polarized along three dimensions. First, countries are globally divided: North American, European, and South Asian remote platform workers attract most jobs, while many Global South countries participate only marginally. Secondly, remote jobs are pulled to large cities; rural areas fall behind. Thirdly, remote work is polarized along the skill axis. The workers with in-demand skills attract profitable jobs, while others face intense competition and obtain low wages These findings suggest that agglomerative forces linked to the unequal spatial distribution of skills, human capital, and opportunities shape the global geography of remote work. These forces pull remote work to places with institutions that foster specialization and complex economic activities, i.e. metropolitan areas focused on information and communication technologies. Locations without access to these enabling institutions—in many cases, rural areas—fall behind.

In addition to addressing management challenges, research also focuses on understanding the impact of remote work on employee dynamics and well-being. By surveying employees and conducting qualitative interviews, researchers gain insights into the factors influencing job satisfaction, work-life balance, and career advancement in a remote work environment. Research plays a crucial role in identifying strategies for supporting remote employees and fostering a sense of belonging and community. Whether through virtual team-building activities, mentorship programs, or wellness initiatives, businesses can leverage research findings to create a supportive and inclusive remote work culture.

Hybrid Work

With the end of the pandemic, the hybrid work is becoming a popular choice as it combines the benefits of telework for the employees like flexibility and lack of commute and in-office work that promotes collaboration, innovation and a sense of loyalty for the company. The hybrid work reduces the disadvantages of both in-person and remote work. The hybrid work has created new opportunities, benefits and *challenges* for each of the three stakeholders -employers, managers and employees. Hybrid work may become the new normal.

Addressing Technological and Security Considerations

Another key aspect of remote work research revolves around technological considerations and infrastructure requirements. As businesses rely increasingly on digital tools and platforms to facilitate remote collaboration and communication, it is essential to understand the implications of these technologies on organizational efficiency and security.

Research helps businesses evaluate the effectiveness of remote work technologies and identify areas for improvement. By studying user experiences, adoption rates, and performance metrics, researchers can inform the development of user-friendly interfaces, robust cybersecurity protocols, and seamless integration with existing systems.

Moreover, research enables businesses to anticipate future technological trends and stay ahead of the curve in a rapidly evolving digital landscape. Whether through investments in artificial intelligence, virtual reality, or cloud computing, businesses can leverage research insights to enhance their remote work capabilities and maintain a competitive edge in the market.

CONCLUSION

In conclusion, ongoing research in understanding the impact of teleworking on businesses and organizations is helping telework and hybrid work. Through systematic inquiry and analysis, researchers provide invaluable insights into the opportunities and challenges presented by remote work.

From navigating management challenges to empowering employees and addressing technological considerations, research serves as a guiding light for businesses seeking to thrive in a remote-first world. By embracing evidence-based practices and staying abreast of emerging trends, businesses can unlock the full potential of remote work and build a more resilient, inclusive, and prosperous future workforce.

As we continue to navigate the complexities of the remote work landscape through a commitment to research and innovation. we have to leverage the transformative power of teleworking to create a better, brighter tomorrow for businesses, employees, and society as a whole. The ongoing research will help us to overcome some of the disadvantages of remote work including Isolation, loss of work-life balance, increased need for meetings, cybersecurity concerns, difficulty maintaining confidentiality, unstable/inconsistent internet access and challenges for new hires. In business, human connections and relationships at work are extremely important to employees' sense of belonging that creates a successful organization. Fostering relationships and human connections drive employee productivity. The co-workers are the top influencers, followed by their managers, leadership, and then HR to create human connections. Managers can create a sense of connection but their passivity can lead to disconnects. Organizations create better human connections across their team, inviting engagement and building alignment along every step of the employee experience—from onboarding to learning and development, performance management, career growth, organizational change, and every moment and milestone along the way. Employers have to focus on connections, engagement, and growth of employees. The corporate office can be a place for relationship and community building. The top benefit of going into the workplace is fostering stronger relationships. Skill sharing with colleagues and impromptu in-office interactions create connections.

This book, *Impact of Teleworking and Remote Work on Business: Productivity, Retention, Advancement, and Bottom Line*, has 10 chapters.

Chapter 1

Enhancing Work-Life Balance in Remote Working via Good Health to Enhance Organizational Performance

Chapter1 deals with enhancing work-life balance in remote working by keeping good health and well-being. There are many advantages to working remotely, like flexibility and work-life balance. One of the drawbacks is a potential for blurring work and personal life boundaries. Working remotely from home can cause loneliness, stress and burnout. Prolonged sitting and lack of physical activity can lead to musculoskeletal problems. Maintaining good employee health is essential to increasing employee performance. Employees are more likely to have a pleasant work attitude and be more productive when they are feel well. A good work-life balance of the employee enhances the organizational performance. Implementing flexible work hours, clear communication, remote work policies, and employee support programs play a crucial role in achieving a more productive and balanced work environment.

Chapter 2

The Influence of Teleworking on Work-Life Balance and Employee Psychological Health: Evidence From Mauritius, a Small Island Economy

Chapter 2 deals with the influence of teleworking on work- life balance, and employee psychological health. The study was conducted in a small island Mauritius in East Africa.

This chapter focuses on the psychological health of employees Adopting a positivist, quantitative approach, this study investigates the employees' perception of their psychological health resulting from the drastic change in the work sphere. This chapter adopted a positivist approach and conducted a survey in the post-pandemic world, to assess the employees' perception of their psychological health resulting from the drastic change in the work sphere. This study used the social exchange, the psychological contract and the job characteristic theories to analyze the impact of teleworking on employees' psychological health. The variables assessed could be extended to psychosocial dimensions. The research investigates the impact of job design, job satisfaction, work-life balance, and burnout on employee psychological health. The sample population for the survey was limited to online survey, sourced via social media, and mailing list, due to the lockdown. Home based teleworking could have significant influence on psychological health depending on other moderators such as home environment, family status, support from the organization.

Chapter 3

Remote Work and Job Satisfaction: Navigating the Complex Landscape of Modern Employment

This chapter analyzes various factors that with the remote Work and affect employee job satisfaction, relationships, environment, structure, and growth opportunities. The paper and provides strategies to enhance job satisfaction in remote work settings.

It discusses the future of work and the shift towards decentralized structures. This paper reveals the complex relationship between remote work and job satisfaction shedding light on how they affect each other.

Chapter 4

Understanding Rational Decision-Making in Virtual Collaborations

This chapter analyzes the knowledge gap in relation to how virtual collaboration might influence rational decision-making. Based on eight interviews with practitioners who work in virtual teams. three factors were identified that influence rational decision-making. First, efficient information integration via the usage of ICT tools. Second, reduced emotional impact on the decision-maker. Third, better control and enhanced confidence of the decision-maker.

Chapter 5

Strategies to Overcome Isolation in Remote Work

This chapter deals with insights on the role of employer and also self-help by the employees to ward off isolation feeling due to remote work. Hybrid work, remote work and telework has become part of the new normal era. Remote work has added its own measure of woes to the trail of negative effects in terms of bonding within family left as residue by the overuse of mobile phones. Isolation and in turn depression are two undisputable negative impacts of remote work. This chapter intends to list out an array of strategies that can be inculcated by the corporate sector to minimize and help their employees overcome the ill effects of isolation. The chapter starts off with the meaning and conceptual background of remote work. The strategies to overcome isolation due to remote work form the focal point of this chapter and these strategies have been listed point wise and explained.

Chapter 6

Relationship Between Remote Working and Work Outcomes of Constantly Connected BYOD Knowledge Workers

This chapter deals with the work outcomes of constantly connected BYOD (Bring Your Own Device) knowledge workers. It chooses the Job Demands-Resources theory. A research model is proposed by analyzing the key constructs related to the impacts of remote working for BYOD users. The strategies and recommendations tailored to enhance the remote work experience are presented.

Chapter 7

Arduousness in Technological Aspects of Remote Work

This chapter deals with concerns over cyber security issues for secured transaction of work in remote environment. The necessity of implementing a thorough cybersecurity framework that includes encrypted data transmission, secure communication channels, and reliable access control is examined. The chapter also explores the importance of continuous monitoring systems and endpoint security solutions in identifying and addressing possible threats quickly. It highlights the challenges and safeguard measures to be taken to prevent the cyber threats in near future and eventualities of remote working in the years ahead.

Chapter 8

Workplace Transformation Navigating the Transition from Remote to Hybrid Work

This chapter chapter deals with the transition from remote to hybrid work. It discusses the significant effects hybrid work has on businesses and the employees they hire, as well as the prospects and obstacles that may follow. In addition, this chapter discusses implications for the path of work, taking into consideration the profound and complex impacts of this fundamental shift on businesses, workers, and society at large. Recommendations are provided to organizations striving to make their way to adopt hybrid work to enable them to establish a more adaptable, long-lasting, favorable work environment.

Chapter 9

Fostering Manager-Employee Relationships in the Era of Remote Work

This chapter deals with the critical need to cultivate robust manager-employee relationships in remote work to deal with the challenges of communication, teamwork, and sustaining employee engagement. The authors emphasize the importance of nurturing connections, assessing the evolving managerial role, and recognizing the central role of technology. This chapter provides insights and strategies to fortify manager-employee relationships to enhance productivity, job satisfaction, and overall team performance.

Chapter 10

Telework Perspectives of Employers, Managers, and Employees

This chapter examines the perspectives of employers, managers and employees on telework and summarizes the most recent work trend of Hybrid work. With the end of the pandemic, the hybrid work is becoming a popular choice as it combines the benefits of telework for the employees like flexibility and lack of commute and in-office work that promotes collaboration, innovation and a sense of loyalty for the company.

Harish Chandra Chandan
Independent Researcher, USA

Chapter 1
Enhancing Work-Life Balance in Remote Working via Good Health to Enhance Organizational Performance

Bhupinder Singh
https://orcid.org/0009-0006-4779-2553
Sharda University, India

Christian Kaunert
https://orcid.org/0000-0002-4493-2235
Dublin City University, Ireland

Komal Vig
https://orcid.org/0009-0008-8771-0481
Sharda University, India

ABSTRACT

Technology is advancing so quickly that it has completely changed the way we work, with remote work growing in popularity. While there are many advantages to working remotely, like flexibility and work-life balance, there are drawbacks as well, such a potential for blurring work and personal life boundaries. Improving worker health and well-being is essential to raising output and performance levels. Employees are more likely to have a pleasant work attitude and be more productive when they are well. Remote working can exacerbate existing health issues, such as stress, burnout, and musculoskeletal problems, due to prolonged sitting and lack of physical activity.

DOI: 10.4018/979-8-3693-1314-5.ch001

Implementing strategies such as flexible work hours, clear communication, remote work policies, and employee support programs can play a crucial role in achieving a more productive and balanced work environment. This chapter comprehensively explores the various dimensions impact of teleworking and remote work on business and how to uplift the productivity, retention, and business advancement in this technological era.

1. INTRODUCTION

Work-life balance has developed into a comprehensive idea that includes physical and mental health, general well-being, and the caliber of restorative sleep. It is no longer limited to simply dividing work and personal time. In order to better understand the complex relationships between employee well-being, sleep quality, and the resulting effects on organizational productivity, it explore the various facets of work-life balance in the context of remote work. Understanding how these components are interdependent is essential to developing methods that improve remote workers' lives while also making a major contribution to corporate performance (Blankson, 2017). This can have a detrimental impact on the quantity and quality of sleep, which might have a domino effect that hurts the individual as well as the company. Developing successful tactics that promote a positive work environment requires an understanding of how sleep patterns are impacted by remote work and, in turn, how this affects an individual's cognitive function and job performance (Rubin, 2022).

The paper explores the more general facets of worker health and wellbeing, acknowledging them as essential elements of the work-life balance. There are certain difficulties that come with working remotely, such as social isolation and the blending of work and personal life. An in-depth grasp of the variables affecting work-life balance may be obtained by looking at how these difficulties affect remote workers' general health and well-being (Singh, 2023).

1.1 Background of Study

The search for work-life balance has progressed beyond temporal division to a comprehensive comprehension that includes the mental, emotional, and physical aspects of wellbeing. Organizations are faced with the challenge of comprehending and resolving the complex interactions between work and life in the context of remote work as a result of their recognition of the need to adapt to this shifting environment. The success of such endeavors is contingent upon a nuanced comprehension of the factors influencing employee well-being and, by extension, organizational productivity (Darouei & Pluut, 2021).

Sound sleep, as a foundational element of well-being, emerges as a linchpin in the work-life balance equation. The transition to remote work often disrupts established routines and introduces new challenges to achieving a restful night's sleep. The implications of sleep quality on cognitive function, emotional resilience, and overall work performance are profound, necessitating a closer examination of sleep patterns in the remote work setting (Budhiraja & Kant, 2020). In the context of remote work, the larger picture of employee health and well-being becomes crucial. Mental health may be impacted by social isolation, a decrease in in-person connections, and being physically far from coworkers. The overall health of remote workers includes things like stress, job satisfaction, and a feeling of purpose in addition to the absence of disease. Employers who put employee well-being first understand the inextricable relationship between it and engagement, productivity, and, eventually, the success of the company as a whole.

1.2 Significance of the Paper

In light of the present state of professional dynamics, it is imperative that we explore the nuances of work-life balance in the context of remote work, with a particular emphasis on healthy sleep patterns, overall wellbeing, and overall health. Understanding and resolving the issues faced by remote workers becomes critical as more and more companies adopt remote work as a long-term operational strategy. This study has a number of significant ramifications for the commercial and academic communities (Mullins et al., 2022).

First off, by illuminating how work-life balance is changing in this paradigm, the study adds to the growing corpus of research on remote work. The conventional division between work and personal life has become increasingly brittle in the context of remote work, calling for a sophisticated comprehension of the variables impacting worker well-being (Ninaus et al., 2021). The goal of this research is to provide a thorough examination of the various aspects of work-life balance, acknowledging its revolutionary effects on how businesses organize their operations and provide for their workforce.

Second, the study acknowledges the critical role that restful sleep plays in remote workers' general health and productivity. Understanding how sleep affects cognitive function, decision-making, and job performance becomes critical as firms struggle to retain a cohesive and engaged staff across diverse locations (Singh, 2024). The paper is even more significant because it focuses on health and well-being in the context of remote employment. For enterprises, ensuring the entire well-being of its workforce is not only morally required but also strategically vital. Organizations may create focused treatments that improve employee resilience, work happiness, and overall organizational performance by recognizing the relationship between physical and mental health and job performance (Kroemer & Kroemer, 2016).

Figure 1. Objectives of the paper

1.3 Objectives of the Paper

The paper has the following objectives to:

- explore the impact of remote work on work-life balance.
- investigate the relationship between sound sleep, health, and well-being with employee productivity.

- identify strategies for organizations to support remote employees in achieving work-life balance.
- assess the influence of work-life balance on organizational performance.

1.4 Structure of the Paper

This paper comprehensively explores the various dimensions of enhancing work-life balance of employees considering the multifarious factors of sound sleep, health (both physical-mental) which contribute for boosting performance of the organizations. Section 2 deals with Remote Working Trends and Its Impact on Work-Life Balance. Section 3 specifies Relationship between Work-Life Balance and Employee Health. Section 4 explains the Influence of Work-Life Balance on Employee Productivity. Section 5 discusses the Health and Well-Being in Remote Working. Section 6 explores the Uplifting Productivity through Work-Life Balance. Section 7 put stress on Strategies to Enhance Work-Life Balance in Remote Working. Section 8 discovers Organization Performance. And finally Section 9 deals with Conclusion and Future Scope.

1.5 Definition and Concept of Work-Life Balance

Work-life balance, a term that has gained increasing prominence in contemporary discourse on employment and well-being, refers to the equilibrium individuals strive to achieve between their professional commitments and personal lives. It embodies the pursuit of harmony between the demands of work, often characterized by job responsibilities, deadlines, and professional expectations, and the equally crucial aspects of life beyond the workplace, including family, health, leisure, and personal development (Hamlin, 2018). The concept recognizes that individuals are not merely employees but multifaceted beings with diverse roles and responsibilities outside the confines of their professional domains. The notion has acquired importance as a standard for evaluating corporate cultures that prioritize the overall welfare of their employees. The idea of work-life balance captures people's continuous search for balance in their lives, acknowledging that the lines between work and personal life are hazy and open to personal interpretation. It expresses a comprehensive view of human existence, realizing that success and contentment go beyond career accomplishments to include the diverse range of experiences that make up a meaningful and complete life (Sigeman & Adolfsson, 2021).

2. REMOTE WORKING TRENDS: IMPACT ON WORK-LIFE BALANCE

The advent and swift uptake of remote work have sparked a revolutionary change in the employment scene, bringing with it a plethora of developments that have a substantial influence on the delicate balance between work and life (Onyeka, 2022). This paradigm shift has given rise to various noticeable trends that redefine the interaction between work and personal life. It has been driven by technical improvements and highlighted by the world events of the previous ten years. The breaking down of geographical barriers is one of the key trends in remote working. Professionals are no longer restricted to certain places, allowing companies to create inclusive and varied teams and accessing a worldwide talent pool. Although this trend increases accessibility and flexibility, it also brings with it difficulties because of time zone variations and the possibility of blending professional and personal life borders (Onder, 2016). A more complex strategy for work-life balance that takes into account both temporal and geographical factors is required since remote workers may have to manage a 24/7 work cycle. The way that digital technologies are woven into remote work has an impact on how professional and personal lives are defined, making it necessary to develop healthy boundaries and manage digital weariness (Ireland, 2022).

2.1 Challenges Faced by Remote Workers in Maintaining Work-Life Balance

While working remotely offers previously unheard-of freedom and opportunity, it has also brought with it a number of new difficulties that remote workers must overcome in order to maintain a positive work-life balance. The lines between personal and professional life are becoming increasingly blurred, which is one of the biggest obstacles. Once thought to be a haven for personal life, the physical division between the house and job has grown more illusive (ten Brummelhuis, ter Hoeven, & Toniolo-Barrios, 2021). Distinguishing between work and personal time can be difficult for remote workers as they frequently have to deal with the ongoing presence of work in their living areas

The conventional notions of work-life balance are seriously challenged by this lack of geographical demarcation, necessitating a reevaluation of how people organize and divide their personal and professional obligations.

The difficulties that remote workers encounter in preserving a healthy work-life balance are numerous and call for an all-encompassing strategy. Both individuals and organizations need to manage the intricacies of indistinct borders, digital expectations, social distancing, adaptability, and family obligations to create a work environment where working remotely is beneficial to employees' general health as well as productivity. Creating sustainable remote work practices that prioritize individuals' overall needs in the ever-evolving modern workplace requires acknowledging and resolving these obstacles (ten Brummelhuis, ter Hoeven, & Toniolo-Barrios, 2021).

2.2 Strategies for Promoting Work-Life Balance in Remote Settings

Encouraging work-life balance in distant environments necessitates a multidimensional strategy that tackles the particular difficulties brought about by the decentralization of labor. An essential tactic is to set and communicate boundaries that are unambiguous. Because there is no physical barrier between the home and the office, remote workers frequently find it difficult to draw boundaries between their personal and professional lives. Organizations that establish regulations that specify working hours, promote breaks, and prohibit after-hours interactions can play a critical role in creating work-life balance. Additionally, leaders can model and promote healthy work practices by respecting these boundaries and emphasizing the importance of downtime (Ruderman et al., 2017). Flexibility, while a hallmark of remote work, can be both a blessing and a challenge. The organizations can strategically leverage flexibility by empowering employees to structure their work hours around personal needs and preferences. This may involve adopting alternative work schedules, allowing for compressed workweeks, or providing options for part-time remote work. With emphasizing clear boundaries, embracing flexibility, fostering a sense of community, promoting wellness initiatives, leveraging technology judiciously, and investing in training and education, organizations can create a remote work environment that not only enhances productivity but also prioritizes the overall well-being of their workforce. Ultimately, these strategies contribute to the cultivation of a sustainable and harmonious balance between professional and personal life in the remote work landscape (Weber & De Fino, 2022).

2.3 Impact of Work-Life Balance on Employee Well-Being

The impact of work-life balance on employee well-being is profound and far-reaching, influencing not only the individual's health but also their overall satisfaction and performance within the workplace. Fundamentally, work-life balance is an important factor in determining an individual's overall mental and physical health. Employees' overall life satisfaction is greatly increased when they are able to balance the demands of their work obligations with the needs of their personal lives. Maintaining this balance is crucial to avoiding burnout, health problems brought on by stress, and the loss of one's sense of purpose (Pradhan & Kumar, 2021).

There exists a direct association between work-life balance and both job happiness and total workplace engagement. Workers are more likely to be content in their employment and be more motivated, committed, and productive if they believe that their personal and professional life are harmoniously balanced. On the other hand, a lack of work-life balance can lead to employee disengagement, low morale, and higher turnover rates as workers look for workplaces that put their overall wellbeing first. Work-life balance affects employees' well-being not just on an individual basis but also on team dynamics and company culture. Businesses that place a high priority on work-life balance send a message to their staff that they care about their health and well-being, which promotes a happy and encouraging work atmosphere (Singh, 2023).

3. RELATIONSHIP BETWEEN WORK-LIFE BALANCE AND EMPLOYEE HEALTH

The relationship between work-life balance and employee health extends beyond the individual to impact interpersonal relationships and the broader community. A workforce that experiences a harmonious balance is likely to exhibit positive relationships with colleagues, reducing workplace tension and fostering a collaborative environment. Social support and a sense of community contribute to both mental and physical health, acting as buffers against the negative effects of stress. The organizations that prioritize work-life balance as a component of their workplace culture reap the benefits of a healthier and more engaged workforce. Employee health programs, flexible work arrangements, and policies that encourage time off contribute to a supportive environment that values the holistic well-being of employees (Ochoa, 2023). Such organizations often experience lower rates of absenteeism, reduced turnover, and higher levels of job satisfaction, creating a positive feedback loop that further enhances employee health.

Mental health stands as a prominent dimension where the relationship between work-life balance and well-being is starkly evident. A balanced work-life dynamic acts as a protective factor against stress, anxiety, and burnout. When employees can disengage from work during non-working hours and enjoy sufficient downtime, it contributes significantly to mental resilience. Physical health is equally influenced by the dynamics of work-life balance. The stress associated with an imbalanced work-life relationship can manifest in various physical health issues, such as cardiovascular problems, digestive disorders, and compromised immune function. Adequate rest, relaxation, and time spent on personal pursuits outside of work contribute to physical well-being. Individuals with a balanced work-life dynamic are more likely to engage in health-promoting activities, such as regular exercise and proper nutrition, further enhancing their overall health. Conversely, the neglect of work-life balance may result in poor lifestyle choices, contributing to a higher risk of chronic health conditions (Fallica, 2022).

4. INFLUENCE OF WORK-LIFE BALANCE ON EMPLOYEE PRODUCTIVITY

The influence of work-life balance on employee productivity is a critical and reciprocal relationship that lies at the heart of organizational success. The key to maximizing worker performance is striking a healthy work-life balance, which promotes more productivity, innovation, and general job happiness. Employees are more likely to bring their best selves to work, which has a favorable influence on productivity, when they can balance their personal and professional obligations. The avoidance of burnout is the main way that work-life balance affects productivity. The flexibility to transition between professional and personal pursuits encourages a wide variety of experiences, which sparks creativity and problem-solving abilities (Amendola, Weisburd,

Hamilton et al, 2011). Workers who successfully manage their personal and professional lives are more likely to approach their job with new insights and creative ideas, which makes the workplace more vibrant and effective.

The most important components of work-life balance is flexible work arrangements, which have a significant impact on productivity. Businesses that give their workers the freedom to pick when and where they work frequently see increases in productivity. Employees that have this flexibility are better able to match their work hours to their times of highest production, which results in more focused and productive work. It further enhances productivity by exhibiting a degree of autonomy and trust that supports a healthy workplace culture (Kollmann et al., 2019).

5. HEALTH AND WELL-BEING IN REMOTE WORKING

The paradigm shift toward remote working has brought forth a renewed focus on the health and well-being of employees as they navigate the challenges and opportunities of decentralized work environments. The relationship between health and remote working is multifaceted, intertwining physical, mental, and emotional aspects of well-being. The paramount considerations in this context is the ergonomic setup of remote workspaces. Employees working from home must contend with issues related to posture, prolonged screen time, and inadequate seating, which can contribute to musculoskeletal problems. As a result, businesses are realizing how crucial it is to advise staff members on ergonomic procedures and assist them in setting up comfortable and physically healthy home offices (Clark & Sousa, 2018).

In the context of remote work, mental health is a crucial aspect of wellbeing that has received more attention. Negative effects on mental health might arise from social isolation, a lack of regular face-to-face encounters, and a blurring of the lines between personal and work life. Stress, worry, and loneliness are common among remote workers. In response, businesses are putting in place programs like mental health services, virtual wellness initiatives, and frequent check-ins to assist remote workers' emotional resilience. Seeing how these initiatives affect people's general well-being, a lot of companies are making mental health a top priority and including it into their remote work plans. The remote work environment poses distinct obstacles to sustaining a salubrious work-life equilibrium (Lorenzen, 2008).

The role of leadership in promoting health and well-being in remote working cannot be overstated. Leaders who prioritize employee well-being, model healthy work practices, and demonstrate empathy create a positive and inclusive work culture. Regular communication, transparent leadership, and providing the necessary resources for employee support contribute to a thriving remote work environment where well-being is at the forefront of organizational priorities (Singh, 2023).

5.1 Physical Health Implications of Remote Work

The shift towards remote work, while offering numerous advantages, has significant implications for the physical health of employees. The primary concern revolves around the ergonomic challenges associated with makeshift home offices. Many remote workers may not have access to proper office furniture and equipment, leading to suboptimal workstations that contribute to musculoskeletal issues. Extended durations of sitting in non-ergonomic chairs, improper positioning of computer monitors, and repeated movements can cause soreness, stiffness, and eventually more severe ailments including carpal tunnel syndrome and back pain. Remote work's sedentary nature is another significant element affecting physical health. Because of the increased screen time, working remotely may potentially have an impact on eye health. Workers may spend a lot of time in front of computers, which can cause eye strain, weariness, and discomfort (Bhende, 2020).

This extended screen time may be a factor in digital eye strain, which manifests as headaches, impaired vision, and dry eyes. In remote work situations, putting into practice techniques like the 20-20-20 rule taking a 20-second break to stare at something 20 feet away every 20 minutes—can help reduce eye strain and improve eye health. The consequences of remote work on physical health draw attention to how critical it is to take preventative action to solve ergonomic issues, encourage physical exercise, and give overall wellbeing first priority (Kelly, 2011). Both people and organizations need to be aware of these consequences and take action to provide a remote work environment that promotes employees' physical well-being. To lessen the possible negative impacts of remote work on physical well-being, this may entail offering ergonomic tools, increasing knowledge of healthy behaviors, encouraging frequent breaks and exercise, and cultivating a work-life balance culture.

5.2 Mental Health Considerations for Remote Workers

While remote work offers flexibility and autonomy, there are particular mental health issues for people who are adjusting to working from home. The risk of social isolation is one of the main worries. The regular social connections that come with working in standard office environments may not be present for remote workers (Chung et al., 2022).

Lack of in-person interactions, informal talks with coworkers, and the supportive environment of the job can all lead to feelings of isolation and loneliness that can negatively affect mental health. In order to solve this, organizations could promote virtual social relationships among team members through scheduled team meetings, virtual coffee breaks, and collaborative platforms that allow for casual conversations. The blurring of the lines between work and personal life is another important factor to consider when it comes to mental health. Digital technology's ability to link people constantly might result in a "always-on" mindset, making it difficult for remote workers to draw boundaries between work and personal time.. This blurring of the lines can lead to more stress, burnout, and a feeling of being attached to one's job all the time. Encouraging workers to establish defined work schedules, take breaks, and unplug during non-working hours is crucial for fostering a positive work-life balance and protecting mental health (Adler et al., 2022).

5.3 Coping Mechanisms and Wellness Programs

A comprehensive strategy to assisting people must include both these coping strategies and health initiatives, especially when confronted with difficult situations like the rigors of remote work or other pressures. These techniques are meant to build mental health, increase resilience, and cultivate general well-being (Roth et al., 2023).

Coping Mechanisms

Activities that promote mindfulness and meditation can be effective coping strategies. These methods support people in managing their stress, staying in the now, and growing in self-awareness.

Frequent Exercise: Research has shown that exercise improves mental wellness. Frequent exercise improves mental and physical health as well as lower levels of stress, anxiety, and depression. It might be as easy as going for a stroll, doing yoga, or working out at home.

Creating and Following Routines: Having a daily schedule may give life structure and a sense of regularity. Establishing a consistent routine aids in efficient time management, diminishing ambiguity, and fostering a harmonious work-life equilibrium.

Digital detox: It's important to take deliberate vacations from digital gadgets since remote work requires ongoing contact. Setting aside certain times to avoid social media, emails, or other notifications might help you mentally recharge.

Social Connection: Emotional health depends on actively sustaining social ties, especially in online environments. Frequent check-ins with loved ones, coworkers, or friends support a sense of community and fight feelings of isolation.

Setting limits: Burnout may be avoided by clearly defining limits between one's personal and professional lives. A positive and long-lasting work environment may be established by clearly communicating these limits to managers and coworkers.

Wellness Programs

Employee Assistance Programs (EAPs): EAPs offer mental health resources, private therapy, and assistance with a range of personal and professional issues. These programs are intended to help staff members deal with stress, mental health challenges, and other problems that impact their overall wellbeing.

Wellness Seminars and Workshops: Businesses can host seminars covering subjects like work-life balance, stress management, and constructing resilience. These training programs provide staff members with the skills and information they need to improve their general well-being.

Programs for Fitness and Nutrition: Holistic approaches to well-being are encouraged by wellness efforts that emphasize physical health. Examples of these include gym memberships, nutrition seminars, and fitness challenges.

Flexible Work Arrangements: Offering reduced workweeks, allowing for remote work, and granting flexibility in work schedules all help to promote a better work-life balance. A crucial component of wellness programs is flexibility, which recognizes the wide range of demands and preferences among staff members.

Mental Health Days: Several organizations include mental health days in their wellness programs because they understand how important mental health is. With these days off, workers may put their mental health first without feeling the shame that comes with taking sick time off.

Financial Wellness Programs: Stress related to money may have a big influence on general health. A comprehensive approach to employee health includes wellness programs that address financial literacy, give tools for budgeting, and provide assistance with financial planning.

Through the integration of complete wellness programs and coping methods, people and organizations may establish a nurturing atmosphere that promotes mental health, resilience, and general well-being. These programs may help create a happy and motivated staff in any type of business, in addition to helping those who are struggling with the demands of working remotely (Sharma & Singh, 2022).

5.4 Sound Sleep and Sleeping Patterns: Relevance in Work-Life Balance

Achieving and sustaining work-life balance is heavily reliant on getting enough sleep and adopting healthy sleeping habits, which have a positive impact on a person's personal and professional well-being. Sleep is an essential biological requirement that supports mental clarity, emotional stability, and physical well-being (Lu et al., 2022). It is impossible to overestimate the significance of getting a good night's sleep in the context of work-life balance since it has a direct influence on one's capacity to handle the demands of both their personal and professional lives with vigor and clarity. The main effects of getting enough sleep on productivity and cognitive performance have a significant impact on work-life balance. For cognitive functions like attention, memory consolidation, and problem-solving, one must get enough good sleep. People who regularly get enough sleep are better able to handle challenging activities, make wise judgments, and maintain focus during the workday. In the personal sphere, this cognitive resilience helps with improved communication, social or familial problem-solving skills, and general clarity in handling a variety of obligations (Kondo & Sonenshein, 2020).

Prioritizing sufficient and quality sleep not only enhances cognitive function and emotional well-being but also contributes to overall physical health. Individuals who recognize the importance of sleep in the delicate dance between professional and personal responsibilities are better equipped to navigate the complexities of modern life with resilience, focus, and a holistic sense of well-being. As organizations increasingly recognize the value of employee well-being, fostering a culture that acknowledges and supports healthy sleep habits is essential for creating workplaces that prioritize work-life balance and sustained individual and collective success (Siganidou, 2020).

6. UPLIFTING PRODUCTIVITY THROUGH WORK-LIFE BALANCE

Uplifting productivity through work-life balance is a strategic imperative that recognizes the symbiotic relationship between the well-being of individuals and their professional performance. In a world characterized by constant connectivity and evolving work structures, organizations are increasingly realizing that a harmonious balance between professional and personal life is not only beneficial for the workforce but also directly contributes to enhanced productivity and organizational success (Tenorio & Vu, 2021). When employees are consistently overburdened with work and lack the necessary time for recuperation and personal pursuits, burnout becomes a significant risk. Burnout not only erodes individual well-being but also hampers productivity and creativity. Organizations that prioritize work-life balance recognize that preventing burnout is a strategic investment in sustaining high levels of productivity and retaining a motivated and healthy workforce.

The flexible work arrangements are instrumental in uplifting productivity through work-life balance. Allowing employees the autonomy to structure their work schedules and choose locations that best suit their needs empowers them to optimize their time and energy. This flexibility not only accommodates diverse personal circumstances but also acknowledges that individuals have varying peak productivity periods (Cheese, 2021). Whether it involves working from home, adjusting working hours, or implementing compressed workweeks, the flexibility provided by organizations fosters an environment where employees can align their professional responsibilities with their individual preferences, ultimately enhancing productivity. The organizations that prioritize work-life balance often invest in wellness programs that address physical and mental health.

These initiatives include resources for stress management, fitness programs, mental health support, and educational workshops. By nurturing the holistic well-being of employees, organizations contribute to the creation of a resilient and healthy workforce, laying the foundation for sustained productivity. Employees who feel physically and mentally well are better equipped to tackle challenges, think creatively, and contribute meaningfully to the organization's objectives (Pearman, 2023).

6.1 Factors Influencing Productivity in Remote Working

Remote working has become a prevalent mode of work, bringing about both challenges and opportunities. Several factors influence productivity in remote working environments, and understanding these dynamics is essential for organizations and individuals striving for optimal performance.

1. **Technological Infrastructure**: The availability and reliability of technology are fundamental to remote work productivity. High-speed internet, secure communication tools, and efficient project management platforms enable seamless collaboration and task execution. Organizations that invest in robust technological infrastructure contribute significantly to remote workers' ability to perform tasks efficiently and stay connected with their teams.
2. **Work Environment:** The physical space in which remote work takes place has a direct impact on productivity. A well-designed home office, free from distractions, and equipped with ergonomic furniture and technology, promotes focus and efficiency. Factors such as lighting, noise levels, and comfort contribute to creating a conducive work environment that positively influences productivity.
3. **Communication and Collaboration**: Effective communication is paramount in remote work settings. Clear and frequent communication channels, along with collaboration tools, facilitate information exchange and teamwork. Organizations that establish transparent communication norms and encourage collaboration create an environment where remote workers can stay aligned with goals and work cohesively.
4. **Time Management**: Remote workers must navigate the challenge of setting and adhering to their own schedules. Effective time management strategies, such as creating daily routines, setting priorities, and using productivity techniques like the Pomodoro Technique, play a crucial role

in optimizing work hours. Establishing boundaries between work and personal life is equally vital for sustaining productivity over the long term.

5. **Flexibility and Autonomy**: The flexibility to manage one's work schedule and autonomy in decision-making contribute to increased motivation and productivity. Remote workers who have the freedom to structure their work in a way that suits their individual preferences and peak productivity periods often exhibit higher levels of engagement and efficiency.
6. **Goal Setting and Clarity**: Clear expectations and well-defined goals are essential for remote work productivity. Organizations that provide clarity regarding tasks, deadlines, and performance expectations empower remote workers to stay focused and aligned with overarching objectives. Regular check-ins and performance feedback further contribute to goal clarity.
7. **Employee Well-Being**: The well-being of remote workers significantly influences productivity. Organizations that prioritize employee health, both physical and mental, contribute to a workforce that is more resilient, focused, and energetic. Well-being initiatives, mental health services, and community-building projects all contribute to a happy and effective remote work environment.
8. **Training and Opportunity for Skill Development**: Ensuring that remote workers have access to sufficient training and chances for skill development helps to guarantee that they are prepared to do their jobs effectively. In addition to improving individual skills, ongoing learning helps create a workforce that is flexible and competent in a changing work environment.
9. **Organizational Culture**: The work atmosphere and productivity are shaped by the organizational culture, especially in distant settings. A motivated and engaged remote workforce is a result of an inclusive, pleasant work environment that values teamwork, acknowledges accomplishments, and encourages a good work-life balance.
10. **Security Precautions**: Preserving productivity in remote work contexts requires ensuring security. A safe and stable work environment lowers the likelihood of interruptions. Strong cybersecurity protections, data protection procedures, and secure access to organizational resources all contribute to this.

6.2 Role of Work-Life Balance and Employee Well-Being in Enhancing Productivity

In the modern workplace, improving employee well-being and work-life balance are critical to increasing productivity. Organizations are becoming more aware of the connection between worker well-being and overall productivity as they work to maximize performance and preserve a competitive advantage (Wan, 2022). Maintaining work-life balance is essential to maintaining employee well-being, and when it is given priority, it may lead to increased motivation, engagement, and productivity. An important component of this synergy is how work-life balance affects mental health. Workers who can successfully manage their personal and work lives in harmony report feeling less stressed, burned out, and having better mental health overall. When employees feel encouraged to balance work and personal obligations, they are more likely to remain resilient, creative, and focused when faced with obstacles.

This positive mental state directly translates into increased productivity, as employees are more likely to approach tasks with clarity, enthusiasm, and sustained energy. Another aspect that is impacted by work-life balance is physical health, which is crucial for increasing productivity. People with robust physical health are more likely to have balanced lives that include enough time for exercise, rest, and healthy eating. Thus, there is an increase in energy, a decrease in absenteeism, and an improvement in cognitive function. Well-being makes workers more resilient to the demands of the workplace, which boosts output and maintains high standards of performance (Battah, 2020).

Beyond an individual's health, employee well-being affects team dynamics and company culture. Businesses that proactively encourage employee well-being and work-life balance foster a culture of trust, camaraderie, and support. An environment at work where workers feel appreciated and encouraged to strike a good work-life balance promotes a favorable work culture that improves communication, cooperation, and general job satisfaction. Increased productivity and organizational success are the results of a contented and motivated staff, which is more inclined to work together toward shared objectives. Improving work-life balance helps to avoid burnout, which is a common problem in today's hectic workplaces (Singh, 2022).

7. STRATEGIES TO ENHANCE WORK-LIFE BALANCE IN REMOTE WORKING

The necessity of techniques to improve work-life balance has become more apparent as remote work continues to develop as a popular method of employment. Both people and organizations are realizing that deliberate strategies are required to guarantee that remote work continues to be a viable and satisfying part of contemporary employment. A good work-life balance may be fostered in remote work situations through a number of important measures (Grenville-Cleave et al., 2021).

First and foremost, it's critical to draw distinct lines between one's personal and professional lives. One of the biggest challenges faced by remote workers is the "always-on" mindset that causes the lines between home and work to blur. A more disciplined routine may be achieved by establishing set work hours, allocating a designated workstation inside the house, and intentionally "closing" the virtual office at the end of the workday. People may reduce the danger of overworking and burnout by setting clear boundaries for when work starts and ends (Moore, 2017). This helps people feel more closure and smoothly transition into their leisure time. Putting in place flexible work schedules is another essential tactic for improving work-life balance in remote environments. Employees can arrange their work hours to coincide with their times of highest production when choices like flexible working hours, compressed workweeks, or part-time schedules are provided. This adaptability allows for a range of unique situations and preferences, enabling people to make the most use of their time and energy. It also recognizes that productive work may be done without strict adherence to regular office hours (Madan et al., 2022).

In remote work contexts, it is essential to promote frequent breaks and downtime in order to maintain a healthy work-life balance. It might be difficult for remote workers to put in a day's work when their home office doubles as their place of employment. But taking short breaks whether it's to go for a quick stroll, practice meditation, or just put the device down is crucial to staying focused and recharging. Organizations can foster a culture that values breaks and emphasizes the importance of regular rest to sustain productivity and overall well-being (Silvester & Konstantinou, 2010). Communication plays a pivotal role in remote work, and establishing effective communication norms contributes to a healthier work-life balance. Clearly articulating expectations regarding response times, setting guidelines for after-hours communication, and utilizing asynchronous communication tools help manage the constant influx of messages. By fostering a culture that respects individual boundaries and emphasizes the importance of downtime, organizations contribute to a more balanced and sustainable remote work environment (Singh, 2022).

8. ORGANIZATION PERFORMANCE

An entity's ability to produce value, accomplish its objectives, and adjust to its operational environment is all measured by its organizational performance. It includes a wide range of elements, including as the capacity to adapt to change, operational effectiveness, financial success, and strategic alignment. An organization's ability to successfully integrate its people, processes, and systems is typically reflected in its high level of organizational performance (Combs, 2017). When assessing the effectiveness of a business, financial measures are frequently a crucial element. Key indications that stakeholders, including investors and shareholders, use to assess an organization's financial health are profitability, revenue growth, and return on investment. Not only is financial success a goal unto itself, but it also shows how well an organization can raise funds for ongoing projects, new ideas, and strategic planning.

The other critical component of organizational effectiveness is operational efficiency. Cost-efficiency and overall effectiveness are influenced by simplified workflows, effective procedures, and optimal resource use. Businesses that put an emphasis on operational excellence are better able to meet customer requests quickly, supply goods and services more effectively, and stay ahead of the competition in their particular markets. The capacity of an organization to match its actions with its long-term aims and objectives is referred to as strategic alignment. An company that performs successfully makes sure that its strategies are clearly stated, effectively conveyed, and accepted at all levels. A cohesive strategy helps guide decision-making, prioritize initiatives, and unify the efforts of the entire workforce toward common objectives (Singh, 2020). The adaptability and responsiveness to change are increasingly recognized as critical elements of organizational performance. In a dynamic business environment, the ability to navigate uncertainties, embrace innovation, and adjust strategies in response to market shifts is essential. Agile organizations can seize opportunities and mitigate risks more effectively, positioning themselves for sustained success in a rapidly evolving landscape (Hubble, 2017).

8.1 Impact of Employee Well-Being on Overall Organizational Performance

The impact of employee well-being on overall organizational performance is profound and multifaceted, extending far beyond individual job satisfaction. A workforce that experiences high levels of well-being is not only more engaged and motivated but also contributes significantly to the collective success of the organization (Amendola, Weisburd, Heitmann et al, 2011). The key to developing a positive workplace culture, which in turn affects many facets of organizational performance, is employee well-being. Productivity is one of the main ways that employee well-being affects corporate performance. People who are in good physical and mental health are more likely to be efficient, creative, and focused in their work. High well-being among employees makes them more capable of meeting workplace expectations, which boosts output, improves quality of work, and boosts productivity as a whole.

Employee well-being is positively correlated with decreased absenteeism and turnover rates, which minimizes interruptions and related expenses for the company. The essential to developing a creative and cooperative workplace culture is the well-being of the workforce. Being in a good mental state helps you be creative, solve problems and be open to trying new things (Molek-Winiarska, 2022). The effects of staff well-being also extend to client loyalty and satisfaction. Positive attitudes and behaviors are more likely to be displayed by employees who report high levels of well-being while interacting with consumers. This kind of interaction improves client satisfaction and encourages word-of-mouth and loyalty, all of which help the company's standing in the marketplace. Talent acquisition and retention are clear indicators of the relationship between employee well-being and business effectiveness. Top talent finds companies that put employee well-being first more appealing when applying for jobs. Higher levels of job satisfaction and employee loyalty are a result of a healthy work environment and a dedication to the well-being of employees. This lowers employee turnover and the related expenses of hiring new employees and providing training. In this sense, wellbeing turns into a tactical advantage for creating and retaining a highly productive workforce (Ilmola-Sheppard, Strelkovskii, Rovenskaya, Abramzon, & Bar, 2020).

8.2 Key Performance Indicators (KPIs) Related to Remote Working and Productivity

Organizations utilize key performance indicators (KPIs) pertaining to productivity and remote work to evaluate the efficacy and efficiency of their activities. One critical KPI is task completion or project milestones, which measures the ability of remote workers to meet deadlines and deliverables. Monitoring project timelines and assessing the achievement of key milestones provides insights into the overall productivity of remote teams.

The other important KPI is the utilization of collaboration tools and communication platforms. Metrics related to the frequency and effectiveness of virtual meetings, the use of project management tools, and communication responsiveness help gauge the level of collaboration and coordination among remote team members. Employee engagement is a significant KPI linked to remote working productivity. Surveys, feedback mechanisms, and other indicators of employee satisfaction and motivation provide valuable insights into the overall well-being and commitment of remote workers. Keeping an eye on and resolving engagement-influencing issues helps create a happy work environment, which increases productivity. Furthermore, KPIs that track the timely completion of specific tasks—like the quantity of support tickets closed or sales conversations completed—provide detailed insights into the daily productivity of remote workers (Singh, 2019).

KPIs pertaining to output might include the quantity of tasks finished, the amount of code produced, or the amount of sales objectives met. These indicators assist companies in evaluating the real benefits that remote workers bring to the table and pinpointing areas that require further development or assistance. By allowing people to take brief breaks from work, the usage of break and downtime features may be measured. This can help prevent burnout and promote a healthy work-life balance. It can also provide information about the well-being of remote workers. In conclusion, a thorough set of KPIs about productivity and remote working includes a variety of factors, such as employee engagement, task completion on an individual basis, project milestones, and the caliber of work production.

For firms to evaluate the efficacy and impact of remote work efforts, Key Performance Indicators (KPIs) pertaining to productivity and remote working are essential. These key performance indicators (KPIs) offer insightful information on many facets of remote job performance, supporting well-informed decision-making and ongoing development. The following are important metrics that businesses typically use:

Task Completion and Project Milestones

KPIs: On-time work completion, project milestone accomplishment, and deadline observance.

Justification: Evaluates remote workers' capacity to oversee and complete tasks within the time constraints, offering valuable information about their total productivity.

Utilization of Collaboration Tools

KPIs: Utilization data for communication channels, project management software, and collaboration platforms.

Justification: Shows the degree of cooperation and involvement between distant teams, which is necessary for efficient communication and information exchange.

Employee Engagement and Satisfaction

KPIs: Feedback systems, employee surveys, and involvement in virtual team activities.

Justification: Evaluates remote workers' contentment, drive, and general involvement, which affects their well-being and output.

Communication Effectiveness

KPIs: It includes communication responsiveness, participation in online meetings, and email response times.

Justification: Evaluates the efficacy and efficiency of routes for communication, making sure that distant teams remain informed and connected.

Individual Task Metrics

KPIs: The quantity of tasks finished, code lines produced, or sales goals met.

Justification: Gives businesses detailed information on how productive remote workers are on a daily basis, enabling them to monitor individual contributions.

Quality and Quantity of Work Output

KPIs: There are output measures that are particular to the type of work being done, such the quantity of closed support tickets or finished sales calls.

Justification: Assesses the observable benefits of remote work, making sure that productivity and quality requirements are upheld.

Well-Being and Work-Life Balance

KPIs: Employee well-being surveys and the use of the break and downtime features.

Justification: Tracks well-being indicators for staff members, assisting in preventing burnout and encouraging a balanced attitude to work.

Remote Access and Security Compliance

KPIs: Security compliance metrics, secure access to company resources.

Rationale: Assesses the security and compliance of remote work setups, ensuring data protection and mitigating risks.

Training and Skill Development Participation

KPIs: Participation rates in virtual training sessions, completion of online courses.

Rationale: Measures the engagement of remote employees in ongoing learning and development opportunities.

Meeting Effectiveness

KPIs: Metrics related to virtual meeting attendance, engagement during meetings, and post-meeting actions.

Rationale: Evaluates the efficiency and impact of virtual meetings, ensuring that collaboration time is utilized effectively.

These KPIs collectively offer a comprehensive view of remote work performance, guiding organizations in optimizing processes, supporting employee well-being, and enhancing overall productivity in the remote work landscape.

8.3 Comparative Analysis of Organizations With Different Approaches

A comparative analysis of organizations with different approaches reveals the diverse strategies and methodologies employed within the business landscape. Organizations often adopt distinct approaches based on their industry, culture, leadership philosophy, and overarching goals. One common point of comparison lies in organizational structures, where some enterprises embrace hierarchical structures with clearly defined roles and reporting lines, while others opt for flat or matrix structures, emphasizing collaboration and flexibility. Strategic approaches also vary widely among organizations. Some adopt a market-driven strategy, focusing on customer needs and competitive positioning, while others may prioritize innovation, seeking to disrupt markets with groundbreaking products or services. Comparative analyses frequently explore corporate cultures, with some organizations emphasizing stability, consistency, and obedience to established processes, while others foster an environment of creativity, risk-taking, and adaptation (Ilmola-Sheppard, Strelkovskii, Rovenskaya, Abramzon, & Bar, 2020).

The crucial variable for comparison is the use of technology. The digital transformation of modern businesses is ongoing, with differing degrees of dependence on technology to improve overall performance, efficiency, and communication. While some businesses completely embrace new technology, others could take a more measured or gradual approach, taking industry rules or financial limitations into account. The ways that firms manage their workforces also vary. While some prioritize a results-driven strategy and provide their workers flexibility and autonomy, others may take a more conventional, time-centric approach with set working hours (Saez, 2023). There are considerable differences in the focus that different firms place on work-life balance, professional growth, and employee well-being, all of which have an impact on employee happiness and total productivity.

9. CONCLUSION AND FUTURE SCOPE

A crucial requirement in today's organizational environments is the investigation of improving work-life balance in remote working through a holistic strategy that promotes healthy sleep, general well-being, and health. It is clear that worker productivity and physical and mental health are related, which emphasizes the necessity for businesses to implement all-encompassing solutions that go beyond conventional work paradigms. The emphasis on sound sleep recognizes its pivotal role in cognitive function, emotional well-being, and physical health, creating a foundation for sustained productivity and creativity. Through tackling issues like possible isolation, indistinct work-life boundaries, and psychological effects, companies may establish work cultures that promote resilience, engagement, and a feeling of community among workers who work remotely. Consequently, this fosters a favorable company culture that increases output and drives all employees toward common objectives.

The potential for improving work-life balance in remote employment appears to be endless, given the continuous progress in technology, data analytics, and a better comprehension of human welfare. Employers are likely to monitor and promote worker well-being by utilizing advanced platforms and tools, putting customized interventions in place depending on each worker's requirements and preferences. By combining AI and machine learning, employers might be able to offer more individualized services to their staff, such as scheduling optimization and wellness recommendations based on individual needs. Performance measurements may be reevaluated in the future, expanding beyond measures of traditional productivity to include indicators of holistic well-being. Businesses may be realizing more and more that having a healthy, balanced staff makes them not just more productive but also more resilient to setbacks. The growing popularity of remote and flexible work arrangements among workers worldwide will make creative solutions and a dedication to employees' changing requirements necessary to maintain work-life balance. The comprehensive approach to improving work-life balance in this changing environment becomes a proactive strategy for organizational success rather than only a reaction to present issues. It puts companies in a position to draw in and hold on to top personnel, adapt quickly to changes in the business landscape, and foster an innovative and healthy culture. The journey towards a balanced and productive remote work model is ongoing, promising a future where organizations and employees mutually thrive in a dynamic and harmonious work ecosystem.

REFERENCES

Adler, D. A., Tseng, E., Moon, K. C., Young, J. Q., Kane, J. M., Moss, E., & Choudhury, T. (2022). Burnout and the Quantified Workplace: Tensions around Personal Sensing Interventions for Stress in Resident Physicians. *Proceedings of the ACM on Human-computer Interaction, 6*(CSCW2), 1-48. 10.1145/3555531

Amendola, K. L., Weisburd, D., Hamilton, E., Jones, G., Slipka, M., Heitmann, A., & Tarkghen, E. (2011). *The impact of shift length in policing on performance, health, quality of life, sleep, fatigue, and extra-duty employment*. Police Foundation.

Battah, P. (2020). *Humanity at work: Leading for better relationships and results*. LifeTree Media.

Bhende, P. M. (2020). *Quality of Work Life of Employees in Banking Sector-A Study of Bank Managers and Staff in Goa* [Doctoral dissertation]. Goa University.

Blankson, A. (2017). *The Future of Happiness: 5 Modern Strategies for Balancing Productivity and Well-Being in the Digital Era*. BenBella Books, Inc.

Budhiraja, S., & Kant, S. (2020). Challenges associated with work-life balance: A meta-analysis. *J. Strat. Human Res. Manag, 9*, 11–16.

Cheese, P. (2021). *The New World of Work: Shaping a Future that Helps People, Organizations and Our Societies to Thrive*. Kogan Page Publishers.

Chung, H., Jaga, A., & Lambert, S. (2022). Possibilities for change and new frontiers: Introduction to the work and family researchers network special issue on advancing equality at work and home. *Community Work & Family, 25*(1), 1–12. doi:10.1080/13668803.2022.2008057

Clark, A., & Sousa, B. (2018). How to be a happy academic: A guide to being effective in research, writing and teaching. *Sage (Atlanta, Ga.).*

Combs, K. M. (2017). *Strategies for retaining employees for call centers* [Doctoral dissertation]. Walden University.

Darouei, M., & Pluut, H. (2021). Work from home today for a better tomorrow! How working from home influences work-family conflict and employees' start of the next workday. *Stress and Health*, *37*(5), 986–999. doi:10.1002/smi.3053 PMID:33887802

Fallica, A. (2022). *Smart working: Measurement of sustainability impact*. Academic Press.

Grenville-Cleave, B., Guðmundsdóttir, D., Huppert, F., King, V., Roffey, D., Roffey, S., & de Vries, M. (2021). *Creating the world we want to live in: How positive psychology can build a brighter future*. Routledge. doi:10.4324/9781003031789

Hamlin, C. R. (2018). *ICT and Diminished Work-life Balance among Civilian Law Enforcement Professionals: A Multiple Case Study* [Doctoral dissertation]. Northcentral University.

Hubble, R. (2017). *Re-Engineer Your Workday: Create Time for Life*. Balboa Press.

Ilmola-Sheppard, L., Strelkovskii, N., Rovenskaya, E., Abramzon, S., & Bar, R. (2020). *A systems description of the national well-being system*. Version 1.0.

Ireland, S. G. (2022). A Study on the Need for Investment in Employee Wellbeing in Small and Medium Enterprises (SMEs) In Bangalore, India. Academic Press.

Kelly, M. (2011). *Off Balance: Getting Beyond the Work-Life Balance Myth to Personal and Professional Satisfaction*. Penguin.

Kollmann, T., Stöckmann, C., & Kensbock, J. M. (2019). I can't get no sleep—The differential impact of entrepreneurial stressors on work-home interference and insomnia among experienced versus novice entrepreneurs. *Journal of Business Venturing*, *34*(4), 692–708. doi:10.1016/j.jbusvent.2018.08.001

Kondo, M., & Sonenshein, S. (2020). *Joy at work: Organizing your professional life*. Pan Macmillan.

Kroemer, A. D., & Kroemer, K. H. (2016). *Office ergonomics: Ease and efficiency at work*. CRC Press. doi:10.1201/9781315368603

Lorenzen, E. M. (2008). *Work/life Issues in America: A Public Policy Analysis* [Doctoral dissertation]. California State University, Northridge.

Lu, G., Du, R. Y., & Peng, X. (2022). The impact of schedule consistency on shift worker productivity: An empirical investigation. *Manufacturing & Service Operations Management*, *24*(5), 2780–2796. doi:10.1287/msom.2022.1132

Madan, P., Tripathi, S., Khalique, F., & Puri, G. (Eds.). (2022). *Re-envisioning Organizations Through Transformational Change: A Practitioners Guide to Work, Workforce, and Workplace*. CRC Press. doi:10.4324/9781003267751

Molek-Winiarska, D. (Ed.). (2022). Shaping Employee Experience in the Changing Social and Organisation Conditions. Wydawnictwo Uniwersytetu Ekonomicznego we Wrocławiu. doi:10.15611/2022.989.4

Moore, K. A. (2017). 21. Mindfulness at work. *Research handbook on work and well-being*, 453.

Mullins, L. B., Scutelnicu, G., & Charbonneau, É. (2022). A Qualitative Study of Pandemic-Induced Telework: Federal Workers Thrive, Working Parents Struggle. *Public Administration Quarterly*, *46*(3), 258–281. doi:10.37808/paq.46.3.4

Ninaus, K., Diehl, S., & Terlutter, R. (2021). Employee perceptions of information and communication technologies in work life, perceived burnout, job satisfaction and the role of work-family balance. *Journal of Business Research*, *136*, 652–666. doi:10.1016/j.jbusres.2021.08.007

Ochoa, J. J. (2023). *Occupational Burnout and Work-Life Balance Strategies Among Chaplains* [Doctoral dissertation]. University of Arizona Global Campus.

Onder, C. C. (2016). *Unfolding of telecommuting's effects in organizations: Performance, commitment, and mechanisms of action* [Doctoral dissertation]. Colorado State University.

Onyeka, N. F. (2022). *Exploring Female Perspectives of Work-Life Balance and the Role of Organizational Culture* [Doctoral dissertation]. Capella University.

Pearman, C. D. (2023). *Tai Chi Enhanced Workplace Health and Wellness: A Systematic Literature Review for a Program Design, Implementation, and Evaluation* [Doctoral dissertation]. University of Arizona Global Campus.

Pradhan, R. K., & Kumar, U. (Eds.). (2021). *Emotion, well-being, and resilience: Theoretical perspectives and practical applications*. CRC Press.

Roth, S. C., Hinton, E. G., & Jivanelli, B. (2023). Impact of COVID-19 on Parents/Guardians in the Library Profession: A Narrative Review and Shared Experiences. *Library Leadership & Management (Online)*, *37*(2), 1–72.

Rubin, K. I. (2022). *The Great Balancing Act: Women Seeking Work-Life Balance during COVID-19* [Doctoral dissertation]. University of Southern California.

Ruderman, M. N., Clerkin, C., & Deal, J. J. (2017). The long-hours culture. *The Routledge Companion to Wellbeing at Work: Taylor & Francis*, 207-221.

Saez, K. (2023). Problematizing workplace learning for the work-from-home and hybrid workforce following. *COVID*, 19.

Sharma, A., & Singh, B. (2022). Measuring Impact of E-commerce on Small Scale Business: A Systematic Review. *Journal of Corporate Governance and International Business Law*, *5*(1).

Siganidou, S. (2020). *Work-related employee stress in 4* & 5* hotels in Greece. A study about the city of Thessaloniki*. Academic Press.

Sigeman, H., & Adolfsson, M. (2021). *Maintaining well-being when working remotely: Work habits of employees at a geographical IT office in Sweden during the COVID-19 pandemic, their perceived well-being and productivity*. Academic Press.

Silvester, J., & Konstantinou, E. (2010). *Lighting, well-being and performance at work*. City University.

Singh, B. (2019). Profiling Public Healthcare: A Comparative Analysis Based on the Multidimensional Healthcare Management and Legal Approach. *Indian Journal of Health and Medical Law*, *2*(2), 1–5.

Singh, B. (2020). Global science and jurisprudential approach concerning healthcare and illness. *Indian Journal of Health and Medical Law*, *3*(1), 7–13.

Singh, B. (2022a). Relevance of Agriculture-Nutrition Linkage for Human Healthcare: A Conceptual Legal Framework of Implication and Pathways. *Justice and Law Bulletin*, *1*(1), 44–49.

Singh, B. (2022b). COVID-19 Pandemic and Public Healthcare: Endless Downward Spiral or Solution via Rapid Legal and Health Services Implementation with Patient Monitoring Program. *Justice and Law Bulletin*, *1*(1), 1–7.

Singh, B. (2023c). Tele-Health Monitoring Lensing Deep Neural Learning Structure: Ambient Patient Wellness via Wearable Devices for Real-Time Alerts and Interventions. *Indian Journal of Health and Medical Law*, *6*(2), 12–16.

Singh, B. (2023a). Blockchain Technology in Renovating Healthcare: Legal and Future Perspectives. In Revolutionizing Healthcare Through Artificial Intelligence and Internet of Things Applications (pp. 177-186). IGI Global.

Singh, B. (2023b). Federated Learning for Envision Future Trajectory Smart Transport System for Climate Preservation and Smart Green Planet: Insights into Global Governance and SDG-9 (Industry, Innovation and Infrastructure). *National Journal of Environmental Law*, *6*(2), 6–17.

Singh, B. (2024). Legal Dynamics Lensing Metaverse Crafted for Videogame Industry and E-Sports: Phenomenological Exploration Catalyst Complexity and Future. *Journal of Intellectual Property Rights Law*, *7*(1), 8–14.

ten Brummelhuis, L. L., ter Hoeven, C. L., & Toniolo-Barrios, M. (2021). Staying in the loop: Is constant connectivity to work good or bad for work performance? *Journal of Vocational Behavior*, *128*, 103589. doi:10.1016/j.jvb.2021.103589

Tenorio, J. L., & Vu, H. T. (2021). *A Cross-sectional Study during the COVID-19 Pandemic: Work Events and Affective Reactions as Predictors of Loss of Productivity and Intention to Leave* [Master's thesis]. Handelshøyskolen BI.

Wan, W. (2022). *Making Kindness Our Business*. Marshall Cavendish International Asia Pte Ltd.

Weber, M. B., & De Fino, M. (2022). *Virtual technical services: a handbook.* Rowman & Littlefield. doi:10.5771/9781538152645

Chapter 2
The Influence of Teleworking on Work–Life Balance and Employee Psychological Health:
Evidence From Mauritius, a Small Island Economy

Zuberia Aminah Hosanoo
https://orcid.org/0000-0002-9583-9268
Edinburgh Napier University, UK

Soujata Rughoobur-Seetah
Curtin Mauritius, Mauritius

ABSTRACT

This chapter investigates the impact of teleworking on the psychological health of employees in the new era of work shaped by the pandemic. Adopting a positivist approach in the post-pandemic world, this study aims to assess the employees' perception of their psychological health resulting from the drastic change in the work sphere in the small island economy of Mauritius. Teleworking has been observed as a core behaviour emerging from the global pandemic has become enduring and gradually institutionalised. Widespread continuation and advocacy of teleworking may fundamentally change job characteristics and outcomes. This research assesses the impact of teleworking on the psychological health of employees in the new era of work shaped by

DOI: 10.4018/979-8-3693-1314-5.ch002

the pandemic. This chapter adopted a positivist approach and conducted a survey in the post-pandemic world to assess the employees' perception of their psychological health resulting from the drastic change in the work sphere in the small island economy of Mauritius.

INTRODUCTION

The world has changed post-pandemic, and the popularity of remote work or teleworking is increasing worldwide (El Hajal, 2022). The COVID-19, which originated in Wuhan, became a global pandemic (Tomasi, 2020). The propagation of the virus was so fast and intense that forced countries to re-think on their security and how to sustain the economy. Countries are facing unprecedented economic crisis and even contraction. The economic climate has also been compared to that of the Spanish Flu pandemic which was then the world's largest recession and now the COVID-19 pandemic is being acknowledged to be the world's largest recession since the second world war. The global outbreak of the virus led to countrywide lockdown, to ensure the safety of the population in various countries. Teleworking has experienced a positive rebound, facilitated by government initiatives and driven by the need for business continuity. Remote working became the contingency plans of both governments and the companies globally (Belzunegui-Eraso & Erro-Garcés, 2020).

Flexible work practices allude to the mutual agreements between employers and employees concerning the hours and location of work, aimed at meeting organisational requirements and enhancing employees' Work-Life Balance (WLB) (Thompson, et al., 2015). Telework, also referred to as telecommuting, e-working, electronic working or remote working (Farrell, 2017); is a flexible work practice that involves working away from the office setting, at the employee's preferable working location, usually home or satellite office (Allen et al., 2015). The evolution of teleworking has been associated with the fast progression of information and communication technology and changes in the economy (Hill et al., 2023). Today, with the rising capabilities of home computing technologies, bolstered by the pandemic, employees have been empowered to work beyond office boundaries (Hopkins, & Bardoel, 2023). Similarly, the transition to an information economy has boosted the quantum of jobs lending themselves to teleworking (Kizza, 2013; Betchoo, 2020).

Teleworking has been studied from broad disciplines: human resources management, industrial relations, management, psychology, sociology, family studies, information systems, operations and logistics (Beauregard et al., 2019). E-working provides multiple benefits to employees such as more control over working hours; improved productivity and creativity; lesser stress and distractions; closer to family members and home; improved WLB (Timsal & Awais, 2016). Using the social exchange, the psychological contract and the job characteristic theories, this study aims at analysing the impact of teleworking on employees' psychological health of employees in Mauritius.

Mauritius

Mauritius, a high-income emerging Small Island Economy (SIE) (Haughton & Ivey, 2023). has political stability, a streamlined tax administration and ultramodern infrastructure analogous to some East Asian emerging economies (Sannassee et al., 2014). Before the pandemic, teleworking has been an offbeat futuristic work practice in Mauritius, whereby the deep-rooted bureaucratic work routines like fingerprint attendance clocking system which are in place to warrant stringent adherence to attendance, serve to curb work flexibility. Braesemann, et.al. (2022) research on the global polarisation of remote work found that states are globally divided, with global south countries partaking marginally on remote workers. Betchoo (2020) explains that teleworking is not customary practice in Mauritius, whereby remote working was commonly undertaken by employees in the informal sector who work from their residence, while in the formal economy, workforces commute to strategic locations/offices to work. Flexible working arrangement such as flexitime or teleworking have time and again surfaced in both private and public companies in Mauritius, but its praxis has been insignificant. Beebeejaun & Gunputh (2022) further enlighten that prior to the global pandemic, there was no legal code in Mauritius regulating teleworking practices, but the labour law was subsequently amended in emergency during the lockdown in Mauritius to include the Workers' Rights (Working from Home) Regulations 2020. Hence, this study aims at empirically testing the influence of teleworking on employees' psychological health and is founded on the triangulation of three theories, namely, the social exchange theory, psychological contract theory and the job characteristic model.

THEORETICAL SUPPORT: THE TRIANGULATION OF THEORIES

Social Exchange Theory (SET)

The Social Exchange Theory (Emerson 1976) is a useful theoretical perspective in the study of teleworking. SET concerns the obligations spawned through successive transactions between parties (Emerson, 1976). SET proposes that individuals pursue social situations maximising their beliefs while minimising personal disadvantages (Even, 2020). More so, Organ (1988) believed that the exchange relationship is mostly geared towards the economic exchange, tending to be quid pro quo, where lesser trust is involved. Hence to produce and sustain favourable work outcomes, firms must identify the best practices to prevent teleworks' feeling of social and organisational isolation (Even, 2020). SET being a robust theory has often been criticized by various researchers (Cropanzano et al., 2017) where it has been acknowledged that the true perception of individuals is not well captured when using SET as the theory has its own limitations. Therefore, for this chapter SET will be accompanied by PCT and JCM as the most plausible and accurate theories to assess and understand the proposed framework.

Psychological Contract Theory (PCT)

PCT refers to the implicit expectations that employers and employees have of the employment relationship, such as the aspirations, values, and the mutual expectations and obligations, that are beyond the formal contract of employment binding both parties (Argyris, 1960; Rousseau,1989). Psychological contracts have often been studied when it comes to employment relationships. Factors like teleworking, burnout, work life balance have been reported to significantly impact employee performance (e.g. Chan, 2021; Sui, Yu & Zhou; 2023); Jayaweera et al., 2021). Yet, it has to be acknowledged some studies have concluded that teleworking sometimes leads to isolation and work life conflict (Nemţeanu & Dabija, 2023). PCT is a useful theoretical framework to assess the influence of working from home and WLB. Since the mutual expectations are implicit, it infers fairness, and good faith of both employers and employees. The use of PCT will enable better understanding of the emotions of employees and how they perceive teleworking. Besides, the JCM also is an integral part in understanding whether the employees can manage their working remotely.

Job Characteristics Model (JCM)

JCM, brought forward by Hackman & Oldham (1975), is a one of the significant contributions (Fried & Ferris, 1987; Ali et al., 2014), as it tends to forecast job performance based on skill variety, task identity, task significance, autonomy, and feedback. These five factors can be depicted to be important to help an employee in performing their tasks. Skill variety helps an employee to get heavily engaged in the job by using a diverse set of skills in a wide range of activities, thus enhancing their performance. Task identity enables the employee to identify and familiarise themself with the tasks which result in a better understanding of the overall work system and the importance and connectedness of all the tasks. Task significance helps employees to recognise the meaningfulness of their work and how their work contributes to the overall wellbeing of others. Therefore, it can be said that skill variety, task identity and task significance contribute towards the employees in connecting and making sense of their work. Job autonomy provides more leeway to employees to adapt and adopt any work-related processes to facilitate and enhance their current work. Finally, feedback provides employees with relevant work information to enhance their performance and work systems.

The Proposed Framework

Various studies are now emerging and assessing the impact of COVID-19 on employees' psychological health (Carnevale & Hatak, 2020; Tuzovic & Kabadayi, 2020) or the relationship between teleworking and employee's health and well-being in the covid and post pandemic era (Njaramba, 2023; Parent-Lamarche & Boulet 2021) These studies adopted various methods and assessed several factors with different theories. Nevertheless, research has been few in developing and assessing a conceptual framework with factors like WLB, job design, burnout, job satisfaction and their relationship with the employee's psychological health. The triangulation of the three will contribute to assessing the relationships between the independent and dependent variables and also among independent variables of the proposed model (See Figure 1).

Figure 1. The proposed theoretical framework

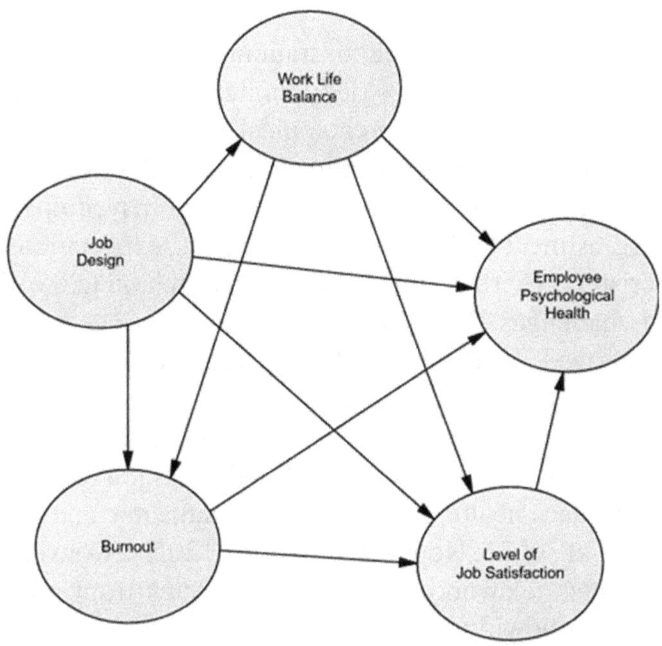

Development of the Proposed Relationships

Job Design, Psychological Health, Work-Life Balance, Burnout, and Satisfaction

Job design is the structure of jobs that employees perform in an organisation. The focus of job design is on the tasks and activities employees undertake daily (Oldham & Fried, 2016). Job specialisation and simplification contribute positively to employee's efficiency at work (Babbage, 1835; Smith, 1850). Job design leaves a deep imprint on the employees' psychological health, since the way work activities are organised is lived and experienced daily by employees (Hackman & Oldham, 1976; Pee & Lee, 2015). Job design and characteristics were linked to WLB (Jindal et al., 2013) and burnout of employees. Research shows that burnout was negatively related to the job characteristics such as autonomy and skill variety. Job satisfaction was positively associated with the job design variables such as autonomy (Mahoney et al., 2020). It can also be stipulated that job design depicts the exchange relationship between the employer and employee where employees are expected to deliver certain tasks for a pay (Organ, 1988).

In most organisations, the time for training, re-training or preparing the employees for teleworking were strictly limited to most critical job tasks to ensure business continuity. The psychological contract between the employer and employee must be re-negotiated to reflect the changes. However, such evolution takes time, and the implicit nature of psychological contract complicates adjustment to the new working norms, expectations, conditions and characteristics. There is ample existing research on teleworking and its benefits and challenges. Teleworking has been associated with decreased stress and improved WLB and wellbeing (Song & Gao, 2020). However, most were generated from a pandemic free context, where teleworking was an occasional or optional practice (Sinclair et al., 2020), and not linked to job characteristics or design. Several studies discuss job redesign in view of improving workplace health and safety in the pandemic and post pandemic era (e.g Reina et al., 2021; Hopskins & Bardoel, 2023). Conversely, Nguyen (2021) investigates teleworking, job characteristics from a transportation perspective. The following hypotheses are proposed:

H1: There is a positive relationship between job design and employee's psychological health.
H2: There is a positive relationship between job design and employees' WLB.
H3: There is a negative relationship between job design and burnout.
H4: There is a positive relationship between job design and the level of job satisfaction.

Work-Life Balance and Burnout

WLB is a situation whereby the employees can achieve a proper balance between work and personal life or any other important commitments. It is the individual's satisfaction with the compatibility of the latter's multiple roles (Soelton et al., 2020), such as paid work, family and personal responsibilities (Hill et al., 2001). Various sources of work-related stress or role overlap can cause role conflict, hence leading to burnout. In the pre-COVID work setting, though overlapping were bound to exist, there was still the physical separation of work-related tasks being conducted in the physical work premise. Hence, the personal space was almost separated. Teleworking during the lockdown leads to a sudden invasion of home/personal space. The blurred demarcation between work and personal life, the sudden change of the psychological contract, and the lack of mutual support from peers could trigger burnout. Existing literature confirmed the significant relationship between WLB factors with aspects of burnout, such as depersonalisation from work, emotional exhaustion, and reduced personal achievement (Soelton et al., 2020). Some researchers have explained how the life-to-work and work-to-life conflicts arise because of the blurred boundaries, hence causing stress (Parham & Rauf, 2020). Poor work-life balance often demonstrates the capitalist nature of doing more with less which also depicts the social and economic exchange between employers and employees (Blau, 1968). Hence, the following hypothesis is proposed:

H5: There is a negative relationship between WLB and burnout.

Burnout, Employee Psychological Health, and Job Satisfaction

Burnout is defined by Schaufeli & Greenglass (2001) as a degree that an individual experiences emotional, physical and mental fatigue when he/she has been involved for a longtime in work related situations that consumes a lot of emotions. Burnout and its symptoms have a direct impact on job satisfaction of the employees. Job satisfaction is defined '. . . a pleasurable or positive emotional state resulting from the appraisal of one's job or job experience' (Locke, 1976, pp. 1300). There are various studies statistically confirming the relationship between the various dimensions of burnout and job satisfaction (Gómez-García et al., 2020).

Burnout has often been associated with the negative impacts on employees' psychological health. Various research work reported a positive relationship between burnout and employee psychological health (e.g. Schaufeli & Enzmann, 1998; De Beer et al., 2016). Burnout results in physical, psychological and emotional damage. Various studies emerged lately attempting to relate burnout with the impact of COVID-19 on employees (e.g. Matsuo et al., 2020; Launer, 2020). It must be acknowledged burnout has always been an issue in the workplace, however, with the impact of COVID-19 pandemic, employees can experience burnout with teleworking. From a theoretical perspective, employees start experiencing burnout when they find inadequacy in the social exchange processes (Schaufeli et al., 1996) between them and their respective employers. They might be not able to relate themselves to their work and have overall issues with their job features. Besides, employees can also feel burnout due to inadequate social exchange while working remotely. Thus, any imbalance damages their psychological contract, thus resulting in burnout leading to psychological distress at work. Therefore, the below hypotheses are proposed:

H6: There is a negative relationship between burnout and employees' job satisfaction.

H7: There is a positive association between burnout and employee psychological health.

WLB and Job Satisfaction and Employee Psychological Health

According to Mathur et al. (2013), WLB emerged when employees started to experience the difficulties in balancing their work with their personal lives. Employees started neglecting their personal relationships and responsibilities because of the pressure from the workplace. Various research demonstrated a positive relationship between WLB and job satisfaction (e.g. Kaliannan et al., 2016; Kasbuntoro et al., 2020). The study of Hjálmsdóttir & Bjarnadóttir (2020) demonstrated that employees, especially women, were finding difficulties to handle both their family and work responsibilities. Wiradendi-Wolor et al. (2020) noted an interesting link between WLB and job satisfaction with remote working during the pandemic. They noted that teleworking influences positively job satisfaction. However, no significant relationship was found between the introduction or termination of remote work with WLB. Yet, permanent remote work tends to contribute towards work imbalance.

Additionally, poor WLB is acknowledged to lead to poor employee psychological health (Jones et al., 2013; Haider et al., 2018). The relationship between job satisfaction and employee psychological health has often been discussed (Ruotsalainen et al., 2020). Job satisfaction enhances employee psychological health in the workplace. Being dissatisfied with their work often leads to employees' undesirable organisational. Therefore, it becomes important for employers to ensure that their staff are satisfied when they are working from home. Therefore, the below hypotheses are proposed:

H8: There is a positive relationship between a good WLB and level of employees' job satisfaction.

H9: There is a positive relationship between a proper WLB and employee psychological health.
H10: There is a positive relationship between job satisfaction and employee psychological health.

METHODOLOGY

This study has made use of a purely positivist approach to collect and analyse the data. The positivist approach has been utilised by researchers in diverse fields, for instance in accounting (Luft & Shields, 2014) and in science (MacArthur et al., 2012) amongst others. The purpose of this research is to assess the psychological health of teleworkers in the Mauritian context, whereby the remote working is a relatively new design, triggered by the COVID-19 pandemic. Job design, burnout, WLB, job satisfaction and employee psychological health were measured based on established scales. Job design was measured by scale devised by Morgeson and Humphrey (2006) where some items were deleted, and some were modified based on the nature and context of this current research. The WLB scale was borrowed Gropel (2006) and the items were slightly modified and adapted to this study. This study made use and modified the Maslach Burnout Inventory (Maslach et al., 1986) to measure burnout of employees. Job satisfaction was measured by the scale proposed by Cammann et al., (1979) and 1 item was added to the scale. Finally, the psychological health of Employees by the General Health Questionnaire by Goldberg & Williams (1988) and the scale was slightly amended. The Likert scale was used to measure the items where "1" stands for Strongly Disagree, "2" stands for Disagree, "3" stands for Neutral, "4" stands for Agree and "5" stands for Strongly Agree.

A pilot study was conducted where five employees were asked to provide their opinions and responses with regards to the clarity and flow of the questionnaire. This was done for the purpose to ensure the content validity. After the pilot test, some amendments were made to the questionnaire and the final revised questionnaire was sent to around 300 employees using the online survey method. The online survey method was deemed to be the most appropriate tool for the data collection for broader outreach. A response rate of (N= 171) was recorded. The reliability and validity test were conducted using the guidelines set by Nunnally (1978) where a score >0.70 was considered to be appropriate. Following the reliability and validity test, this study made use of the Partial Least Square Structural Equation Modelling (PLS-SEM) through the SmartPLS 3.0 to evaluate both the measurement and structural model. Therefore, this study followed the two steps approach proposed by Anderson and Gerbing (1988), that is, a Confirmatory Factor Analysis (CFA) to evaluate the measurement model and the PLS-SEM to evaluate the structural model.

The CFA confirmed that the proposed model was a multi-dimensional one. Therefore, adding sub-dimensions to construct like job design and psychological health of employees as second-order factors have further justified the use of PLS-SEM as the sub-dimensions acted as formative constructs that allowed a better understanding of the reflective constructs. The use of second-order factor analysis has provided much in-depth analysis of this research as the researcher was able to better understand firstly the construct itself and the relationships among the constructs. The second-order factor analysis has significantly contributed to this research. While conducting the CFA, it is important to take note that factor loading <0.70 was not retained (Hair et al., 2011). The PLS-SEM was the most appropriate method to evaluate the proposed model as this research is dealing with formative indicators and a small size with regards to the respondents has been received. PLS-SEM allows better handling of data when it comes to small sample size compared to the covariance-based SEM (Hair et al., 2011). According to Ringle et al. (2009) and Hwang et al. (2010), by using PLS-SEM more reliable results are obtained and PLS-SEM provides more predictive power to the model. PLS-SEM also allows researchers to contribute towards theory development (Astrachan et al., 2014; Nitzl, 2016) as the researcher has the flexibility to find explanations for results that have not been supported based on past literature. This aspect of PLS-SEM is contributive as it leads to further theory development.

ANALYSIS AND DISCUSSION

This study followed the steps provided by Hair et al., (2017) to evaluate both the measurement and structural framework. While carrying out the CFA, factor loadings <0.70 were eliminated and the researcher ensured the Convergent validity is > 0.50; that is; the Average Variance Extracted (AVE) is > 0.50. The reliability and validity of the constructs were validated where the Cronbach Alpha and the Composite Reliability scores were >0.70 as per (Hair et al.,

2017). The second step, the researcher proceeded with the evaluation of the structural model where a PLS-SEM was conducted.

Discriminant Validity

Discriminant validity enables the researcher to identify multicollinearity issues, if any. Despite numerous benefits of using the Fornell and Larcker method to assess the discriminant validity, the study used the Heterotrait-Monotrait (HTMT) ratio. Thus, for this study a threshold not greater that 0.85 has been chosen (Kline, 2023). Authors like Gold & Malhotra (2001) even suggest a threshold of 0.90. That is, the values obtained should not exceed these figures. Figure 2 is a diagrammatical representation of the tested model.

Figure 2. The tested framework
Notes: TC- Task Characteristics; KC-Knowledge Characteristics; SC- Social Characteristics; CC-Contextual Characteristics; EE: Emotional Exhaustion; PA: Personal Accomplishment

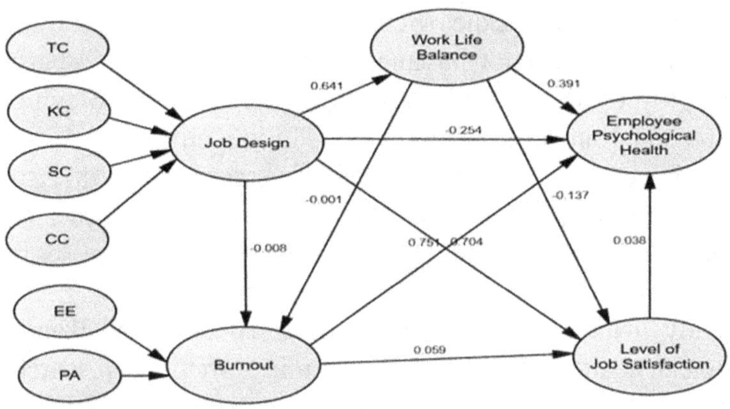

DISCUSSION

Employee Psychological Health

This chapter is based on our study on the inter-relationship between remote working and various factors such as WLB, burnout, job design, job satisfaction and employee psychological health. This study found a non-significant relationship between job design and employee's psychological health while working remotely during this pandemic lockdown (H1). The result contradicts the findings of various research, which stipulate that the way jobs are designed contribute to the psychological wellbeing of employees (Azarbouyeh & Naini 2014; Parker et al., 2019; Song & Gao, 2020). The job structure is essential employees' overall wellbeing. Hence, the tasks and job structure that were shifted to home-based teleworking since the pandemic were conducive to psychological wellbeing of the employees. Most of the jobs that were done remotely since the pandemic were jobs and tasks that could be done using the right technological structure, such as computers, virtual platforms and telematics. Jobs such as intellectual professions with a considerable amount of job autonomy are easier for teleworking (Belzunegui-Eraso & Erro-Garcés, 2020). Therefore, working remotely has not affected positively or negatively the employees' psychological health. It can be concluded working remotely was business as usual for many of them and not influencing the overall psychological health.

The relationship between burnout and employee psychological health was supported in this study (H7). The positive relationship has been empirically proven by many existing studies (Schaufeli & Enzmann, 1998; De Beer et al., 2016). Emotional exhaustion has an impact on the employee's psychological health. Employees are more likely to experience burnout. Staying at home, handling personal and professional responsibilities culminates into emotional exhaustion, hence damaging employees' psychological health. Experiencing burnout is a sign of poor job characteristics where employees are finding difficulties to adjust with the new way of working and symbolises poor social exchanges between the employees and the employers.

This study further supported the relationship between WLB and employee psychological health (H9), hence confirming previous research findings (Jones et al., 2013). Haider et al., (2018) found that psychological health and wellbeing is a mediator of WLB and job performance. Working from home provides the employees with a level of flexibility in handling their work and household responsibilities. Dawson & Golijani-Moghaddam (2020) are of opinion that working from home in the confinement period provides the employees with psychological flexibility in managing their overall wellbeing. This study further adds to this body of knowledge and stipulates that working from home in the COVID-19 pandemic provides both psychological and physical flexibility to the employees as the managers and supervisors' pressure on them are reduced health. The employers and the employees appear to be reasonable with the influence of the job characteristics (flexibility). This results in creating positive psychological contracts among the employees and hence they can experience a good psychological health with WLB.

The findings revealed a negative relationship between the level of job satisfaction and employee psychological health (H10) and rejected the proposed hypothesis of this study where it was assumed that job satisfaction has a positive influence on the employees' psychological health. The results are not in line with existing studies, which propose the spillover effect of job satisfaction on psychological well-being. Nadinloyi et al., (2013) demonstrated a positive relationship between job dissatisfaction and the global mental health index, depression and social action. Aazami et al., (2015) found a relationship between job satisfaction and psychological distress, sleep disorders, gastro-intestinal problems and headaches. However, aligned with this research's findings, there are many studies which only partially supported the influence of job satisfaction on employee's psychological health (Petrovski & Gleeson, 1997). It can be concluded that the employees' level of job satisfaction is mostly related to the economic exchange, that is, they are simply satisfied with their salary and retaining their jobs and the focus is not on the job design. Furthermore, their job satisfaction Since the COVID-19 pandemic, most countries have been experiencing economic and financial crises accompanied by massive job losses. It can be related that this constant fear of job loss somehow cultivates in damaging the psychological wellbeing of employees which justified the negative association among these two variables (Kimbu et al., 2023). There is a contradiction between their pre-COVID and post-COVID psychological contract between the employers and the employees (De Ruiter & Schalk 2023). Since remote working has never been in the agenda and work contract of any of the employees, this sudden shift has shaken the existing psychological contract.

Inter-Relationships Among the Variables

The findings revealed a positive relationship between job design and WLB and between job design and job satisfaction (H2, H4). The results confirmed the studies of Jindal, et al. (2013) and Mahoney et al. (2020). The findings are also in line with the studies of Azarbouyeh & Naini (2014) and Song & Gao (2020) where it was demonstrated that remote working tends to decrease stress and enhance the employees' overall personal and work balance and welfare. Therefore, it can be concluded that despite the remote working in the confinement period where employees had to juggle with handling their work solely through technology and managing both their personal and professional lives, they were overall satisfied and having WLB. The level of job satisfaction and proper WLB of employees demonstrates the proper social exchanges from the employers through the job design, that is, proper job characteristics. The task, knowledge, social and contextual characteristics aided to job satisfaction and proper WLB. It can be assumed that a degree of flexibility was present when they were working from home which contributed to positive psychological contracts of the employees.

On the contrary, the relationship proposed between job design and burnout has been rejected, that is, there is no significant relationship between the two variables (H3). Job design does not contribute in reducing the level of burnout of the employees. It can be assumed that burnout is an ongoing emotion that an employee feels whether when working physically in the organisation or remotely. Due to work pressure and the evolving remote working pattern since the pandemic, job design does not mitigate the level of burnout of the employees. The proposed negative relationship between WLB and burnout and between burnout and job satisfaction have been rejected (H5) and (H6) which contradicts the studies of Mahendran et al. (2019) and Soelton et al. (2020). Burnout is being depicted as a phenomenon which despite the employees might be having a proper job design or WLB, these factors do not significantly contribute to decreasing the level of employee burnout. Burnout has been measured with first order factors like emotional exhaustion and personal accomplishment. Hence, burnout is not being influenced by other independent variables. The relationship between WLB and job satisfaction has also been rejected (H8), that is, having a proper WLB does not significantly contribute to increasing the level of employees' job satisfaction. It can be assumed that WLB is being viewed as an external component of the overall life of the employees and not significantly influencing the other variables apart from the employees' psychological health.

PRACTICAL AND THEORETICAL IMPLICATIONS

The new normal might not be short-lived, but rather well integrated new routine and work set-up. Firms in countries where remote working is not the norm and governed by the law, might consider reviewing their existing policies to cope with the lasting impact of COVID-19 on organisational life and psychological health of employees, precisely the teleworkers. The

Human Resources professionals and departmental managers would need to redesign jobs to suit teleworking practices, while at the same time ensure that most of the tasks and responsibilities are still relevant and meeting the jobs' objectives. To ensure that teleworking does not lead to psychological stress and burnout, employers should provide employee assistance programs such as counselling, coupled by regular e-interaction with colleagues. Most governments have encouraged teleworking, but such employment mode was rarely practiced, or the legislative framework has not evolved to facilitate its operation. Authorities should amend their labour laws to integrate remote working practices and incentivize firms to adopt remote working practices. Hence, employers should implement a policy for remote work. Teleworking should not be reactive to such drastic situations but embedded in the newly designed jobs to ensure a smooth adjustment.

From a theoretical perspective, the three theories have contributed to understanding the complexities of the framework. The Job Characteristics Model was found to be leading the two other theories: SET and PCT. The JCM was the foundation of this study which helps in determining whether there are proper social exchanges between the employers and the employees which helps in culminating positive or negative psychological contracts. JCM can be said to be a fundamental element that employers need to take into consideration if they want their employees to be happy and committed in the workplace. JCM fosters both the social and economic exchanges between the employers and employees which result in creating positive psychological contracts among the employees. Better employee performance and psychological well-being can be expected with positive psychological contracts. The integration of the three theories provided better explanatory powers and allowed more interpretation and understanding of the framework and led to more theory development. The researchers acknowledged that without the triangulation of the theories, it would have been very difficult to understand and interpret the results from the analysis. Using more than one theory has proved to be beneficial as more understanding and exploitation of both the model and the theories have been made.

CONCLUSION AND FUTURE RESEARCH DIRECTION

The research investigates the impact of job design, job satisfaction, WLB, and burnout on employee psychological health. Other variables such as psychosocial safety climate could also be considered by future research. In addition, the sample population for the survey was limited to online survey, sourced via social media, and mailing list, due to the lockdown. Future research can consider having more respondents to have a more general overview of the subject matter. The analysis could focus more on the demographics such as family status, teleworking and employee psychological health: for instance, the presence of young children, dependent requiring special care and gender. This study made use of purely the positivist approach, quantitative research. Future studies can consider using the mixed method, qualitative aspect to capture more insights with respect to employees' psychological health. Further research could investigate the psychosocial aspects of teleworking and employees' psychological health. Home based teleworking could have significant influence on psychological health depending on other moderators such as home environment, family status, support from the organisation.

REFERENCES

Aazami, S., Shamsuddin, K., Akmal, S., & Azami, G. (2015). The relationship between job satisfaction and psychological/physical health among Malaysian working women. *The Malaysian Journal of Medical Sciences : MJMS*, *22*(4), 40. PMID:26715907

Ali, S. A. M., Said, N. A., Abd Kader, S. F., Ab Latif, D. S., & Munap, R. (2014). Hackman and Oldham's job characteristics model to job satisfaction. *Procedia: Social and Behavioral Sciences*, *129*, 46–52. doi:10.1016/j.sbspro.2014.03.646

Anderson, J. C., & Gerbing, D. W. (1988). Structural equation modeling in practice: A review and recommended two-step approach. *Psychological Bulletin*, *103*(3), 411–423. doi:10.1037/0033-2909.103.3.411

Argyris, C. (1960). *Understanding organizational behavior*. Academic Press.

Astrachan, C. B., Patel, V. K., & Wanzenried, G. (2014). *Journal of Family Business Strategy*.

Azarbouyeh, A., & Naini, S. (2014). A study on the effect of teleworking on quality of work life. *Management Science Letters*, *4*(6), 1063–1068. doi:10.5267/j.msl.2014.5.027

Babbage, C. (1832). *On the economy of machinery and manufactures.* Academic Press.

Beauregard, T. A., Basile, K. A., & Canónico, E. (2019). *Telework: Outcomes and facilitators for employees.* Academic Press.

Beebeejaun, A., & Gunputh, R. P. (2022). A Critical Analysis of the Mauritius Workers' Rights (Working from Home) Regulations 2020 in the Wake of COVID-19. *Industrial Law Journal, 51*(1), 174–193. doi:10.1093/indlaw/dwab036

Belzunegui-Eraso, A., & Erro-Garcés, A. (2020). Teleworking in the Context of the Covid-19 Crisis. *Sustainability (Basel), 12*(9), 3662. doi:10.3390/su12093662

Betchoo, N. K. (2020). Evaluating the Prospect of Teleworking in Mauritian Society in the Light of COVID-19 Pandemic. *Middle East Journal of Applied Science & Technology, 3*(4), 69–80.

Braesemann, F., Stephany, F., Teutloff, O., Kässi, O., Graham, M., & Lehdonvirta, V. (2022). The global polarisation of remote work. *PLoS One, 17*(10), e0274630. doi:10.1371/journal.pone.0274630 PMID:36264859

Cammann, C., Fichman, M., Jenkins, D., & Klesh, J. (1979). *The Michigan organizational assessment questionnaire.* Unpublished manuscript, University of Michigan.

Carnevale, J. B., & Hatak, I. (2020). Employee adjustment and well-being in the era of COVID-19: Implications for human resource management. *Journal of Business Research, 116*, 183–187. doi:10.1016/j.jbusres.2020.05.037 PMID:32501303

Chan, S. (2021). The interplay between relational and transactional psychological contracts and burnout and engagement. *Asia Pacific Management Review, 26*(1), 30–38. doi:10.1016/j.apmrv.2020.06.004

Dawson, D. L., & Golijani-Moghaddam, N. (2020). COVID-19: Psychological flexibility, coping, mental health, and wellbeing in the UK during the pandemic. *Journal of Contextual Behavioral Science, 17*, 126–134. doi:10.1016/j.jcbs.2020.07.010 PMID:32834970

De Beer, L. T., Pienaar, J., & Rothmann, S. Jr. (2016). Work overload, burnout, and psychological ill-health symptoms: A three-wave mediation model of the employee health impairment process. *Anxiety, Stress, and Coping*, *29*(4), 387–399. doi:10.1080/10615806.2015.1061123 PMID:26079200

De Ruiter, M., & Schalk, R. (2023). The Employment Relationship Amidst and Beyond the COVID-19 Pandemic: The Role of (Responsible) Inclusive Leadership in Managing Psychological Contracts. In *Virtual Management and the New Normal: New Perspectives on HRM and Leadership since the COVID-19 Pandemic* (pp. 121–139). Springer International Publishing. doi:10.1007/978-3-031-06813-3_7

Emerson, R. M. (1972). Exchange theory, part II: Exchange relations and networks. *Sociological Theories in Progress, 2*, 58-87.

EvenA. (2020). The evolution of work: Best practices for avoiding social and organizational isolation in telework employees. *Available at* SSRN 3543122. doi:10.2139/ssrn.3543122

Farrell, K. (2017). *Working from home: A double-edged sword.* Academic Press.

Fried, Y., & Ferris, G. R. (1987). The validity of the job characteristics model: A review and meta-analysis. *Personnel Psychology*, *40*(2), 287–322. doi:10.1111/j.1744-6570.1987.tb00605.x

Gold, A. H., Malhotra, A., & Segars, A. H. (2001). Knowledge management: An organizational capabilities perspective. *Journal of Management Information Systems*, *18*(1), 185–214. doi:10.1080/07421222.2001.11045669

Goldberg, D. P., & Williams, P. (1988). *A user's guide to the General Health Questionnaire*. Academic Press.

Gómez-García, R., Bayón-Calvo, S., & Lucas-García, J. (2021). The relationship between burnout and job satisfaction in a sample of Spanish social workers. *British Journal of Social Work*, *51*(8), 3115–3134. doi:10.1093/bjsw/bcaa150

Gropel, P. (2006). Balancing work and personal life: Relation to needs and life satisfaction. *Československá Psychologie*, *1*, 71–83.

Hackman, J. R., & Oldham, G. R. (1975). Development of the job diagnostic survey. *The Journal of Applied Psychology*, *60*(2), 159–170. doi:10.1037/h0076546

Hackman, J. R., & Oldham, G. R. (1976). Motivation through the design of work: Test of a theory. *Organizational Behavior and Human Performance*, *16*(2), 250–279. doi:10.1016/0030-5073(76)90016-7

Haider, S., Jabeen, S., & Ahmad, J. (2018). Moderated mediation between work life balance and employee job performance: The role of psychological wellbeing and satisfaction with coworkers. *Journal of Work and Organizational Psychology*, *34*(1), 29–37. doi:10.5093/jwop2018a4

Hair, J. F. Jr, Matthews, L. M., Matthews, R. L., & Sarstedt, M. (2017). PLS-SEM or CB-SEM: Updated guidelines on which method to use. *International Journal of Multivariate Data Analysis*, *1*(2), 107–123. doi:10.1504/IJMDA.2017.087624

Hair, J. F., Ringle, C. M., & Sarstedt, M. (2011). PLS-SEM: Indeed a silver bullet. *Journal of Marketing Theory and Practice*, *19*(2), 139–152. doi:10.2753/MTP1069-6679190202

Hajal, G. E. (2022). Teleworking and the jobs of tomorrow. *Research in Hospitality Management*, *12*(1), 21–27. doi:10.1080/22243534.2022.2080953

Haughton, A., & Ivey, W. (2023). A Growth Comparison of Mauritius and Jamaica. In *Overcoming Productivity Challenges in Small Countries: Lessons from Jamaica* (pp. 229–258). Springer Nature Switzerland. doi:10.1007/978-3-031-23301-2_9

Hill, E. J., Hawkins, A. J., Ferris, M., & Weitzman, M. (2001). Finding an extra day a week: The positive influence of perceived job flexibility on work and family life balance. *Family Relations*, *50*(1), 49–58. doi:10.1111/j.1741-3729.2001.00049.x

Hill, E. J., McClanahan, K. J., & Driggs, T. (2023). Telework. In *Encyclopedia of quality of life and well-being research* (pp. 1–3). Springer International Publishing. doi:10.1007/978-3-031-17299-1_2990

Hjálmsdóttir, A., & Bjarnadóttir, V. S. (2021). "I have turned into a foreman here at home": Families and work–life balance in times of COVID-19 in a gender equality paradise. *Gender, Work and Organization*, *28*(1), 268–283. doi:10.1111/gwao.12552 PMID:33041540

Hopkins, J., & Bardoel, A. (2023). The future is hybrid: How organisations are designing and supporting sustainable hybrid work models in post-pandemic Australia. *Sustainability (Basel)*, *15*(4), 3086. doi:10.3390/su15043086

Hwang, H., Malhotra, N. K., Kim, Y., Tomiuk, M. A., & Hong, S. (2010). A comparative study on parameter recovery of three approaches to structural equation modeling. *JMR, Journal of Marketing Research*, *47*(4), 699–712. doi:10.1509/jmkr.47.4.699

Jayaweera, T., Bal, M., Chudzikowski, K., & De Jong, S. (2021). The impact of economic factors on the relationships between psychological contract breach and work outcomes: A meta-analysis. *Employee Relations*, *43*(3), 667–686. doi:10.1108/ER-03-2020-0095

Jindal, A., Agarwal, S., Garg, P., & Rastogi, R. (2013). Role of job design for achieving work life balance. In *Conference on Excellence in Research and Education* (pp. 203-217).

Jones, F., Burke, R. J., & Westman, M. (Eds.). (2013). *Work-life balance: A psychological perspective*. Psychology Press. doi:10.4324/9780203536810

Kaliannan, M., Perumal, K., & Dorasamy, M. (2016). Developing a work-life balance model towards improving job satisfaction among medical doctors across different generations. *Journal of Developing Areas*, *50*(5), 343–351. doi:10.1353/jda.2016.0035

Kasbuntoro, D. I., Maemunah, S., Mahfud, I., Fahlevi, M., & Parashakti, R. D. (2020). Work-life balance and job satisfaction: A case study of employees on banking companies in Jakarta. *International Journal of Control and Automation*, *13*(4), 439–451.

Kimbu, A. N., Adam, I., Dayour, F., & de Jong, A. (2023). COVID-19-induced redundancy and socio-psychological well-being of tourism employees: Implications for organizational recovery in a resource-scarce context. *Journal of Travel Research*, *62*(1), 55–74. doi:10.1177/00472875211054571 PMID:36605829

Kline, R. B. (2023). *Principles and practice of structural equation modeling*. Guilford publications.

Launer, J. (2020). Burnout in the age of COVID-19. *Postgraduate Medical Journal*, *96*(1136), 367–368. doi:10.1136/postgradmedj-2020-137980 PMID:32457131

Locke, E. A. (1976). The nature and causes of job satisfaction. Handbook of Industrial and Organizational Psychology.

Luft, J., & Shields, M. D. (2014). Subjectivity in developing and validating causal explanations in positivist accounting research. *Accounting, Organizations and Society*, *39*(7), 550–558. doi:10.1016/j.aos.2013.09.001

MacArthur, D. G., Balasubramanian, S., Frankish, A., Huang, N., Morris, J., Walter, K., Jostins, L., Habegger, L., Pickrell, J. K., Montgomery, S. B., Albers, C. A., Zhang, Z. D., Conrad, D. F., Lunter, G., Zheng, H., Ayub, Q., DePristo, M. A., Banks, E., Hu, M., ... Tyler-Smith, C. (2012). A systematic survey of loss-of-function variants in human protein-coding genes. *Science*, *335*(6070), 823–828. doi:10.1126/science.1215040 PMID:22344438

Mahendran, A. V., Panatik, S. A., Rajab, A., & Nordin, N. (2019). The Influence of Work-life Balance on Burnout among Nurses. *Indian Journal of Public Health Research & Development*, *10*(11), 3338. doi:10.5958/0976-5506.2019.04097.X

Mahoney, C. B., Lea, J., Schumann, P. L., & Jillson, I. A. (2020). Turnover, burnout, and job satisfaction of certified registered nurse anesthetists in the United States: Role of job characteristics and personality. *AANA Journal*, *88*(1), 39–48. PMID:32008617

Maslach, C., Jackson, S. E., & Leiter, M. P. (1997). *Maslach burnout inventory*. Scarecrow Education.

Mathur, A., Purohit, A., & Mathur, R. (2013). Camparative Analysis Of Overall Work Life Balance Of Medical Professionals. *International Journal of Management Research and Development*, *3*(2).

Matsuo, T., Kobayashi, D., Taki, F., Sakamoto, F., Uehara, Y., Mori, N., & Fukui, T. (2020). Prevalence of health care worker burnout during the coronavirus disease 2019 (COVID-19) pandemic in Japan. *JAMA Network Open*, *3*(8), e2017271–e2017271. doi:10.1001/jamanetworkopen.2020.17271 PMID:32749466

Morgeson, F. P., & Humphrey, S. E. (2006). The Work Design Questionnaire (WDQ): Developing and validating a comprehensive measure for assessing job design and the nature of work. *The Journal of Applied Psychology*, *91*(6), 1321–1339. doi:10.1037/0021-9010.91.6.1321 PMID:17100487

Nadinloyi, K. B., Sadeghi, H., & Hajloo, N. (2013). Relationship between job satisfaction and employees mental health. *Procedia: Social and Behavioral Sciences*, *84*, 293–297. doi:10.1016/j.sbspro.2013.06.554

Nemțeanu, M. S., & Dabija, D. C. (2023). Negative impact of telework, job insecurity, and work–life conflict on employee behaviour. *International Journal of Environmental Research and Public Health*, *20*(5), 4182. doi:10.3390/ijerph20054182 PMID:36901192

Nguyen, M. H. (2021). Factors influencing home-based telework in Hanoi (Vietnam) during and after the COVID-19 era. *Transportation*, *48*(6), 3207–3238. doi:10.1007/s11116-021-10169-5 PMID:33518829

Nitzl, C. (2016). The use of partial least squares structural equation modelling (PLS-SEM) in management accounting research: Directions for future theory development. *Journal of Accounting Literature*, *37*(1), 19–35. doi:10.1016/j.acclit.2016.09.003

Njaramba, L. (2023). *New Ways of Working and Burnout in the Post-Pandemic Era: A Case of National Health Insurance Fund (NHIF), Kenya* [Master's thesis].

Oldham, G. R., & Fried, Y. (2016). Job design research and theory: Past, present and future. *Organizational Behavior and Human Decision Processes*, *136*, 20–35. doi:10.1016/j.obhdp.2016.05.002

Organ, D. W. (1988). *Organizational citizenship behavior: The good soldier syndrome*. Lexington Books/DC Heath and Com.

Parham, S., & Rauf, M. A. (2020). COVID-19 and obligatory remote working in HEIs: An exploratory study of faculties' work-life balance, well-being and productivity during the pandemic. *International Journal of Economics. Commerce and Management*, *8*(12), 384–400.

Parker, S. K., Andrei, D. M., & Van den Broeck, A. (2019). Poor work design begets poor work design: Capacity and willingness antecedents of individual work design behavior. *The Journal of Applied Psychology*, *104*(7), 907–928. doi:10.1037/apl0000383 PMID:30640488

Parker, S. K., Andrei, D. M., & Van den Broeck, A. (2019). Poor work design begets poor work design: Capacity and willingness antecedents of individual work design behavior. *The Journal of Applied Psychology*, *104*(7), 907–928. doi:10.1037/apl0000383 PMID:30640488

Pee, L. G., & Lee, J. (2015). Intrinsically motivating employees' online knowledge sharing: Understanding the effects of job design. *International Journal of Information Management*, *35*(6), 679–690. doi:10.1016/j.ijinfomgt.2015.08.002

Petrovski, P., & Gleeson, G. (1997). The relationship between job satisfaction and psychological health in people with an intellectual disability in competitive employment. *Journal of Intellectual & Developmental Disability*, *22*(3), 199–211. doi:10.1080/13668259700033411

Reina, R., Ventura, M., Cristofaro, C. L., & Vesperi, W. (2021). Digitalize work in pandemic time: Practices for remote working and job redesign in public organization. *HR Analytics and Digital HR Practices: Digitalization post COVID-19*, 73–99.

Ringle, C. M., Wende, S., & Will, A. (2009). The finite mixture partial least squares approach: Methodology and application. Handbook of partial least squares: Concepts, methods, and applications. Berlin: Springer.

Rousseau, D. M. (1989). Psychological and implied contracts in organizations. *Employee Responsibilities and Rights Journal*, *2*(2), 121–139. doi:10.1007/BF01384942

Ruotsalainen, S., Jantunen, S., & Sinervo, T. (2020). Which factors are related to Finnish home care workers' job satisfaction, stress, psychological distress and perceived quality of care?-a mixed method study. *BMC Health Services Research*, *20*(1), 1–13. doi:10.1186/s12913-020-05733-1 PMID:32988396

Sannassee, R. V., Seetanah, B., & Lamport, M. J. (2014). Export diversification and economic growth: the case of Mauritius. In Connecting to Global Markets (pp. 11-23). WTO iLibrary.

Schaufeli, W., & Enzmann, D. (2020). *The burnout companion to study and practice: A critical analysis*. CRC Press. doi:10.1201/9781003062745

Schaufeli, W. B., Dierendonck, D. V., & Gorp, K. V. (1996). Burnout and reciprocity: Towards a dual-level social exchange model. *Work and Stress*, *10*(3), 225–237. doi:10.1080/02678379608256802

Smith, A. (1976). *An inquiry into the nature and causes of the wealth of nations* (R. H. Campbell, A. S. Skinner, & W. B. Todd, Eds.).

Soelton, M., Hardianti, D., Kuncoro, S., & Jumadi, J. (2020, February). Factors affecting burnout in manufacturing industries. In *4th International Conference on Management, Economics and Business (ICMEB 2019)* (pp. 46-52). Atlantis Press.

Song, Y., & Gao, J. (2020). Does telework stress employees out? A study on working at home and subjective well-being for wage/salary workers. *Journal of Happiness Studies*, *21*(7), 2649–2668. doi:10.1007/s10902-019-00196-6

Sui, M., Yu, Z., & Zhou, M. (2023). The Impact of Psychological Contract, Physical and Mental Health on Burnout in Grassroots Civil Servants: Evidence from China. *Psychology Research and Behavior Management*, *16*, 3461–3476. doi:10.2147/PRBM.S423011 PMID:37664138

Thompson, R. J., Payne, S. C., & Taylor, A. B. (2015). Applicant attraction to flexible work arrangements: Separating the influence of flextime and flexplace. *Journal of Occupational and Organizational Psychology*, *88*(4), 726–749. doi:10.1111/joop.12095

Timsal, A., & Awais, M. (2016). Flexibility or ethical dilemma: An overview of the work from home policies in modern organizations around the world. *Human Resource Management International Digest*, *24*(7), 12–15. doi:10.1108/HRMID-03-2016-0027

Tomasi, D. (2020). Coronavirus disease (COVID-19). A socioepidemiological analysis. Bennington, VT: Vermont Academy of Arts and Sciences LV.

Tuzovic, S., & Kabadayi, S. (2021). The influence of social distancing on employee well-being: A conceptual framework and research agenda. *Journal of Service Management*, *32*(2), 145–160. doi:10.1108/JOSM-05-2020-0140

Wiradendi Wolor, C., Solikhah, S., Fidhyallah, N. F., & Lestari, D. P. (2020). Effectiveness of e-training, e-leadership, and work life balance on employee performance during COVID-19. *Journal of Asian Finance, Economics and Business, 7*(10).

Chapter 3
Remote Work and Job Satisfaction:
Navigating the Complex Landscape of Modern Employment

Likitha Venkatesh
PES University, India

Zidan Kachhi
https://orcid.org/0000-0002-8317-6356
PES University, India

ABSTRACT

This chapter delves into the impact of technology and COVID-19 on the modern workplace, with a focus on remote work. It analyzes various factors that affect job satisfaction, such as relationships, environment, structure, and growth opportunities. The chapter explains remote work, highlighting its benefits and challenges, and provides strategies to enhance job satisfaction in remote work settings. Furthermore, it discusses the future of work and the shift towards decentralized structures. Ultimately, the chapter reveals the complex relationship between remote work and job satisfaction, shedding light on how they affect each other.

DOI: 10.4018/979-8-3693-1314-5.ch003

INTRODUCTION

The world of work has undergone a significant transformation in the 21st century as a result of technological advancements, changing social norms, and increasing global interconnectedness. One of the most notable developments in this evolution is the rise of remote work. The COVID-19 pandemic has further accelerated the adoption of flexible work arrangements, such as remote working and teleworking, which has sparked a crucial conversation about its impact on job satisfaction.

The term "remote working," refers to work arrangements in which people carry out their duties and activities outside of the typical office setting, such as from home.

According to E.A. Locke, job satisfaction is defined as "the pleasurable emotional state resulting from the appraisal of one's job as achieving or facilitating the achievement of one's job values." Job satisfaction simply means an individual's overall perception of their job. It comprises various factors like interpersonal relationships at work, work environment, organization structure, and growth opportunities.

In recent years, telecommuting has become increasingly common in today's job market. As organizations around the world adopt flexible working arrangements, it has become important to examine the impact of telecommuting on job satisfaction. The research problem this chapter seeks to address is to identify the various factors affecting job satisfaction in a telecommuting system and to navigate the challenges of modern employment. We reviewed the existing literature to address the research problem and provide valuable insights. One key area the research will study is the role of technology in facilitating or hindering remote work experiences. This includes communication tools, project management systems, and virtual workflows that affect employee satisfaction. In addition, the study will explore psychological and social factors peripheral to remote work, such as feelings of isolation, the lack of boundaries between work and personal life, and the importance of social relationships to status in a virtual reality.

Further, the research will examine the impact of organizational policies, leadership styles, and support systems on remote job satisfaction. The aim is to identify best practices that foster a positive work experience for remote workers. With the ongoing global shift towards remote work, the research problem seeks to address the nuanced challenges and opportunities associated with this paradigm shift. It aims to provide valuable insights for organizations, policymakers, and individuals navigating the evolving landscape of modern employment.

By understanding the intricacies of remote job satisfaction, this research aims to contribute to the development of effective strategies that enhance employee well-being, productivity, and overall satisfaction in the dynamic realm of remote work. It is a crucial step towards creating a better work environment for remote workers and improving the future of work.

THE INTRICATE LANDSCAPE OF MODERN EMPLOYMENT

The landscape of employment practices has gone through an incredible evolution throughout history, influenced by societal, technological, and economic changes. From the beginning of organized labour to the current era of remote work and gig economies, the way we work has continually adjusted to the prevailing conditions of each time. This evolution demonstrates the dynamic nature of human endeavours and the constant pursuit of improved efficiency and flexibility in the world of work.

The start of new job traditions sprang from the Industrial Revolution. This big change switched farming societies to ones focused on industry and factories. This switch is what started the boss-worker relationship. Later on, during the 1800s and 1900s, groups fighting for workers' rights made it possible to have rules about jobs and pay. After World War II, people began to want jobs that they could keep their whole lives. Employers gave benefits like health coverage and plans for when workers retired. This time placed importance on workers being loyal and feeling secure. Those major changes lead the way for how we work today. It showed the changes in society and economy, from focusing on factory jobs to putting more value on workers and how long they stay with a company.

In the second half of the 20th century, technology changed how people work. Things like computers, automated processes, and the internet made a huge difference in work. They didn't just make operations more efficient but also promoted creative work setups. This progress influenced an increase in remote working, which altered the old-school view of an office space. Thanks to tech, individuals can now contribute their skills to many tasks from different places. This shift ushers in a fresh wave of flexible work arrangements.

In the early 2000s, job practices made a swift turn. A key shift was the rise of the gig economy, marked by short-term, self-employed, or contract-based roles. Now, companies can tap into expert skills when they require, and employees have chances for diverse tasks thanks to a spirited and adaptable work model. The gig economy also aligns with an evolution in labor tendencies, with individuals seeking autonomy and a range of work adventures.

The Covid-19 outbreak influenced the work-from-home culture a lot. It sped up trends that were already surfacing and made businesses reassess their normal work methods. Working remotely, first, was just a quick fix, but the pandemic made it a lasting and common practice. This switch to off-site work allowed businesses to be more bendable and resilient. Communication tools like video calling, task management software, and online talk platforms emerged as vital for keeping up productivity and connectivity in remote work scenes.

In the last hundred years or so, work routines have dramatically shifted. Technology advancements, worldwide trade, and the popular freelancing trend are leading the charge towards more adaptable work culture. The idea of what work is, is being altered even now. The old rules of what a job means are being questioned, revealing thrilling opportunities and hurdles in the ever-changing work climate.

UNDERSTANDING THE CONCEPT OF REMOTE WORK

Information technology has enabled alternative ways of organizing work that challenges traditional management paradigms. Evolved organizational structures now offer a range of options to meet the growing demand for greater work flexibility. Telework is rapidly expanding as the right tools emerge, allowing for remote work from individual homes or satellite offices to reduce commuting (Boselie, 2010).

Before COVID-19, most workers had little experience with remote work, and their organizations were unprepared to support it. However, the unprecedented outbreak of the COVID-19 pandemic in 2020 has forced millions of people around the world to work remotely, leading to a de facto global experiment of remote working (Kniffin et al., 2020).

The COVID-19 pandemic has accelerated the transition away from traditional office work. Even companies that do not intend to continue full-time remote work policies beyond the pandemic are unlikely to return entirely to their pre-COVID work arrangements (Lund et al., 2020). Instead, they are likely to adopt a hybrid work model, which includes occasional telecommuting or a mixed-mode model.

In a completely virtual work setup, workers carry out their tasks completely from non-office locations. There's no need for them to be physically in a main office. Teamwork and talking mostly happen via online platforms and digital tools. The company's setup must be built to back up working remotely, with solid rules, tech, and communication methods in place. This way of working remotely was mostly in practice during the health crisis.

Hybrid work blends at-home and in-office tasks. Workers split their time between house and workspace. Different parts of a week are spent remotely or on-site. Plans might be different. Some may choose set days at the office, while others may pick total home-based work. This method aims for flexibility but keeps some face-to-face teamwork. Hybrid work needs equal weight on remote policies and an office layout encouraging teamwork for office days. After the pandemic, most groups used this remote work style.

Occasional telecommuting refers to a flexible work arrangement where employees have the option to work remotely on an irregular or occasional basis. It is not a full-time remote arrangement but allows for periodic work from a location other than the office. Employees may choose to telecommute on specific days or when personal circumstances warrant it. Occasional telecommuting provides employees with flexibility while maintaining a connection to the traditional office environment. In the mixed-mode model, firms are comprised of a mixture of full-time remote employees and full-time office employees. The mixed-mode model of remote work is a flexible and adaptive approach that combines various forms of remote work to create a customized and dynamic working environment. In this model, organizations allow employees to choose and switch between different modes of work, such as fully remote, hybrid, or in-office, based on their job responsibilities, personal preferences, and the needs of the team. The mixed mode model recognizes that a one-size-fits-all approach to remote work may not be suitable for all employees or all tasks within an organization. The mixed mode model aims to strike a balance between the benefits of remote work (flexibility, autonomy) and the advantages of in-person collaboration (team building, spontaneous interactions). This balance may vary for different teams and projects.

EXPLORING JOB SATISFACTION IN THE CONTEXT OF REMOTE WORK

Job satisfaction is often associated with job performance and productivity. According to Locke E.A. (1969), job satisfaction is the pleasant emotional state resulting from the appraisal of one's job as achieving or facilitating the achievement of one's job values. It refers to the value responses or fulfillment an individual experiences when engaging in a task compared to their perceived standard of measure. In human resource management (HRM), job satisfaction is a crucial component of one's motivation to perform (Culibrk et al., 2018). Performance is defined as the total expected value to the organization of the discrete behavioral episodes that an individual carries out over a standard period. It pertains to results that influence one's behavior over time, either contributing to or deterring from expected organizational goals. When employees feel a sense of achievement related to work tasks (i.e., job satisfaction), their organizational behaviors (or performance) tend to be positive (Siengthai S, Pila-Ngarm, 2016).

The importance of on-the-job interactions on employee job satisfaction has been recognized by organizational theorists. Early scholars such as Mayo (1949) and Maslow (1954) suggested that workplace connections and social interactions could lead to employee motivation and fulfillment of needs. Later, Sims, Szilagyi, & Keller (1976) noted that jobs that offer opportunities for feedback, friendship, and dealing with others could improve employee satisfaction. Some scholars believe that face-to-face communication is the most effective way to interact meaningfully as it allows for the exchange of multiple and rich cues (Short, Williams, & Christie, 1976). It is widely acknowledged that face-to-face interaction is fundamental in western interpersonal and relational scholarship, and this assumption extends to organizational scholarship as well.

Employees who interact with each other regularly have unrestricted channels of communication, which enhances psychological closeness and social presence (Short et al., 1976). Working in a co-located environment fosters mutuality, connection, and common ground, improving communication quality and increasing levels of information exchange. However, employees with limited presence in the office and increased reliance on technology may experience lower levels of appropriate communication, accurate information sharing, discussion quality, and communication richness (Lowry et al., 2006). Teleworkers who feel isolated from the office network may experience lower job satisfaction (Cooper & Kurland, 2002), further emphasizing the importance of social interaction in the workplace.

Organizations rely on their employees as an internal resource to remain competitive in today's knowledge-driven economy. The concept of job satisfaction has been studied for over a century, as it is believed that satisfied employees perform better and remain longer with the organization (Izvercian et al., 2016). Even today, it remains an essential indicator of employee experience, particularly in remote work environments (Pinsonneault & Boisvert, 2001).

Despite popular belief, early studies failed to provide strong evidence that teleworkers are more satisfied with their jobs (Bailey & Kurland, 2002; Morganson et al., 2010). To address the inconsistencies in previous research, Golden and Veiga (2005) investigated the impact of telework intensity on job satisfaction among professional-level employees. Their findings indicated a curvilinear relationship between telework intensity and job satisfaction, with satisfaction plateauing at higher levels of telework. Specifically, teleworkers who work more than 15.1 hours per week tend to have lower job satisfaction, while those who work less are more likely to experience higher job satisfaction. This may be due to the fact that remote work reduces interpersonal interactions and weakens connections with colleagues, leading to feelings of social isolation (Allen and Golden, 2015). This is important because job satisfaction is closely linked to the need for relatedness, which is necessary for a positive work experience. Considering that telework has become the new normal during the pandemic, with most employees and supervisors working remotely for more than 15.1 hours a week (Fonner and Roloff, 2010), it is essential that we find ways to address issues related to social isolation and relatedness to improve job satisfaction among remote workers.

In recent years, teleworking has emerged as a popular topic of research. Early studies on the subject were inconclusive, but more recent research has confirmed that telework has a positive impact on job satisfaction. For example, a meta-analysis of 46 studies, involving more than 12,000 employees by Gajendran and Harrison (2007), as well as research conducted by Bae and Kim (2016), have found a strong positive association between telework and job satisfaction.

Schall suggests that there is a linear, positive relationship between remote work and job satisfaction, stating that employees who work remotely are generally more satisfied. A study highlighted the benefits of remote work, including better work-life balance, increased flexibility, autonomy, reduced commuting time, and increased productivity (Staniulienė & Leonavičiūtė, 2022). Furthermore, Allen et al. have pointed out the advantages of remote work, such as higher job satisfaction, organizational commitment, reduced stress, better job performance, pay, and career potential.

As happens with the relationship between remote work and job satisfaction, the relation with job engagement is ambiguous at best. The positive effect found between one and the other is rather small (Martin, MacDonnell, 2012). Moreover, in a study by Delanoeije and Verbruggen in 2020, they discovered that allowing employees to work remotely didn't affect their overall engagement levels compared to when they were working in the office. However, one downside they found was that remote work tends to reduce employee job engagement. They further noted that the demands and resources of a job can influence this relationship.

Back in 1991, Andrew conducted a study comparing levels of job satisfaction and productivity among telecommuters and in-house employees. Participants included 34 in-house staff and 34 teleworkers responsible for data entry and coding. Job satisfaction was measured using the Minnesota Job Satisfaction Questionnaire and the five-item Work Plan subscale. Input per hour was used as a measure of team performance. Results indicated that there were no significant differences in overall job satisfaction and work organization scores between the two groups. However, the study identified seven individual factors related to working from home that distinguished the two groups. In addition, 12 of the 25 items of the questionnaire had significantly different importance between telecommuters and in-house employees. It should be noted that the experimental design had a few limitations, but the findings still shed light on possible differences between teleworkers and in-house workers.

According to a study conducted by Smith, Patmos and Pitts (2018) personality traits can offer insights, into practices particularly for individuals who work remotely. The researchers examined how the Big Five personality traits (openness, introversion, detachment, agreeableness and depression) relate to satisfaction with communication styles (such as email, instant messaging, phone calls and video conferencing) and overall satisfaction with work. The findings revealed that while neuroticism did not have an impact on job satisfaction there was a correlation between communication channel satisfaction (including email, video technology, instant messaging and telecommunications) and job satisfaction. This suggests that telecommuters who are content, with their chosen communication methods tend to experience levels of job satisfaction.

Telework has often been seen as an opportunity for employees to have more autonomy. When working remotely, they have the freedom to choose how, when, and where to work. For instance, they can start their workday earlier or later, or even work from a nearby coffee shop. According to Morganson et al. (2010), telework allows workers to have a sense of freedom and discretion, since they are physically and psychologically removed from the office environment and away from face-to-face supervision.

However, this study also highlights a tension that comes with telework. While employees may feel more autonomous when working remotely, using technology at home may also limit their behavior in ways that were previously beyond the reach of managerial control. Therefore, it is important to remember that although autonomy is considered one of the main benefits of telework, it cannot be taken for granted and may not always contribute to teleworkers' satisfaction.

Despite several attempts to understand teleworkers' experiences, there is still a need for a solid theoretical model that can explain how remote work can enhance job satisfaction. Gajendran and Harrison's (Golden & Veiga, 2008) and Fonner and Roloff's (2010) mediation models have contributed to a better understanding of teleworking outcomes. However, their models are complex and include nine different psychological mediators, making it difficult for practitioners to determine which levers to use to ensure teleworkers' satisfaction. Therefore, it is necessary to gain a clear insight into teleworkers' psychological experiences and compare them with those of office workers.

It's crucial to remember that the link between remote work and job satisfaction varies significantly across people and different organizations. Effective communication, careful planning, and consideration of each employee's particular requirements and preferences are frequently necessary for remote work arrangements to be successful. It is also worth to note that job satisfaction while remotely working is influenced by various subjective factors, indicating that remote work has its own benefits and challenges (discussed below in detail) but studies have also found certain factors that can make remote work more beneficial and less challenging. These factors include the ability to choose whether to work remotely or not, as well as having flexible scheduling (Allen et al. 2015). The application of few theories in remote work setting to enhance the job satisfaction is discussed in 'strategies to enhance job satisfaction in remote work' section.

BENEFITS OF REMOTE WORK

Remote work offers employees unparalleled flexibility, which allows them to structure their work hours according to their most productive times. This autonomy empowers employees to achieve a harmonious work-life balance, as they are no longer bound by the constraints of commuting. The elimination of the daily commute reduces stress and enhances overall well-being. Employees can seamlessly integrate personal and professional responsibilities, thereby reclaiming precious hours to redirect towards personal pursuits, family time, or even self-care. This contributes to a more fulfilling and balanced lifestyle and accommodates diverse personal circumstances, catering to the needs of caregivers, parents, and individuals with unique scheduling requirements.

Contrary to initial skepticism, remote work has demonstrated its potential to boost productivity and enhance overall performance. The absence of workplace distractions combined with the ability to create a personalized and focused work environment allows employees to delve deeply into their tasks. Many individuals find that they can achieve a higher level of concentration and efficiency when working remotely, leading to a tangible increase in output. Moreover, employees can choose a conducive workspace, whether it be a quiet home office, a co-working space, or a local coffee shop, which enables them to tailor their surroundings to optimize productivity. The result is a workforce that is not only more productive but also more engaged and motivated. The shift towards outcome-based performance metrics, rather than mere hours clocked in an office, becomes more feasible and aligns with the principles of remote work.

Remote work has the advantage of transcending geographical boundaries, allowing organizations to access a vast and diverse talent pool. Companies are no longer confined to recruiting talent within their physical offices' commutable radius. Instead, they can tap into a global network of skilled professionals, which brings a rich tapestry of perspectives and experiences to the workforce.

This advantage is especially useful in areas where specific skills are highly sought after. Organizations can carefully pick out skilled individuals from various places, nurturing a broad and welcoming work environment. Plus, getting rid of location boundaries makes it easier to hire people with distinct cultural understanding, language skills, and a worldwide viewpoint. The outcome is a team that mirrors the wide range of the international society, enhancing creativity and flexibility.

Working remotely saves a lot of money for both companies and their workers. Firms save by not needing as much office space. This means they spend less on things like rent, utilities, and upkeep. With less overhead, they can use money more wisely.

Employees, on the other hand, experience direct financial benefits. The elimination of daily commutes translates into savings on transportation costs, including fuel, public transportation fees, or parking expenses. Additionally, remote work often results in reduced spending on work attire, lunches, and other incidental expenses associated with an office-centric work model. This financial relief contributes to employees' overall job satisfaction and economic well-being.

Companies globally are starting to recognize the environmental benefits of remote work. By reducing commuting, remote work helps decrease carbon emissions and traffic congestion. Additionally, remote work reduces energy consumption and waste generation by decreasing the reliance on physical office spaces.

Working from home does more than just saving the environment. It's in step with a company's ethical duty and green ways. Taking on work-from-home rules shows a company's promise to a greener future and lures in workers who care about the Earth. With global warming on the rise, working remotely serves as a thoughtful and accountable method for conducting business activities. In essence, working from home greatly boosts employee contentment. It focuses on adaptability, independence, and a sound work-life equilibrium, making remote work amplify job pleasure and overall joy. Workers value the confidence placed in them to organize their tasks autonomously and the chance to customize their work area and timeline to match personal choices.

This job satisfaction translates into increased employee retention, reducing turnover costs and contributing to long-term organizational success. Companies that offer remote work options are often better positioned to attract and retain top talent, as employees are more likely to remain loyal to organizations that prioritize their well-being and work preferences.

COVID-19 has highlighted the value of remote work for company sustainability and resilience. Firms with thorough remote-capabilities easily moved to complete remote operation. This action lessened the effect of sudden shutdowns and interruptions. This flexibility branches to unforeseeable circumstances like natural disasters or political troubles. Businesses with strong remote set-up are more prepared to steer through uncertainties and protect their operations. The pandemic has cultivated a new understanding;

remote work is not a bonus. It's a strategic necessity for continuous business success in a world full of surprises.

Remote work is good for promoting inclusion and diversity in the workforce. It removes the usual obstacles related to being in a certain place or fitting into an office culture. This allows people from a variety of backgrounds to do really well. Others see benefits too, like people with disabilities, those caring for others, or people often overlooked. Companies believing in remote work show they're committed to being inclusive. The adaptability that comes with remote work means employees can work wherever suits them best. This suits their needs, their situation. The result? A more diverse and creative place to work. So, the upside to remote work isn't just skipping the daily commute.

Working remotely changes, a lot. It touches everything from how we work together, to how we feel, to how companies do business. With remote work, workers can better balance their lives against their jobs. Companies can hire from anywhere, cutting costs and building strategic advantages at the same time. As the business world changes, it's really important to see just how much potential remote work offers. The benefits go beyond just money, though. Remote work also affects worker happiness, it's better for our planet, and it makes businesses stronger over time. Companies that notice and take advantage of these benefits are not just set up for success today, but are also helping to form a long-term, inclusive, adaptable work environment for the future.

CHALLENGES OF REMOTE WORK

The landscape of work has transformed significantly with the rise of remote work. Firms adapting to this change must grasp different elements that affect job contentment in a home-based work environment. Remote work offers freedom and self-direction, yet it also presents specific hurdles which could influence workers' job happiness. Top enjoyment in working from home comes from good communication. In-person chats are filed down to the bare bones, relying primarily on online methods. As a result, talking must be clear, open and frequent to cover the distance made by being online. Employees at home must remain looped in, linked and part of the broader business picture. The challenge is sorting out the problems that can come up when body language isn't there to help with understanding.

Companies can solve this problem by developing an environment filled with communication. Routine online meetings, video calls, and various collaborating equipments can promote smooth communication. Management has an important part in setting the mood for strong communication. They do this by highlighting being open and making sure news travels easily across the team that works remotely. When workers know they are listened to, kept up-to-date, and involved, this reflects well in their job happiness. This helps create a sense of community despite being in different locations.

Tech advancements have provided countless perks to groups. Now, we can structure conversations in four distinct methods, depending on time and place. Digital communication occurs, not requiring in-person gatherings. It could be managed through video, audio, or text. There are synchronized online meetings where everyone participates at the same time, or asynchronous meetings where the meeting is recorded and stored for employees with different time shifts. Participants share the same virtual space and web during these meetings. Each of these settings can be applied based on the situation, purpose, availability, resources, and context of the transmitted message. It is critical for management to determine which media is most appropriate to use in each situation.

Remote work, while offering flexibility, can blur the boundaries between work and personal life. It is challenging to maintain a healthy work-life balance as the physical separation between office and home diminishes. The risk of overworking and burning out increases, which can negatively affect employees' well-being and job satisfaction.

Organizations can turn this challenge into an opportunity by promoting and supporting a balanced work-life dynamic. Flexible work hours, clear work expectations, and encouragement for breaks and time off can contribute to a healthier equilibrium. Leaders can set an example by respecting boundaries, discouraging excessive working hours, and fostering a culture that values the well-being of remote employees. When individuals have the autonomy to manage their time effectively, it not only improves their work-life balance but also enhances their job satisfaction.

Kirchmeyer (2000) says a balanced life needs satisfactory experiences at work and at home. To get this, you need time and energy. You need to share these resources fairly between work and home. Clark (2000) agrees. He says to avoid conflict between roles, you need to be satisfied and function well at work and home (Greenhaus, Collins & Shaw, 2003).

Trusting technology is essential when working remotely as it directly affects job contentment. Problems like internet breakdowns, software errors, or insubstantial digital resources can interrupt work and lead to annoyance among remote workers. The real test is to offer an uninterrupted tech experience that helps work activities rather than blocking them.

Companies can tackle this problem by putting money into a strong online system. This involves making sure there's fast internet, giving dependable team tools, and offering help for tech issues. Education programs can help workers use digital systems well, reducing tech problems at their job. A solid tech setup not only makes jobs more enjoyable by taking away issues. It also helps create an atmosphere of smooth work and skills.

The belief in management backing and its noticeable presence deeply impacts job happiness in virtual work environments. Workers-based remotely may experience feelings of isolation from leaders, particularly when the dialogue is largely online. The difficulty arises in trying to create a feeling of leadership existence and assistance that overcomes any geographical limitations.

Leading can transform difficulties into chances by showing active support and visibility. By doing check-ins virtually on a regular basis, holding town hall sessions, and personal interactions, leadership can be clearly seen. Show empathy, grasp the distinct trials of working remotely, and offer guidance when required to cultivate trust and connectivity. Employees feel valued and realize that their leaders are actively engaged and vested in their success when they receive this kind of support. This promotes job contentment and dedication to the organization.

Headship in an online, off-site, or electronic setting must change to fit the new factors of a digital, global, and adjustable workplace. In this kind of workplace, team members communicate through digital ICTs. Team leaders handle their teams over long distances. These distances could be within and across different time zones. Effective leadership in such a scenario involves the prudent and competent use of ICTs, choosing the best communication channel based on the context, and utilizing physically present communication channels when most appropriate. As Van Wart et al. (2019) suggest, leaders should use ICTs when they are advantageous, choose the most appropriate ICTs available relative to the value of various resources, use physically present communication channels when most appropriate, and use ICTs with competence.

Acknowledgment and praise are core desires of people, playing a big role in job contentment. Working remotely though, brings a challenge. It becomes tough to guarantee that those working from afar get the right amount of praise for their work. Without physical closeness, noticing and valuing individual work can be tough.

For tackling this issue, firms can tap into the possibilities of developing strong appreciation plans for distant workers. Habitually appreciating accomplishments, voicing thankfulness, and commemorating important points can uplift a good working environment. Moreover, introducing a method for coworker recognition, coupled with chances for workers to display their tasks, can increase happiness at work. When people sense they are appreciated and acknowledged for their hard work, it gives them a feeling of achievement and job fulfillment.

One hurdle with working from home is losing the informal chats and social bonds usually happening at the office. This can cause remote employees to feel alone and cut off. It might cause a big problem for job enjoyment. The team spirit and friendship often built in a physical work office may be lost.

Tackling this issue, teams can purposely build social connections in the distance-working setup. Digital group-bonding activities, internet-based social happenings, and casual video discussions can help cultivate a community feeling. Encouraging open communication about non-work-related topics during virtual meetings can replicate the spontaneous interactions that occur in an office setting. Leaders can also play a critical role in creating a virtual culture that emphasizes connection and collaboration. When remote employees feel socially connected, it positively influences their job satisfaction and overall well-being.

Working from home brings its own hurdles to professional improvement. A top issue is giving chances for skills growth and career progress in an online setting. Workers at home might feel cut off from usual paths for development and learning. Facing this hurdle, it's crucial for businesses to dedicate resources to growing their offsite teams. Focused programs like online lessons, virtual training, and mentor guides can all aid in boosting abilities and pushing careers forward. Consistent check-ins and job evaluations serve as platforms for talking over career aspirations and plotting a course for progress. When staff sense a strong backing for their career development, it boosts their work happiness and loyalty to the business.

Working from home can have mixed results. You might love it or hate it. What makes the difference? A whole lot of things! Good communication is key, of course. So is being able to spend time with family or doing hobbies. Sound computer systems and strong support from bosses also help. Online bonding with colleagues is important too. It's nice to feel appreciated, and to be able to shape your workday. Plus, there's always room to learn more and grow in your job! If companies address these aspects well, working from home isn't just manageable. It can make staff really love what they do.

Directing a team is vital in handling the elements that impact creating an upbeat work-from-home culture. Being versatile, emphasizing clear conversation, and offering genuine support for workers' wellness are all key steps. These steps build a remote workspace that not only attains company objectives but also boosts the happiness of its valuable staff. As we continue to adapt to the evolving nature of work, understanding and prioritizing these factors will be crucial in building a resilient and satisfied remote workforce.

STRATEGIES TO ENHANCE JOB SATISFACTION IN REMOTE WORK

The rise of remote work has brought about a new level of flexibility and autonomy, challenging traditional ideas of the workplace. Though working from home gives great advantages, it sure brings a few trials, especially when trying to keep job satisfaction high in remote employees. To build a good work from home experience, businesses should look into extensive plans that cover the many things that affect job satisfaction.

Good communication is key. We need it when we can't meet in person. Clear, easy-to-use ways of talking to each other are important. Workplaces should use strong online tools. These make it easier to work and share information together. Regular online meetings, video chats, and instant messaging will help. It can feel like you are in the same office together. Leaders have a big job. They set how open everyone is with each other. They do this by saying what they expect, listening to worries, and checking on their team. This helps everyone feel like they are part of the conversation at work. If people feel involved, they feel connected. If they feel like they belong, they are happier at work. This helps everyone feel part of a team, even if they are not in the same place.

Trust and the freedom to work are key. For a remote work set-up to work, trust is vital. Companies should promote a laissez-faire culture, where their staff can handle their duties their own way. Over-monitoring harms the remote work atmosphere, lessening trust and work happiness. Proving workers the liberty with their work schedule and duties boosts a sense of duty and ownership. Defining clear goals, rather than setting how to get there, gives workers a chance to tailor their work to their strengths and likes. A culture of trust and liberty doesn't just improve job happiness, but it also brings up a more fired-up and self-reliant team.

Worker happiness is tied to feeling good at work. With remote work, personal time and work time sometimes mix. It's key for firms to make sure their workers are happy. Things like health-focused programs, mental health support, and adaptable working hours help make work life better. Pushing for rest time, getting up and moving around, and taking time off are a big part of wellness plans. Bosses should talk up the company's drive to ensure worker wellness. This builds an environment that cares about all parts of a person's life. When workers are backed in being well, it boosts job happiness and how engaged they are.

Investing in the growth of remote workers is key. A worker's happiness greatly lies in their chances for progress. Due to work from home, employees might feel cut-off from usual growth routes. To tackle this, remote-friendly growth programs should be a big part of a company's plan. Things like virtual classes, internet-based courses, and guidance programs allow for skills to improve and jobs to advance. Regular digital feedback and performance check-ups give chances to talk about goals and growth. By showing a real commitment to help remote workers grow, businesses can boost happiness on the job. It also shapes a workplace where constant learning is a culture.

Encouraging Friendship and Team Unity: Many remote workers feel lonely, it's crucial to boost friendship and team unity. We can use online platforms to create a feeling of togetherness and team spirit. Setting up online team unity activities, digital fun events, and casual video chats can make an imitation of a real office. Promoting team members to talk about their personal stories, pastimes, or things they like during online meetings can make strong bonds beyond just work-related talks. Leaders have a key part in encouraging these activities, shaping the vibe for a cheerful and socially joined remote work environment.

Identifying and Celebrating Achievements: Giving praise and recognition to both the individual and team successes is vital, especially when working remotely. A program that recognizes achievements across online platforms can be effective. This could include praising a team member during a meeting, giving digital awards, or having online parties. All of these actions support a healthy workspace. Encouragement of recognition from coworker to coworker can strengthen the bonds within the team. It is the leaders' task to always show appreciation for the hard work of those working remotely. This helps workers to know that their input is both valuable and noted.

Work Flexibility: Remote work's main allure lies in its flexibility. Job satisfaction can be augmented through companies adopting adaptable work schedules. Providing flexible working hours lets workers sync their schedules with their most efficient times. Moreover, adopting strategies that allow for sporadic telecommuting or adjustable workdays fosters an adaptable atmosphere. Transparent communication about goals and timelines, paired with the ability to arrange work to fit personal necessities, bolsters employees and boosts their contentment with their work set up.

Ensuring Tools for Remote Success: Giving remote workers needed resources is key to their achievement and happiness. Companies must put money into tech that helps smooth communication, teamwork, and task handling. Giving remote workers the proper devices, programs, and learning makes it easier for them to do their jobs effectively. Also, companies must create clear rules for remote work, dealing with stuff like internet safety, data privacy, and online manners.

Continuous Feedback and Performance Management: When working remotely, it's crucial to have constant feedback processes. This includes both official performance assessments and frequent, informal touchpoints. Giving helpful feedback, creating objectives, and celebrating accomplishments, helps remote team members have clear guidance. Performance dialogues should look back at successes, but also set out next steps and career plans. Clear and frequent performance communication builds a workspace that values responsibility and constant betterment. This positively affects job happiness.

Creating Opportunities for Virtual Career Advancement: Staff working from home must never feel boxed in when it comes to moving up. Companies need to open doors for online promotion, making space for advancement, added duties, and managerial posts. Programs for online career growth, mentorship plans, and networking chances all help people feel they're progressing. Bosses should chat about career aims with home workers, offering advice on skills and giving chances to climb higher. If people see their careers doing well online, they'll feel more devoted to the firm and happier in their job.

In addition to these factors, there exist certain theories that can be effectively applied to enhance job satisfaction in remote work. These theories have been extensively studied and analyzed and an provide valuable insights into enhancing job satisfaction in remote work setting. A few of these theories are briefly explained below provide a better understanding of how they can be applied.

Herzberg's two-factor theory breaks job satisfaction down into motivators and hygiene factors. Hygiene factors keep employees from becoming dissatisfied. For remote workers, these could include a good at-home work setup, dependable technology, and clear ways to communicate. However, lack of these factors can lead to unhappiness. On the other hand, motivators boost performance. These might involve chances to learn new skills, facing challenging tasks, and getting rewards and acknowledgment in remote work scenarios.

Social exchange theory highlights the importance of the reciprocal relationship between employees and their organizations. When working remotely, it becomes essential to maintain strong social connections through virtual communication tools. Organizations that focus on building a positive virtual work environment are more likely to promote job satisfaction among their employees.

Equity theory suggests that an individual's job satisfaction depends on the fairness of the exchange between their inputs (effort and time) and the outcomes (rewards). In remote work, ensuring fairness in terms of workload distribution, recognition, and opportunities for career development is crucial for promoting job satisfaction.

There are several other theories which can be applied to enhance job satisfaction. Researchers can further investigate how theories could be applied to enhance job satisfaction in remote work settings. They could also develop theories which could be applied in organizations to enhance productivity and job satisfaction.

FUTURE DIRECTION

The rise of remote work has significant implications for employment and job satisfaction, completely reshaping the traditional paradigm of work. The widespread acceptance of remote work signals a transformative shift towards more flexible and decentralized work structures. This evolution goes beyond just convenience; it impacts how people view their professional lives.

Employment in the future will likely be a hybrid model, blending remote and in-office work to accommodate diverse preferences. This flexibility caters to the changing dynamics of the workforce and opens up opportunities for a more inclusive and diverse employment landscape. Global talent pools become accessible, breaking down geographical barriers and fostering cross-cultural collaborations.

Regarding job satisfaction, remote work prompts a reassessment of the factors that contribute to a fulfilling professional life. The emphasis on autonomy, effective communication, and holistic well-being is likely to endure, influencing organizational cultures. Employers will have to proactively address the challenges posed by remote work to ensure sustained job satisfaction, focusing not only on productivity but also on employee engagement and mental health.

Researchers and scholars could investigate job satisfaction in different types of virtual work environments and find ways to improve strategies and models to enhance job satisfaction in remote work settings. Researchers additionally can investigate how remote work has impacted the lifestyle, psychological well-being, personal life, and social life of employees. Further, there is scope for research in constructing theories as mentioned earlier in the chapter.

In the end, the future of employment will be a dynamic coexistence of remote and traditional work structures, demanding an understanding of individual needs and a commitment to creating environments where job satisfaction thrives in diverse professional settings.

CONCLUSION

The relationship between remote work and job satisfaction reveals a multitude of opportunities and challenges in our constantly evolving professional landscape. We can conclude that modern employment is being redefined, with professionals no longer confined to cubicles or commutes. Remote work offers flexibility and autonomy, making it a powerful force in shaping the job satisfaction narrative. However, it is essential to acknowledge the potential challenges of isolation and communication gaps. By understanding and embracing these nuances, organizations and individuals can work towards a future where work is not just a task but a fulfilling journey, free from physical constraints.

REFERENCES

Albertson, L. A. (1980). Review Essay: Trying To Eat an Elephant The Social Psychology of Telecommunications, by John Short, Ederyn Williams, and Bruce Christie. London: John Wiley, 1976. *Communication Research*, *7*(3), 387–400. doi:10.1177/009365028000700307

Allen, T. D., Golden, T. D., & Shockley, K. M. (2015). How effective is telecommuting? Assessing the status of our scientific findings. *Psychological Science in the Public Interest*, *16*(2), 40–68. doi:10.1177/1529100615593273 PMID:26403188

Bae, K. B., & Kim, D. (2016). The Impact of Decoupling of Telework on Job Satisfaction in U.S. Federal Agencies: Does Gender Matter? *American Review of Public Administration*, *46*(3), 356–371. doi:10.1177/0275074016637183

Bailey, D. E., & Kurland, N. B. (2002). A review of telework research: Findings, new directions and lessons for the study of modern work. *Journal of Organizational Behavior, 23*, 383–400. doi:10.1002/job.144

Belzunegui-Eraso, A., & Erro-Garcés, A. (2020). Teleworking in the context of the covid-19 crisis. *Sustainability (Basel)*, *12*(9), 3662. doi:10.3390/su12093662

Clark, S. C. (2000). Work/family border theory: A new theory of work/family balance. *Human Relations*, *53*(6), 747–770. doi:10.1177/0018726700536001

Cooper, C. D., & Kurland, N. B. (2002). Telecommuting, professional isolation and employee development in public and private organizations. *Journal of Organizational Behavior, 23*, 511–532. doi:10.1002/job.145

Culibrk, J., Delic, M., & Mitrovic, S. (2018). Job Satisfaction, Organizational Commitment and Job Involvement: The Mediating Role of Job Involvement. *Frontiers in Psychology*. https://www.researchgate.net/publication/323223935_Job_Satisfaction_Organizational_Commitment_and_Job_Involvement_The_Mediating_Role_of_Job_Involvement

Darouei, M., Delanoeije, J., & Verbruggen, M. (2022, August). When daily home-to-work transitions are not all bad: A multi-study design on the role of appraisals. In *AoM Proceedings: Annual Meeting of the Academy of Management*. https://lirias.kuleuven.be/retrieve/666508

DuBrin, A. J. (1991). Comparison of the job satisfaction and productivity of telecommuters versus in-house employees: A research note on work in progress. *Psychological Reports*, *68*(3, suppl), 1223–1234. doi:10.2466/pr0.1991.68.3c.1223

Fonner, K. L., & Roloff, M. E. (2010). Why teleworkers are more satisfied with their jobs than are office-based workers: When less contact is beneficial. *Journal of Applied Communication Research*, *38*(4), 336–361. doi:10.1080/00909882.2010.513998

Gajendran, R. S., & Harrison, D. A. (2007). The good, the bad, and the unknown about telecommuting: Meta-analysis of psychological mediators and individual consequences. *The Journal of Applied Psychology*, *92*(6), 1524–1541. doi:10.1037/0021-9010.92.6.1524 PMID:18020794

Golden, T. D., & Veiga, J. F. (2005). The Impact of Extent of Telecommuting on Job Satisfaction: Resolving Inconsistent Findings. *Journal of Management*, *31*(2), 301–318. doi:10.1177/0149206304271768

Greenhaus, J. H., Collins, K. M., & Shaw, J. D. (2003). The relation between work-family balance and quality of life. *Journal of Vocational Behavior*, *63*(3), 510–531. doi:10.1016/S0001-8791(02)00042-8

Izvercian, M., Seran, S., & Ivascu, L. (2016). Job Satisfaction Variables: A Grounded Theory Approach. *Procedia - Social and Behavioral Sciences*. https://www.researchgate.net/publication/304106042_Job_Satisfaction_Variables_A_Grounded_Theory_Approach

Johnson, N., Pinsonneault, A., & Boisvert, M. (2001). *The Impacts of Telecommuting on Organizations and Individuals*. https://www.researchgate.net/publication/315386065_The_Impacts_of_Telecommuting_on_Organizations_and_Individuals

Kirchmeyer, C. (2000). Work-life initiatives: Greed or benevolence regarding workers' time? In C. L. Cooper & D. M. Rousseau (Eds.), *Trends in organizational behavior, Vol. 7. Time in organizational behavior* (pp. 79–93). John Wiley & Sons Ltd. https://psycnet.apa.org/record/2001-00452-005

Kniffin, K. M., Narayanan, J., Anseel, F., Antonakis, J., Ashford, S. P., Bakker, A. B., Bamberger, P., Bapuji, H., Bhave, D. P., Choi, V. K., Creary, S. J., Demerouti, E., Flynn, F. J., Gelfand, M. J., Greer, L. L., Johns, G., Kesebir, S., Klein, P. G., Lee, S. Y., ... Vugt, M. (2021). COVID-19 and the workplace: Implications, issues, and insights for future research and action. *The American Psychologist*, *76*(1), 63–77. doi:10.1037/amp0000716 PMID:32772537

Locke, E. A. (1969). What is job satisfaction? *Organizational Behavior and Human Performance*, *4*(4), 309–336. doi:10.1016/0030-5073(69)90013-0

Lowry, P. B., Roberts, T. L., Romano, N. C. Jr, Cheney, P. D., & Hightower, R. T. (2006). The Impact of Group Size and Social Presence on Small-Group Communication: Does Computer-Mediated Communication Make a Difference? *Small Group Research*, *37*(6), 631–661. doi:10.1177/1046496406294322

Lund, S., Madgavkar, A., Manyika, J., & Smit, S. (2020). *What's next for remote work: An analysis of 2,000 tasks, 800 jobs, and nine countries*. McKinsey & Company. https://www.mckinsey.com/featured-insights/future-of-work/whats-next-for-remote-work-an-analysis-of-2000-tasks-800-jobs-and-nine-countries

Mallett, O., Marks, A., & Skountridaki, L. (2020). Where does work belong anymore? The implications of intensive homebased working. *Gender in Management*, *35*(7/8), 657–665. doi:10.1108/GM-06-2020-0173

Martin, B., & Mesler, R. (2012). Is Telework Effective for Organizations? A Meta-Analysis of Empirical Research on Perceptions of Telework and Organizational Outcomes. *Management Research Review*. https://www.researchgate.net/publication/242340515_Is_Telework_Effective_for_Organizations_A_Meta-Analysis_of_Empirical_Research_on_Perceptions_of_Telework_and_Organizational_Outcomes

Morganson, V., Major, D., Oborn, K., Verive, J., & Heelan, M. (2010). Comparing Telework Locations and Traditional Work Arrangements: Differences in Work-Life Balance Support, Job Satisfaction, and Inclusion. *Journal of Managerial Psychology*, *25*(6), 578–595. Advance online publication. doi:10.1108/02683941011056941

Raišienė, A. G., Rapuano, V., Varkulevičiūtė, K., & Stachová, K. (2020). Working from home—Who is happy? A survey of Lithuania's employees during the COVID-19 quarantine period. *Sustainability (Basel)*, *12*(13), 5332. doi:10.3390/su12135332

Rook, K. S. (1984). The negative side of social interaction: Impact on psychological well-being. *Journal of Personality and Social Psychology*, *46*(5), 1097–1108. doi:10.1037/0022-3514.46.5.1097 PMID:6737206

Schall, M. A. (2019). The Relationship Between Remote Work and Job Satisfaction: The Mediating Roles of Perceived Autonomy, Work-Family Conflict, and Telecommuting Intensity. *Master's Theses., 5017*. Advance online publication. doi:10.31979/etd.2x82-58pg

Shepherd, M. M., & Martz, W. B. (2006). Media Richness Theory and the Distance Education Environment. *Journal of Computer Information Systems*, *47*, 114–122. doi:10.1080/08874417.2006.11645945

Siengthai, S., & Pila-Ngarm, P. (2016). The interaction effect of job redesign and job satisfaction on employee performance. *Evidence-based HRM: a Global Forum for Empirical Scholarship*. 10.1108/EBHRM-01-2015-0001

Sims, H. P. Jr, Szilagyi, A. D., & Keller, R. T. (1976). The measurement of job characteristics. *Academy of Management Journal*, *19*(2), 195–212. doi:10.2307/255772 PMID:1029346

Smith, S. A., Patmos, A., & Pitts, M. J. (2018). Communication and teleworking: A study of communication channel satisfaction, personality, and job satisfaction for teleworking employees. *International Journal of Business Communication*, *55*(1), 44–68. doi:10.1177/2329488415589101

Staniulienė, S., & Leonavičiūtė, A. (2022). Self-organization of remote employees in Lithuanian companies. *Problems and Perspectives in Management*, *20*(2), 486–496. doi:10.21511/ppm.20(2).2022.40

Veld, M., Paauwe, J., & Boselie, P. (2010). HRM and strategic climates in hospitals: Does the message come across at the ward level? *Human Resource Management Journal*, *20*(4), 339–356. doi:10.1111/j.1748-8583.2010.00139.x

Chapter 4
Understanding Rational Decision-Making in Virtual Collaborations

Yan Zou
Nexi Group, Germany

Frederico V. R. Castro
https://orcid.org/0009-0004-0721-5634
Éstudo, Belgium

Petros Chamakiotis
https://orcid.org/0000-0001-7109-310X
ESCP Business School, Spain

ABSTRACT

Despite the wealth of studies on virtual teams, there is a knowledge gap in relation to how virtual collaboration might influence rational decision-making in this context. With the study presented in this chapter, the authors address this gap by presenting the research findings drawing on eight interviews with practitioners who work in virtual teams. They identify three factors that influence rational decision-making: (a) efficient information integration via the usage of ICT tools, (b) reduced emotional impact on the decision-maker, and (c) better control and enhanced confidence of the decision-maker. The chapter explains how the findings contribute to existing literature and what this means for practitioners on the ground.

DOI: 10.4018/979-8-3693-1314-5.ch004

1. INTRODUCTION

Although virtual teams (VTs) have been around for some time (e.g., Chamakiotis, 2020; Gilson et al., 2021), the Covid-19 pandemic has undoubtedly led to a wider adoption of VTs (e.g., Chamakiotis et al., 2021), such that post-pandemic virtual collaborations now form part of business practices for most of us. Understanding how this form of collaboration affects decision-making is therefore essential to ensure organizations can make informed decisions under changing circumstances. The improved performance brought by VT collaboration has been studied by Hosley (2010), suggesting the usage of Information and Communication Technology (ICT) — such as knowledge databases, video conferencing systems (VCS) and collaborative sharing software — could lead to better VT decision-making. According to Dennis et al. (1998), ICTs must be matched with the task in question to ensure that VT members use the features of different ICTs efficiently. The optimized use of ICT, however, is not the sole element that contributes to the impact of VT collaboration on decision-making.

The relation between VT collaboration and decision-making has been previously studied, however mostly indirectly. For example, Wilson et al. (2019) suggested that taking up digital tools can "debias" the decision-making process, but only when users feel the ICT being used is useful and trustworthy. This has been supported by Turel and Connelly (2012), indicating that motivation of VT members is also influenced by their perception towards the capability of the ICTs at hand. Scholars have found that rationality improves team decision-making; for example, a rational decision maker would optimize the outcome and generate the economic profit maximum (Alkaraan, 2013). Bringing rationality to all employees is valuable, as uncertainty could be reduced by rational decisions, providing better solutions to complex problems (Einhorn, 2020).

Despite the above, there is a significant knowledge gap in the existing literature, as, to our knowledge, no studies have explicitly explored rationality and rational decision-making in the VT context. This chapter aims to bridge this gap by empirically exploring decision-making and rationality in the VT context through rich qualitative interview data from a curated pool of decision-makers in the public and private sector. In what follows, we present background information on the topic and then present our empirical material. The chapter then closes with a discussion of our research findings along with practical recommendations and future research directions.

2. BACKGROUND

2.1 Rationality in Decision-Making

Rationality has been a key determinator in decision-making and has been supported by multiple studies, especially the classic "Bounded Rationality" model, which claims that rationality and irrationality co-exist in decision-makers and the allocation of the two can vary (Simon, 1957).

The bounded nature of rationality means that decision-makers can optimize the outcome by deploying more rationality and excluding the irrationality (Alkaraan, 2013). Rationality in this context refers to the utilization of predetermined tools such as financial analysis to justify the decision, while irrationality is the utilization of personal or psychological factors (Alkaraan, 2013). A decision-making process under bounded rationality is dominated by the ability of the decision-maker to process a significant amount of information, and when the information processing has become overwhelming, individuals tend to rely on personal experiences and preferences, thus causing deviation from the optimized outcome (Wall, 1993).

The concept of "rationality" has received extremely limited attention in the existing VT literature. Indeed, the mechanism of bounded rationality indicates a pipeline where rationality can predominate if irrationality is prevented (Simon, 1957). Based on this bounded rationality model, VT members tend to be less emotionally involved and a "neutral" state can be promoted via VT collaboration (Straus, 1996). Nonetheless, does it imply that VT collaboration directly contributes to rational decision-making? Such a relationship is seldom evident in the existing literature.

2.2 Rational Decision-Making in Virtual Teams

Information exchange has been studied as one of the key notions in decision-making. Rooted in using these ICT-based communication tools, VTs can achieve a higher degree of information integration across a vast range of knowledge sources, including different decision-makers (Draghici et al., 2008). However, whether decision-makers can utilize the information rationally to generate the optimized outcome is what determines the final performance (Alkaraan, 2013). Based on the bounded rationality model (Simon, 1957), the processing of information can determine how much rationality is used in decision-making, reducing reliance on irrationality.

In the comparison of decision-making performance between VTs and face-to-face (F2F) collaboration, information exchange or knowledge sharing is at the core of deciding whether VTs can outperform F2F teams (Davidavičienė et al., 2020). Among limited studies on decision-making in VTs, improved VT performance has been attributed to the enhanced communication via ICTs such as knowledge database, online resources, video conference and collaborative sharing software (Hosley, 2010).

Indeed, the use of ICT can influence VTs in the decision-making process (Schmidt et al., 2001). Recent studies emphasize the importance of considering how technology—including ICT tools (e.g. telecommunicating tools such as Zoom)—influences collective performance in VTs (Hardin et al., in press). An example of how the role of technology has been considered by scholars in existing VT research is "Media Synchronicity Theory" (MST) which argues that if an ICT, based on its degree of synchronicity, matches the task at hand, it fosters the integration of information (Dennis, 1998). However, the perception of collective decision-makers towards ICT can also influence the outcome. For example, Turel and Connelly (2012) mentioned that the motivation of VT members is influenced by their perception towards the capability of the ICT in question. If they deem the ICT to be satisfactory, they would be more collaborative in the decision-making process.

Thus, both the utilization and the perception of the used ICT tools have been found to improve the efficiency of information processing, contributing to higher rationality in the decision-making process. However, factors such as the cognitive influence in VT collaboration and the irrationality side of decision-making have not yet been comprehensively studied, and these could contribute to rebalancing the scale of rationality/irrationality in the decision-making process. Further research is needed on these topics.

All in all, the studies above have supported that enhanced information exchange can contribute to higher performance in VT collaboration, but the links between improved performance and increased rationality remain unclear. To make this void even more apparent, other aspects — such as the emotional biases that are hindering rational decision-making — have been scarcely studied in the context of VT collaboration. A summary of the existing literature in this area is visually depicted in Figure 1.

On one hand, the left side of Figure 1 displays the current studied areas on VT collaboration and decision-making, suggesting that VT collaboration is associated with better decisions, and it can debias decision-making. On the other hand, the right side of Figure 1 shows that improved performance in decision-making is brought by either increased rationality or decreased irrationality. Figure 1 illustrates that there is a lack of understanding around the factors behind the deployment of rational decision-making in the VT context, i.e., how VT collaboration can contribute to rational decisions and what are the intermediaries affecting this process. Although there are indications suggesting that VT collaboration can foster more rational decisions, there is no empirical research explicitly studying the underlying mechanism. Consequently, in this chapter we pose the following Research Question (RQ):

How can VT collaboration contribute to rational decision-making?

The overarching RQ serves as direction for the study. It can be broken down into the following more granular sub-questions:

1. How can rationality be improved with VT collaboration, in decision-making?
2. How can ICT impact rationality in decision-making?
3. Can decision-makers have more control over their irrationality in decision-making with VT collaboration?

Figure 1. Existing literature and knowledge gap

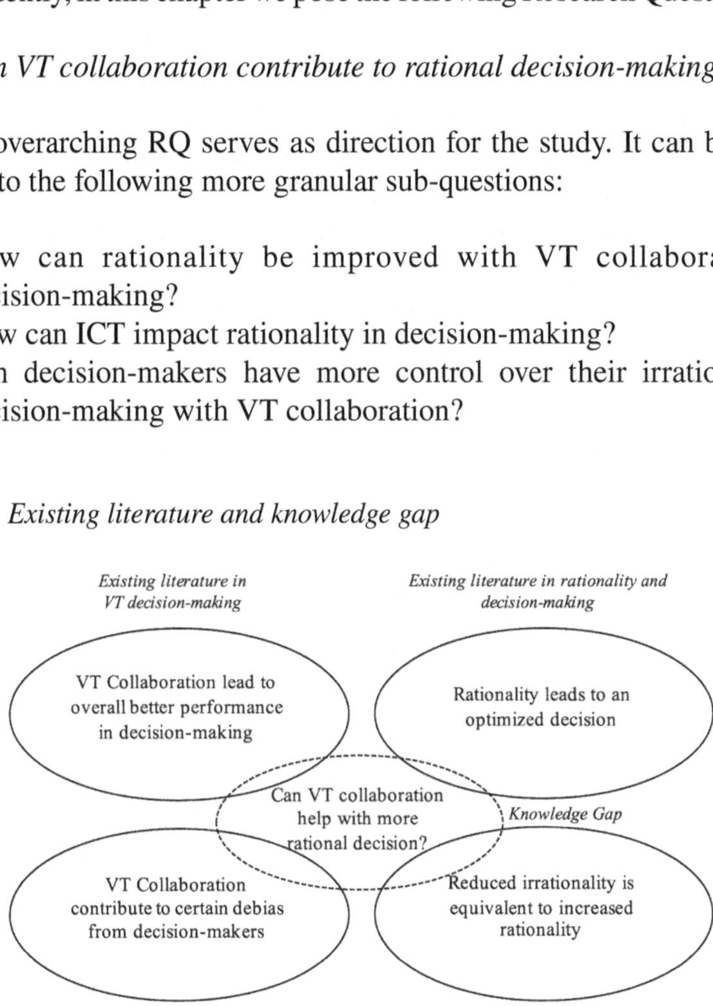

3. RESEARCH DESIGN

3.1 Research Strategy

This research aims to provide a comprehensive understanding of the role of VT collaboration in decision-making rationality, with practical implications for managers, leaders, and organizations seeking to leverage these methods in their operations. To achieve this overarching goal, this study encompasses a cohort of eight distinct organizations, drawn from both the public and private sectors. The researched organizations are different in size, industries and sectors, which can ensure representativeness in addressing the research aim.

Table 1 below presents the participants' background information relevant to the study. While forming a diverse and international cohort, participants have had competency to provide insights on the topic studied, with suitable professional experiences and decision-making involvement.

Table 1. Background of participants

Features	Categories	Participants
Age Group	20-25	3
	25-35	4
	35 and above	1
Biological Gender	Male	3
	Female	5
Native Language	Chinese	3
	Czech	1
	German	2
	Lithuanian	1
	Portuguese	1
Full-time Working Experience from 2020 to 2023	Yes	8
Seniority	More than 5 years	2
	3-5 years	2
	Less than 3 years	4

continues on following page

Table 1 continued

Features	Categories	Participants
Industry of Organization	Consulting	2
	Telecommunication	1
	Chemistry	1
	Finance	1
	Hospitality	1
	IT	2
Geographical Distribution of Organization	Pan Europe	2
	Asia	1
	Worldwide	1
	Western Europe	1
	Northern Europe	2
Sector of Organization	Public sector	2
	Private sector	6
Working mode before COVID-19	On site	2
	Hybrid	6
	Remote	0
Working mode during COVID-19	On site	0
	Hybrid	2
	Remote	6
Working mode after COVID-19	On site	2
	Hybrid	6
	Remote	0
Team Size	Fewer than 3	3
	3 to 10	3
	Above 10	2
Role in decision-making	Only decision maker	3
	Only decision participant	0
	Mix of decision maker and participant	5

continues on following page

Table 1 continued

Features	Categories	Participants
ICT tools used	Instant messengers: MS Teams, WhatsApp, Mobile messengers	6
	Video conference tools: MS Teams, WebEx, Skype, Jabber and others	7
	Non-synchronized messengers (e.g. email)	6
	Non-visible verbal tools (e.g. telephone calls)	3
	Shared document real-time editing (e.g. Google Docs)	3
	Database	3
	Real-time collaborating tools for graphic designing	1
Preferred ICT tools	Video conference tools	1
	Shared document real-time editing (e.g. Google Docs)	5
	Email	2
Self-evaluation on rationality presence	Identified as rational	0
	Identified as irrational	0
	Identified as both rational and irrational	8

3.2 Data Collection and Analysis Process

Eight interviews were conducted with the candidates from January to May 2023. The pre-selection of candidates was based on the following recruitment criteria. Each candidate should have:

4. Worked in both virtual and physical settings, which ensure the data generated can be used to compare both conditions.
5. Worked full-time during this period, which optimizes the ability of the participants to reflect on the topic in the given period.
6. Contributed to either individual or collective decisions, which will make sure the sample contains the decision-making experience.
7. Clearly illustrated their reflection on rationality and irrationality, which will generate relevant data about rational decision-making in different settings.

The interviews were semi-structured and had five central questions (Table 2), focusing on the following topics:

1. Background Information: professional experience of the candidate, especially the working mode.
2. ICT usage: tools utilized by candidates in their work experience.
3. Perception of ICT: the point of view from candidates towards ICT and their experience using ICT in VT collaboration
4. Perception of rationality: the understanding and reflection of candidates on rationality and irrationality.
5. Perception of rationality in decision making: the self-reflection of candidates on their rational approach to decision-making within the context of VT and F2F collaboration.

Table 2. Central question lists for interviews

Question 1	Can you tell me about your working experiences and working mode?
Question 2	What Information and Communication Technologies (ICT) tools did you use?
Question 3	What do you think about these ICT tools?
Question 4	What are your opinions of rationality and irrationality during these experiences?
Question 5	Does it make a difference to you in VT collaboration versus real life in terms of rational decision-making?

Two interviews took place onsite in Germany and six interviews were organized remotely. The duration of interviews ranged from 30 minutes to one hour. All interviewees have signed a consent form regarding the collection and processing of data and read the information sheet about this research study. The content of interviews was audio recorded and transcribed using Artificial Intelligence (AI) tools, then analyzed using ChatGPT before manual analysis using a qualitative, thematic analysis method. Through a comprehensive comparative thematic analysis on the verbal data collected in the eight interviews, a text body of more than 11,000 words was broken down into 25 codes, which are the labels corresponding to a semantic group of text, and further processed into three common themes, contributing to the patterns significant to the researched topic. Proven to reveal hidden patterns behind unstructured large amount of data such as texts (Braun & Clarke, 2006), thematic analysis was considered suitable for exploratory purposes in this study, following six stages of execution: data familiarization, generation of initial codes, themes identification, themes review, themes definition and labelling, and reporting. The reporting of study-relevant findings is strictly in line with data protection regulations and is anonymized as stated in the consent form and information sheet.

4. RESEARCH FINDINGS

Explored in this section are the themes identified in our study. We start with sub-section 4.1 whereby we explore how decision-makers can rely on more rational analysis when facilitated information is integrated into decision making processes thanks to the usage of ICT; then in sub-section 4.2 we further detail the use of VT collaboration for mitigating emotional bias, thus contributing to decision-makers' more analytical and rational decision; and lastly sub-section 4.3 delves into decision-makers' boosted sense of control and confidence in making rational decision making, when given a VT setting, especially with the existence of a favorable projection via the digital self-replica in video conference tools.

4.1 Information Integration Through ICT Facilitates VT Rational Decision-Making

The first theme follows a quite intuitive thread: information is at the core of the decision, and with facilitated information integration into the decision-making process, decision-makers could rely more on rational analysis than irrational judgement.

Participants noted that the judicious selection of the right ICT tool for a specific task significantly facilitates the flow of information. As Fred, for example, stated:

"When I have virtual meetings, I often check the documents for the meetings, such as Excel files, during the meeting with others. The upside of this is that I can have more resources and data."

Efficient ICT usage opens broader access to information and fosters the processing of information, as Michael shared:

"The process of information is much more at ease such that to get some first feedback from email is easy, also I can revise the information based on feedback easily. It is always nice to look at the documents together in VT settings. Additionally, there are more external information at hand in virtual because I can search for certain information anytime."

With confidence in using ICT, decision-makers tend to rely less on irrationality when equipped with sufficient information for rational analysis. This alignment is underscored by unanimous participant agreement (eight out of eight, 100%) that decision-makers operate on both rationality and irrationality.

Having explored in this sub-section the pivotal role of ICT in facilitating VT rational decision-making by efficiently integrating information, we now turn to 4.2 in which we show how VT collaboration serves as a filter to mitigate emotional bias.

4.2 VT Collaboration Filter out the Emotional Bias

Social factors such as emotion are more addressable for most of the participants, and the interplay between emotions and rational decisions was found to play a central role, as we explore here. The data reveal that participants strategically chose VT collaboration for tasks with potentially negative emotional impacts. The perceived "filter function" of VT collaboration has been favored by participants who conducted communication sessions which trigger them negatively. An example was given by Angel:

"It could filter out a lot of very strong negative emotions such as unappreciation, misunderstanding, judgment. And I think maybe that is the reason why on the other hand, I prefer to receive positive feedback in real life is to return those good emotions, good feedback, good feelings being emitted by virtual setting."

With a lack of sight in body language and overall ambiance, participants recognized a holistic effect that aids the "filtering function." The emotion influence is better under control in VT collaboration and thus contributing to a more analytical and rational decision-making process, as Fred explains:

"Virtual meetings have a more analytical view, while the physical meeting has the emotional side. I can feel more the reactions of everyone and focus

on passing the message as the best as I can, in contrast to ensuring the best use of information and data in the virtual meeting."

The orchestration of social factors and technology-driven "filtering functions" collectively shapes the behaviors of decision-makers, where they were seen reducing the involvement of emotional effect and could make more rational decisions.

Having now delved into the influence of emotional factors and the filtering role of VT collaboration, we proceed to the sub-section 4.3 for the final theme: how decision-makers' sense of control and confidence is promoted in VT collaboration and thus adding on to their rationality when making decisions.

4.3 The Power of Control in VT Collaboration

The third theme explored originated from a phenomenon universally experienced by all participants—the small videoconference window that projects oneself. Some shared insights into the impact of having this digital self-replica, expressing how it provides a unique advantage for self-monitoring and awareness. As Hanna shared:

"To be able to see yourself is such an advantage that you can manage your expressions and be more aware of your way of speech, the rhythm of speech. And this part of control is in my power because in real life I have so much more to worry about: what I wear, how do I sound, what image am I delivering to the person in front of me... I feel less hierarchy existing before my superior in virtual setting somehow."

This newfound control over self-presentation—which is unfeasible in F2F interactions—has granted a positive self-perception, enhancing the decision-maker's confidence and consistency in VT decision-making. As Angel put it:

"Yes, I do actually look at the image frequently – to make sure that I look 'presentable', and I do not want to look bad in front of colleagues. I feel empowered because of this control of my appearance. And I am for sure more confident having the camera on with this image presentation."

By having this confidence from the power of control, decision-makers integrate better information and can debias situations where one feels less in control, as Angel adds to the explanation.

Having now presented the three themes that emerged from our analysis of our interview data, we turn to our next section where we discuss our findings and provide recommendations.

5. SOLUTIONS AND RECOMMENDATIONS

The present study has addressed the knowledge gap in the existing literature by qualitatively examining how VT collaboration can contribute to rational decision making. We have addressed our overarching RQ by identifying three factors which contribute to rational decision making in the VT context:

1. Efficient information integration via the usage of ICT tools
2. Reduced emotional impact on the decision-maker
3. Better control and enhanced confidence of the decision-maker

Some of these factors have been individually studied (i.e., the efficiency of information exchange using rich media) in traditional F2F collaboration contexts. However, in this study, we have considered the role of ICTs in the context of VT collaboration which constitutes a widespread form of working worldwide. First, we have shown that with focus on the information exchange aspect in the decision-making process, there is a contribution to the MST (Dennis et al, 1998) with a progression on the ICT adoption, indicating that with VT collaboration, decision-makers can solve data-driven problems in a more rational fashion. This finding has confirmed that matching VT collaboration to data-driven tasks by leveraging ICT has indeed a positive impact on rationality and may thus lead to improved performance, relating back to the bounded rationality model. Furthermore, the second factor has specified the application scenario of the classic bounded rational model in decision-making (Simon, 1957) to a modern setting, where decision-makers can reduce irrationality and deploy more rationality within an organized collaboration setting such as VT. Finally, the third factor has explained how the decision-makers' control can be enhanced via the digital self-replica that only applies to VT collaboration, thus making a more rational decision.

Collectively, the three factors we have identified have helped us to address our RQ and sub-questions. First, we have shown that increased rationality and decreased irrationality in decision-making can be achieved via VT collaboration. Second, the use of ICT facilitates information processing and thus decision-makers can be more rational. Third, the debias of emotional influence and fostered confidence enables decision-makers to rely less on irrationality.

Further to their theoretical significance, our findings may also be used to help management practice. Therefore, we propose specific recommendations for business practitioners, coaches and educators. In particular, we recommend that decision makers select the right working method (virtual vs. F2F) for different types of tasks as follows:

1. Choose VT collaboration for analytical tasks
2. Choose VT collaboration for emotionally-negatively-oriented sessions
3. Choose F2F collaboration for brain-storming or creative idealization

It is also worth noting that the curation of an ideal VT collaboration culture in an organization relies on adequate employee training and coaching, and therefore business practitioners should operate closely with coaches who should implement the appropriate strategy to adapt to the organizational needs and individual team requirements, with well-defined ICTs and organizational processes. Additionally, when it comes to leading in a virtual setting (aka e-leadership), e-leaders should be aware of the uniqueness and limitations surrounding VT collaboration (Chamakiotis & Panteli, 2011). This will enable e-leaders to promote ICT use that contributes to rational decision making depending on the situation. Finally, educators should consider the findings in our chapters in order to provide students with an understanding of the fast-growing digitalization developments and their influence on management practices.

6. FUTURE RESEARCH DIRECTIONS

Our novel findings in this chapter give rise to additional avenues for future research. First among these possibilities is the black-box-like psychological mechanism of decision-makers, which is only on the rationality spectrum in this study. More in-depth studies can attempt to support the underlying psychological factors driving rationality. Future research could consider additional methods to enable a more detailed examination of rational decision-making gin the VT context, including, for example, real-time (daily) diaries so that research participants can reflect on the rationality of their decisions on a daily basis, overcoming this way limitations characterizing research interviews, whereby participants are asked to remember past situations and example. Furthermore, given our small group of participants in this study, future studies could consider larger samples and employ quantitative methods that could also test the statistical generalizability of our findings. Finally, given the fast advancements of ICTs, future studies should consider VTs that use new technologies, including AI chatbots, among others. With the current unprecedented uptake and dominance of digital technologies, which is gradually impacting human interactions, further studies might explore the cognitive impact of increased technological advancements and its influence on the work environment, working practices, and decision-making.

7. CONCLUSION

Our study in this chapter was inspired by an observed demand for further exploration in rational decision-making in VT collaboration. Our main contribution is the identification of factors which might influence rational decisions in VT collaboration: (a) efficient information integration via the usage of ICT tools; (b) reduced emotional impact on the decision-maker; and (c) better control and enhanced confidence of the decision-maker. These factors can be used to provide guidance for the application scenarios to optimize the performance of practitioners who form part of VTs. The limitations of our study mainly come from the complexity of psychological factors interacting with decision-makers' rationality, which could not be fully captured in this study, and other restrictions in the scope of the study. Still, we hope that our study can incentivize further and larger studies, as we discussed in Sections 6 above.

ACKNOWLEDGMENT

We would like to thank the participants in our study for their time and valuable insights; without their input, the study presented in this chapter would not have been possible.

REFERENCES

Alkaraan, F., & Northcott, D. (2013). Strategic investment decision-making processes: The influence of contextual factors. *Meditari Accountancy Research*, *21*(2), 117–143. doi:10.1108/MEDAR-09-2012-0031

Braun, V., & Clarke, V. (2006). Using thematic analysis in psychology. *Qualitative Research in Psychology*, *3*(2), 77–101. doi:10.1191/1478088706qp063oa

Chamakiotis, P. (2020). Virtual Teams as Creative and Agile Work Environments. In C. Grant & E. Russell (Eds.), *Agile Working and Well-Being in the Digital Age* (pp. 133–142). Springer International Publishing. doi:10.1007/978-3-030-60283-3_10

Chamakiotis, P., & Panteli, N. (2011). E-leadership styles for global virtual teams. In *Global business: Concepts, methodologies, tools and applications* (pp. 1688–1706). IGI Global. doi:10.4018/978-1-60960-587-2.ch608

Chamakiotis, P., Panteli, N., & Davison, R. M. (2021). Reimagining e-leadership for reconfigured virtual teams due to Covid-19. *International Journal of Information Management*, *60*, 102381. doi:10.1016/j.ijinfomgt.2021.102381 PMID:34934257

Davidavičienė, V., Al Majzoub, K., & Meidute-Kavaliauskiene, I. (2020). Factors affecting knowledge sharing in virtual teams. *Sustainability (Basel)*, *12*(17), 6917. doi:10.3390/su12176917

Dennis, A. R., Valacich, J. S., Speier, C., & Morris, M. G. (1998). Beyond media richness: An empirical test of media synchronicity theory. *Proceedings of the Thirty-First Hawaii International Conference on System Sciences*. 10.1109/HICSS.1998.653082

Draghici, A. (2008). Influence of multi-culturality in virtual teams. In *Methods and tools for effective knowledge life-cycle-management* (pp. 61–89). Springer Berlin Heidelberg. doi:10.1007/978-3-540-78431-9_4

Einhorn, C. S. (2020). How to make rational decisions in the face of uncertainty. *Harvard Business Review*. https://hbr.org/2020/08/how-to-make-rational-decisions-in-the-face-of-uncertainty

Gilson, L. L., Costa, P., O'Neill, T. A., & Maynard, M. T. (2021). Putting the "TEAM" back into virtual teams. *Organizational Dynamics*, *50*(1), 100847. doi:10.1016/j.orgdyn.2021.100847

Hardin, A., Davison, R. M., Schneider, C., Looney, C. A., & Sarker, S. (in press). Contextualising collective efficacy in virtual team research: The essential role of collaborative technologies in the virtual team efficacy conceptual framework. *Information Systems Journal*, 1–30. doi:10.1111/isj.12484

Hosley, C. F. (2010). *The perceived effects of technology on product management team collaboration* (Publication No. 3425728) [Doctoral dissertation, University of Phoenix]. University of Phoenix ProQuest Dissertations Publishing.

Schmidt, J. B., Montoya-Weiss, M. M., & Massey, A. P. (2001). New product development decision-making effectiveness: Comparing individuals, face-to-face teams, and virtual teams. *Decision Sciences*, *32*(4), 575–600. doi:10.1111/j.1540-5915.2001.tb00973.x

Simon, H. A. (1957). *Models of man: Social and rational; mathematical essays on rational human behavior in society setting*. Wiley.

Straus, S. G. (1996). Getting a clue: The effects of communication media and information distribution on participation and performance in computer-mediated and face-to-face groups. *Small Group Research*, *27*(1), 115–142. doi:10.1177/1046496496271006

Turel, O., & Connelly, C. E. (2012). Team spirit: The influence of psychological collectivism on the usage of E-Collaboration Tools. *Group Decision and Negotiation*, *21*(5), 703–725. doi:10.1007/s10726-011-9245-7

Wall, K. D. (1993). A model of decision making under bounded rationality. *Journal of Economic Behavior & Organization*, *20*(3), 331–352. doi:10.1016/0167-2681(93)90030-S

Wilson, C. G., Bond, C. E., & Shipley, T. F. (2019). How can geologic decision-making under uncertainty be improved? *Solid Earth*, *10*(5), 1469–1488. doi:10.5194/se-10-1469-2019

LIST OF ABBREVIATIONS

AI: Artificial Intelligence
COVID-19: Coronavirus disease 2019 (COVID-19) - Pandemic
F2F: Face-to-Face
ICT: Information and Communication Technology
MST: Media Synchronicity Theory
VT: Virtual Teams

Chapter 5
Strategies to Overcome Isolation in Remote Work

Nisha Thundiyil
https://orcid.org/0009-0008-4211-7545
St. Joseph's College, India

ABSTRACT

The chapter provides insight on the role of employers and also self-help by the employees to ward off isolation due to remote work. Hybrid work, remote work, and telework have become part of the new normal era. Remote work has added its own measure of woes to the trail of negative effects in terms of bonding within family left as residue by the overuse of mobile phones. Isolation and in turn depression are two undisputable negative impacts of remote work. This chapter lists an array of strategies that can be inculcated by the corporate sector to minimize and help their employees overcome the ill effects of isolation. The chapter starts with the meaning and conceptual background of remote work. The strategies to overcome isolation due to remote work form the focal point of this chapter, and these strategies have been listed point wise and explained.

DOI: 10.4018/979-8-3693-1314-5.ch005

REMOTE WORK: AN OVERVIEW

Human beings are social animals. Several motivational studies such as Maslow, Herzberg and so on have portrayed the role of social need as a major factor of motivation. It is an undisputable fact that the era of new normal has crept in with the advent of covid. The practice of telework and hybrid work has evolved as a part of the new normal era. This has led to a metamorphic change in work culture and work environment as well. Cropping up of unimaginable dimensions in the interaction between coworkers, changes in superior subordinate relationship, employee motivational factors, concept of office and work timing, changes in life style, sleeping patterns, virtual meetings, dress code and so on are few aftermaths of covid in the corporate sector. This chapter touches upon just a tiny aspect – isolation due to remote work and its impact and strategies to meet the challenge.

CONCEPTUAL BACKGROUND

Olson and Primps (1984, p. 97), refers working from home to "organizational arrangements which are enabled by technological advancements that allow employees to work at home on a regular basis, as a substitute for attendance at the normal workplace". Remote working is defined as "a flexible work arrangement whereby workers work in locations, remote from their central offices or production facilities, the worker has no personal contact with co-workers there, but is able to communicate with them using technology" (Di Martino & Wirth, 1990, p. 530).

Remote work is the practice of working outside of the traditional office, both in a temporal (time) and a spatial (location) sense (Olson, 1983).. Eurofound publications (Eurofound, 2015; Eurofound and the International Labour Organization, 2017; Eurofound, 2020, p. V), define remote work and telework thus: Telework and ICT-based mobile work (TICTM) is any type of work arrangement where workers work remotely, away from an employer's premises or fixed location, using digital technologies such as networks, laptops, mobile phones and the internet.

The definition of remote work, as quoted by M.H. Olson, is as follows (Olson, 1983, p. 182): "remote work refers to organizational work that is performed outside the normal organizational confines of space and time". Remote working has been defined by Collins et al. (2016) as work that is completed outside the employee's organizational office and using various technologies to communicate with colleagues and customers. For Julsrud (1998) telework is defined as working in the home for 5 or more hours, and mobile work as working outside the home and main office site for 5 hours of more.

The term 'telework' was originally coined in 1973 by Jack Nilles who defined telework as an activity which "includes all work-related substitutions of telecommunications and related information technologies for travel" (Collins, 2005, p. 115). The terms 'e-Work' and 'home-anchored work' have also been suggested as an alternative to 'telework' (Nunes, 2005; Whittle and Mueller, 2009).Hybrid work as a new concept that synthesizes most of the characteristics of the above definitions (Bloometal.,2022; Gratton,2022). It has been defined as "the combination of working in the office and working from home. The idea is to break an employee's working week into tasks, distinguishing between tasks that are typically best [done] in person, like meetings, training events, or mentoring on office days, and those that are best [done] individually, like reading, writing, or coding on home days" (Torre,2022; p.5).

Remote work is a work arrangement in which an employee resides and works at a location beyond the local commuting area of the employing organization's worksite (e.g., Mokhtarian, 1991). A remote worker can be self-employed or dependent on an employer. Telework is fully or partially carried out at an alternative location rather than the default place of work, and personal electric devices (i.e., telecommunications) are used to perform the work (e.g., International Labor Organization, 2020).

A teleworker can be a self-employed or a dependent worker. Home-based telework occurs at or from home using electronic devices. "Permanent teleworkers" spend more than 90% of their working time from home. "Supplementary teleworkers" or "regular teleworkers" spend one full day per week working at home. "Occasional teleworkers" have worked at/from home at least once in the last 4 weeks (e.g., International Labor Organization, 2020). Mokhtarian (1991, p. 3) suggested one reasonable definition of remote work: "work done by an individual while at a different location than the person(s) directly supervising and/or paying for it."

Telework and remote work provide the viewpoint of an individual working remotely using or not using ICT to interact and work with others. The use of technologies also links telework to virtual teamwork. Digital online telework is a form of employment that uses online platforms to enable individuals, teams, and organizations to access other individuals or organizations to solve problems or to provide services in exchange for payment (e.g., Berg et al., 2018).

'Telework' and a host of other terms, such as 'homeworking', 'tele homeworking', 'telecommuting', 'remote working', 'virtual work', 'electronic homeworking' and 'distributed work' have been used interchangeably (Golden and Eddleston, 2018; Haddon and Brynin, 2005; Illegems and Verbeke, 2004; Lautsch et al., 2009; Nunes, 2005).Different types of telework have also been discussed and scholars usually agree on three main categories: home-based work or homeworking, group-based teleworking including satellite-office and neighborhood office centers and mobile telework or otherwise called nomadic (Nunes, 2005; Pérez et al., 2002; Taskin and Devos, 2005; Wilson and Greenhil, 2004; Morganson et al., 2010).

According to the International Labor Organization (2020, p. 6), the basic difference between telework and remote work is that a teleworker uses personal electronic devices in addition to working remotely.

Forms of Remote Work

The term remote work though strictly refers to the employee residing in a different location takes on various forms. Usually in the event of remote work, the employee shall be residing in his/her own native place whereas the office shall be located in a totally different place. Apps and social media and the internet play a major role in ensuring efficient and effective functioning of remote work

Hybrid work is another form of remote work. As the term indicates, it is a blend of remote work and office work. Few days in a week, the employee is required to be present in the office and the remaining days, he may work from home. The hassle in this type of work is that if the residence of the employee is located in a far off place, he cannot commute between office and home in a week. He will have to relocate yet not be able to enjoy the privileges of full time office work.

Telework is very much similar to remote work, the major difference being that as far as a remote worker is concerned, he may be working from anywhere, usually his home, which may be quite far off. In the case of teleworker, he shall be working from outside the office but comparatively nearer to the office. A remote worker may meet his seniors once in a blue moon whereas a teleworker shall have the opportunity to meet the colleagues and superiors more often.

Table 1. Top industries for remote workers in 2023

Computer and IT
Marketing
Accounting and Finance
Project Management
Medical and Health
HR and Recruiting
Customer Service

Source: https://www.forbes.com/advisor/business/remote-work-statistics/

BENEFITS OF REMOTE WORK

There are end number of benefits and luxuries that a remote worker enjoys.

Less Travelling

One of the most enjoyable features in remote work is the leisure of not having to commute to office and back home. This indisputably leads to less fatigue and saving of time.

Less Stress

The stress associated with travelling, probably staying away from family, time for personal work and quality me time, work life balance and so on are expected to be eliminated via remote work

Less Absenteeism

The absenteeism among employees is to a great extent avoided via remote work as the employee feels obliged to work even if he is sick due to being able to work from the comfort of his home and take intermittent rest. Similarly, remote work provides the option to employees to work even while traveling or attending social functions or gatherings.

Convenient Resources and Infrastructure

An employee working from home finds it easy to purchase customized furniture for and other resources for work rather than depend on the standardized furniture and resources provided by the company. There are several employees facing back pain and neck pain issues, more so in the IT sector. Remote work provides them the opportunity to work in a more comfortable environment that suits their comfort level.

Comfortable Dress Code

Remote work enables the employees not to undergo the necessity of adhering to specific dress code. Rather they have the option to dress in the attire that they are more comfortable in as they are working from their home or the most comfortable place that they are in.

Less Gossips and Rumors

Remote work, in a way, eliminates gossips and rumors as the avenues for face-to-face conversations among employees are restricted. Moreover, the bonding and trust between the remote workers is comparatively less and this put a mental block on them not to gossips due to mistrust.

Enjoy Home Cooked Food

Food is the basic need of man. Inspite of restaurants being available, people often prefer to eat home cooked food for health purposes as well as avoid expenses. The opportunity to work from home enables the workers to eat hot, cooked food during work and also eat at their choice of time.

Less Work-related Miscellaneous Expenditure

It is a fact that workers often need to shell out money for miscellaneous expenditures in relation to work. These are generally unavoidable s they are necessary to maintain social relationships, personal expenses during travel and so on. But remote work gives an option to the workers to cut down such types of expenses.

Better Work-Life Balance

The benefit of time available to spend with the family is one of the major advantages of remote work. It provides ample opportunities and time for the employee to be present with family, take cate of their needs and provide moral support with their presence.

CHALLENGES OF REMOTE WORK

There are several challenges to the employer and the employee as well in the event of remote work. An attempt has been made to trace few challenges in relation to remote work.

Less Control Over Employees

Along with remote work comes the hassles of not able to keep a tab on the employees. Moonlighting is one the most recurrent problems associated with remote work.

Excess Workload

Though it may seem that remote work offers less work load for the employees, in many cases it is just the opposite. Unscheduled meetings, expectation that the employee is available for work 24/7, pushing off excess work to be completed with free time in between and son are the grievances often voiced out by remote workers.

Higher Targets

It is generally felt by the remote workers that highest targets are often pushed on to them, probably due to the lackage of face-to-face interaction between the boss and the employee.

Minimized Social Interaction

Remote work is conspicuous due to its lackage of social interaction. The workers seldom get a chance to mingle with their colleagues. They often find it difficult to attend social gatherings, entertainment activities and so on if they shuffle the office work and their personal work together.

Less Friends

Having a smaller number of friends is another negative aspect associated with remote work. These isolated lot get very minimum opportunities to mix wholeheartedly with people be it colleagues or others.

Health Issues

Physical as well as mental depression are two negative impacts of any sort of sedentary work, more so in the case of remote work. Poor posture, long hours of siting, poor neck posture while using laptop, eye problems due to prolonged screen time are part and parcel of the package of remote work.

More Stress and Tension

On one hand, remote wok is associate with less stress. But it is noticed that many remote workers lament about more tension and stress especially due to excess work load, no companion to share their stress with or pass a comment and have a good laugh.

Risk of Outdated Skill Set

Remote workers face the danger of not able to upgrade their skill set as they are mostly out of circulation from the field. This poses the danger of being outdated in their knowledge and skill set.

Communication Barriers

The distance, time, no face-to-face interaction, lack of emotional bonding between employees and also with the company paves way for many problems, the major issue being communication barrier.

Higher Expenses

Remote workers spend more in terms of electricity bills, inter connection, personal furniture for work, laptops and so on which in turn eliminates the myth that remote work leads to less expenses.

Less Systematic Work

An office goer has fixed work timing but he same cannot be said of a remote worker as most companies make it a point to call on duty the remote workers any time of the night or day, especially if the time zones differ for the employer and the employee.

Erratic Sleeping and Eating Patterns

Erratic sleeping and eating patterns are noticeable in remote worker s as they do not have a specific work time. Remote workers are noticed to eat and sleep when they are free off work which leads to minor and later serious health hazards.

Lethargy

An accusation faced by the remote workers is that they tend to lethargic in their duties and play truant often due to the lack of direct supervision and physical submission of work.

Less Competitive

The isolated environment far away from competitors may create feeling of contented as they may not be aware of the competition in the real work scenario. Moreover, the spirit of team work and team spirit may seem to be missing.

Performance Appraisal Difficulties

The superiors may find it a little difficulty to appraise the remote worker as the full essence and, magnitude of their work load and effort may seem a bit blurry due to the lackage of direct interaction.

Depression

Increased depression with no outlet to vent out the emotions, no shoulder to cry on or no one to share the office burden leads to acute depression. It is not possible for everyone to share office stress with family members. In such situations, colleagues offer solace. Deprivation of social needs leads to depression and poor mental health.

Irritation

Seeing the same face of the family members and hearing the same voice full day and night may create irritation in the mind of people.

Stress

Stress is yet another negative impact of isolation. Isolation leads to over thinking, no sharing of burden and in turn creates stress.

Poor Physical and Mental Health

The prolonged sitting posture and too much screen time leads to several chronic diseases such as eye problems, headache, back pain, blood pressure, diabetics and so on. It poses serious mental issues such as depression too.

Poor Creativity and Dull Mind

The absence of direct contact with peers and professionals leads to absence of sharing of ideas, unaware of latest trends, less creativity and a dull mind.

ISOLATION: AN AGENT OF DEPRESSION

When an employee opts for remote work one of the unexpected problems that he faces is the challenge of isolation. From the bubbly environment of campus as a college student or from the enthusiastic environment of office, the sudden push to a situation confined to the 4 walls of his house is the major havoc that covid initiated and in a way, corporate sector continues the trend via remote work. Initially this situation may be found as welcoming as there is seemingly no need to wake up early, dress up or travel to work and has ample time on his hands for family and friends. The comfort of the home may seem inviting at the initial stages as the employee sits comfortably in his chair or maybe even in his bed with a lap top in front of him, tea and breakfast probably served in his bed by the family members. As days roll by, monotony creeps in and the need to crack a joke or the urge to steal a laugh or the need to clear a doubt arises. After the initial days of enjoying the luxury of home, the weariness of routine and no direct mingling with colleagues lifts its neck.

The realization that remote work translates as 24/7 work, less friends, narrow social circle, less laughter, tight targets and so on acts as kerosene poured on to burning fire of isolation already omnipotent with the increased usage of mobile phones that has created the culture of family members living in the virtual world, though residing in the same house.

STRATEGIES TO OVERCOME ISOLATION

It is imperative that adequate steps ought to be taken from the side of the employer to ensure that the remote worker remain in an active and happy state of mind in order to extract maximum productivity. It is necessary that the employer puts effort to drive away isolation so that the employee is at ease. Below mentioned are few strategies to bring down isolation among remote workers.

Daily Huddles at a Scheduled Time

These brief online meetings keep the employees aligned with the expected role and duties they are expected to execute for the day. The brief opportunity to interact makes the remote worker active and feel part of the team. These interactions create a feeling of team spirit and relieve monotony of work amongst the remote workers. They improve the feeling of job accountability and also serves as a platform for problem solving. These meetings facilitate immediate feedback and help them to plan their day effectively based on input received during the huddle. It makes the employee more connected to the mission and vision of the organization and provide mechanism for updates leading to adaptation to organizational changes.

Weekly Meetings

Weekly online meeting to plan and discuss and appreciate the work of the employees create a feeling of something to look forward to and being part of the team feeling and gives a loop for social interaction. It provides a regular and structured platform to discuss projects, challenges and opportunities. It ensures the involvement of the remote workers also in planning and help them focus on priorities. Progress reporting during these meetings keeps them focused and makes them accountable. It ensures that they are able to plan their week ahead and facilitate good work life balance.

Staff Tours and Get -Together

Staff tours and periodic get-togethers often prove o be a source to relieve isolation and work monotony and provide a golden opportunity to the remote workers to meet and mingle with both regular and remote workers. The fun and enjoyment during these programs help eliminate isolation to a large extent. It leads to strengthening of team bonding and stronger relationships. It promotes the feeling of belongingness, improved communication and networking opportunities. It serves as a platform for celebrating achievements, knowledge sharing and reinforcement of company culture. It provides a break from the monotony of routine work and relives stress.

It serves the indirect impact of contributing towards improved communication, better bonding among employees, more networking opportunities, reinforcing company culture, exchange of ideas which may lead to innovations and boost morale and mental health of workers.

Remote Worker's Get-Together

Exclusive get together for the remote workers create opportunities for them to meet each other, have better bonding and be familiar with each other leading to social connection and better friendship. Such get-togethers result in stress reduction, team building and team work.

Training Programs

If the companies arrange training programs for the employees, it gives them an opportunity to update their knowledge and skills and also a way to meet their peers. This ensures that the remote workers are also in tune with new technologies paving way for professional development, increased productivity and improved self confidence among the remote workers. Training programmes ensure that the remote workers are molded to draw boundaries and differentiate their personal life from professional life. It provides them input to carve better time management and sketch out routines. It trains them to develop the flexibility required to handle changing tasks, priorities and work environments.

Pairing of Work

Pairing of work is also referred to as job sharing or collaborative work arrangement. When remote workers are put in groups and group work is allotted or when employees are put in pairs and allotted duties, it goes a long way to bring down feelings of isolation. This results in sharing of knowledge and expertise, reduced workload, flexible scheduling, improved accountability, peer support, resource sharing, efficient problem solving and sense of team work.

Dress Code

Dress code may not directly influence feeling or isolation but insisting on adherence to a specific dress code while at work and insisting to switch on the video during meetings and interactions gives a feeling of professionalism to the employees and the remote worker may get a feeling of being at work. It contributes towards stabling a feeling of professional identity, boost self-esteem and positive attitude. It helps the remote worker to mentally switch on to work mode. Remote workers may interact with clients, customers or team members through video calls, email or other channels of communication. Maintaining a professional dress code helps to create a positive and professional image.

Surprise Gifts During Festivals

Surprise gifts sent by the company to employees makes the employee feel wanted and he is induced to feel team spirit and belongingness. It gives an emotional boost to a person down with the feeling of loneliness. It is an effective method to remind the remote worker that he is valued and appreciated by the organization. It creates a sense of connection and reinforce the idea that remote workers are part of a large community. It helps to promote loyalty towards the company and makes the remote worker feel special. It enhances relationship of the remote worker with his colleagues and superiors. It may also result in encouraging the feeling of reciprocity leading to active participation of remote workers in team activities thus reducing isolation. It enhances the festive mood and make the remote workers feel included in the celebration. It reinforces the feeling that remote workers are a valued part of the organization family.

Online Games

Conducting online games periodically among employees help the remote worker be relieved of loneliness and monotony. This social interaction provides a sense of connection. It encourages workers to collaborate and build strong relationship with their colleagues. It serves as a form of stress relief and reduce emotional strain. It indirectly boos self-esteem and provide a sense of accomplishment. It helps them to forma bonding with colleagues who share their same passion. These activities keep the remote worker engaged and prevent feelings of boredom and isolation. Online games bring together people from diverse backgrounds and locations which fosters a sense of inclusion, helping the remote worker appreciate different perspectives and experiences. It offers a break from work related stressors and serves as a positive distraction when kept under control. It acts as a counter effect against isolation and over seriousness. It is to be noted that gaming shall not become an addiction and a balance should be kept between games and work.

Acknowledging Online Presence During Meetings

During staff meetings, if the online presence of the remote worker is acknowledged and included, it makes the employee feel accepted and boosts his self-esteem and confidence. It sends a clear message that they are acknowledged as part of the team. It fosters a sense of belongingness though separated by distance. It conveys respect for the remote worker's time and efforts and motivates them to actively participate in meetings. It ensures that they don't feel like invisible participants. It establishes a feeling of team work and unity. It brings down feelings of bias, if any. It promotes job satisfaction and positive approach towards work and also boost the morale of the employees. It leads to the remote workers taking ownership of their work and be more accountable.it also has a positive impact on the emotional wellbeing of the individual.

Freedom to Visit the Company Anytime

An open invitation and freedom for the remote worker to visit the organization anytime motivates the remote worker and instills in him a feeling of togetherness and bonding and organizational loyalty. These personal connections can help the remote worker to build stronger relationships and improve team work. These visits shall help the remote worker to imbibe the organization culture and enhance their understanding and alignment with the organization.it also leads to improved networking opportunities which can be valuable for career growth and knowledge sharing. They also get an opportunity to be mentored by more experienced employees.it paves a way for them to participate in problem solving discussions that may be challenging to achieve through virtual communication alone. certain asks and projects require hand on experience which the worker can gain through these visits. Being able to attend on site trainings, workshops and other programs shall help in enhancing the skills and knowledge of the remote workers. It provides then a choice and control over their work environment.

Frequent Video Meetings and Regular Communication

Incorporating frequent video meetings and regular communication into the work routine create a sense of presence and connection.it encourages both work related and social conversations. It makes the remote worker feel wanted and feel instills a sense of work environment, even while at remote work.

Systems

Having a scheduled check in time, check out time, routine huddles, meetings and so on instills a systematic mindset in the mind frame of the remote workers. They get the feeling of being at work, though physically being at home.

CONCLUDING NOTE

The new normal is here to stay. Changing work environments and changing work culture is part and parcel of the new normal. Remote work has offered numerous opportunities to the corporate sector. The corporate world has taken remote work in its stride. It is essential that the workforce also follow suit. The challenges faced by remote work has to be tackled with various strategies so that the disadvantages of remote work may be turned to advantages for the individual and organization as well.

REFERENCES

Becker, W. J., Belkin, L. Y., & Tuskey, S. E., & Conroy, S. A. (2022). Surviving remotely: How job control and loneliness during a forced shift to remote work impacted employee work behaviors and well-being. *Human Resource Management*, *61*(4), 449–464.

Beland, L. P., Brodeur, A., Haddad, J., & Mikola, D. (2020). *Family stress and domestic violence: Remote work, isolation and bargaining power*. Academic Press.

Bell, C. J. (2020). *Feeling remote: Factors influencing isolation in remote workers* [Doctoral dissertation]. Middle Tennessee State University.

Bollestad, V., Amland, J. S., & Olsen, E. (2022). The pros and cons of remote work in relation to bullying, loneliness and work engagement: A representative study among Norwegian workers during COVID-19. *Frontiers in Psychology*, *13*, 1016368.

Bulut, S., & Maimaiti, R. (2021). Remote working in the period of the COVİD- 19. *The Japanese Psychological Research*, *3*(1), 60–65.

Colbert, C. J. (2011). *Telework as part of a business continuity strategy: A path toward organizational resilience*. University of Maryland University College.

Di Martino, V., & Wirth, L. (1990). Telework: A new way of working and living. *Int'l Lab. Rev.*, *129*, 529.

Flores, M. F. (2019). Understanding the challenges of remote working and it's impact to workers. *International Journal of Business Marketing and Management*, *4*(11), 40–44.

Ha, N. (2021). Workplace isolation in the growth trend of remote working: a literature review. *Rev economic Bus Stud, 14*(1), 97-113.

Jalagat, R., & Jalagat, A. (2019). Rationalizing remote working concept and its implications on employee productivity. *Global Journal of Advanced Research, 6*(3), 95-100.

Jämsen, R., Sivunen, A., & Blomqvist, K. (2022). Employees' perceptions of relational communication in full-time remote work in the public sector. *Computers in Human Behavior*, *132*, 107240.

Jones, Y. (2022). *Factors Contributing to Workplace Isolation in Remote Workers* [Doctoral dissertation]. Trevecca Nazarene University.

Julsrud, T. E. (1996). Teleworking–the vision. *Feature Status, 16*.

Kłopotek, M. (2017). The advantages and disadvantages of remote working from the perspective of young employees. *Organizacja i Zarządzanie: kwartalnik naukowy,* (4), 39-49.

Miyake, F., Odgerel, C. O., Hino, A., Ikegami, K., Nagata, T., Tateishi, S., ... Ishimaru, T. (2022). Job stress and loneliness among desk workers during the COVID-19 pandemic in Japan: Focus on remote working. *Environmental Health and Preventive Medicine*, *27*, 33–33.

Nickson, A., Gair, S., & Miles, D. (2016). Supporting isolated workers in their work with families in rural and remote Australia: Exploring peer group supervision. *Children Australia*, *41*(4), 265–274.

Noh, E., & Lee, K. H. (2022). The Empirical Research on Professional Isolation During Remote Work in Context of COVID-19 and Organizational Citizenship Behavior: Sequential Mediation Effect of Psychological Empowerment and Affective Organizational Commitment. *Korean Journal of Industrial and Organizational Psychology*, *35*(3), 513–558.

Olson, M. H. (1983). Remote office work: Changing work patterns in space and time. *Communications of the ACM*, *26*(3), 182–187.

Olson, M. H., & Primps, S. B. (1984). Working at home with computers: Work and non-work issues. *The Journal of Social Issues*, *40*(3), 97–112.

Song, S., & Song, J. (2022). A Study on Workplace Isolation in Remote Working: The Analysis of Workplace Isolation Perception and The Effect on Work Engagement. *Korean Journal of Industrial and Organizational Psychology*, *35*(3), 559–581.

Taser, D., Aydin, E., Torgaloz, A. O., & Rofcanin, Y. (2022). An examination of remote e-working and flow experience: The role of technostress and loneliness. *Computers in Human Behavior*, *127*, 107020.

Toscano, F., & Zappalà, S. (2020). Social isolation and stress as predictors of productivity perception and remote work satisfaction during the COVID-19 pandemic: The role of concern about the virus in a moderated double mediation. *Sustainability*, *12*(23), 9804.

Van Zoonen, W., & Sivunen, A. E. (2022). The impact of remote work and Mediated communication frequency on isolation and psychological distress. *European Journal of Work and Organizational Psychology*, *31*(4), 610–621.

Wang, B., Liu, Y., Qian, J., & Parker, S. K. (2021). Achieving effective remote working during the COVID-19 pandemic: A work design perspective. *Applied Psychology*, *70*(1), 16–59.

Chapter 6
Relationship Between Remote Working and Work Outcomes of Constantly Connected BYOD Knowledge Workers

Melina Doargajudhur
https://orcid.org/0000-0002-0287-9691
Edinburgh Napier University, UK

Zuberia Hosanoo
Edinburgh Napier University, UK

Geshwaree Huzooree
https://orcid.org/0000-0001-8076-6636
Edinburgh Napier University, UK

ABSTRACT

Remote working is growing in popularity globally, and the COVID-19 pandemic has contributed to this trend. While the idea of remote work is not new, the reality is that it is now more common than ever in all industries and organisations where it can be implemented, and it is becoming a preferred method of working globally. The issue of remote work is current, and given how quickly it has spread throughout the globe, it is critical to look at how this new working paradigm impacts work outcomes of constantly connected BYOD (bring your own device) workers. The chapter examines the effects of working remotely for this type of employee. It also presents the remote working trend across the globe, the theoretical lens, and the reason for choosing the

DOI: 10.4018/979-8-3693-1314-5.ch006

job demands-resources theory. A research model is proposed by carefully analysing the key constructs related to the impacts of remote working for BYOD users. The chapter explores strategies and recommendations tailored to enhance the remote work experience for these employees, while also providing insights into future directions.

INTRODUCTION

In the current digital era, working remotely and constant connectivity are closely related concepts. The internet and the development of cutting-edge technology have completely changed how people communicate and work, making the "Bring Your Own Device" (BYOD) practice and remote work more common and accessible than ever. This contributed to people increasingly opting to fulfil their organizational duties using their preferred devices. Hence, it can be contended that the BYOD philosophy, constant connectivity, and remote working all contribute to a contemporary work paradigm that has been increasingly popular in recent years. Together, these three components transform how individuals operate and how businesses handle technology and security (Buonomo et al., 2023; Arunprasad et al., 2022).

Businesses are also increasingly implementing BYOD programs and encouraging remote work. The BYOD phenomenon facilitates constant connectivity to work resources, and these trends are closely connected to remote working, by enabling employees to access work-related applications, emails and communication on their personal devices from anywhere. Workers in contemporary companies are now constantly connected as a result of this strategy, particularly during and after the COVID-19 pandemic. Therefore, it is argued that BYOD and constant connectivity are crucial factors in remote working and have grown in popularity since the pandemic, when businesses were forced to make a hasty switch to remote work arrangements (Buonomo et al., 2023; Arunprasad et al., 2022; Galanti et al., 2021). Since, constant connectivity is a foundational element of remote working, it allows workers to remain connected, collaborate, share materials irrespective of place and location, whether they are in another country or working from a coffee shop or from home (Wang et al., 2021).

Although the BYOD and constant connectivity phenomena are not new study topics, they have gained more popularity in the current teleworking environment as a result of the recent global pandemic across industries. BYOD is a logical organizational response to the surge in work from home arrangements that organizations have enthusiastically welcomed. Due to the recent global pandemic, this topic has gained more attention. IT workarounds and business communications are anticipated to emerge, subjecting employees to nonstop email notifications, a steady stream of task and message reminders, and other work-related activities (Arunprasad et al., 2022; Yang et al., 2022).

Since the culture of work-life integration is becoming more prevalent, the BYOD trend shows no signs of subsiding. Contrarily, businesses that were initially hesitant to using privately owned tools to carry out organizational tasks because of information security risks (Chen & Chen, 2020) have come to understand their importance since the Covid-19 pandemic established the work-from-home culture that has now come to be accepted as the norm.

There is little agreement in the literature regarding the dual effects of constant connectivity on employees' psychological well-being and work-related outcomes such as productivity, organizational commitment, burnout and stress (Hill et al., 2022; Fan & Moen, 2024). Despite the fact that the BYOD trend for work practices has continuously expanded and enables people to perform work duties regardless of the limitations imposed by time and place (Gewald et al., 2017; Köffer et al., 2014; Niehaves et al., 2012; Weeger et al., 2016; Fan & Moen, 2024), this is still the case. It is therefore debatable whether constant connectivity has both beneficial and bad effects on workers. For instance, research frequently links technology use to increased productivity and flexibility. Similar to this, the widespread availability of employees has led to a variety of unusual work arrangements, such telecommuting and flextime (McDaniel et al., 2021). While employees occasionally struggle to strike a balance between their personal and professional lives, technological adoption also contributes to the blurring of the barriers between work and non-work realms. Since work may now be done "anywhere, anytime," leading to work "everywhere, all the time," there is less time currently set aside for non-work activities (Cousins & Robey, 2015; Hill et al., 2022).

In a similar vein, some have even coined the term "Bring Your Own Danger" to describe BYOD and contend that it raises security risks, specifically with regard to data integrity, confidentiality, and authenticity. Organizations are aware of these hazards, but end users frequently do not comprehend or address them, and setting up appropriate security policies for BYOD can be challenging for them. The use of BYOD may potentially give rise to legal and regulatory problems. Diverse jurisdictions may have different privacy and data protection laws, which can present compliance challenges. Additionally, data accuracy and other technical problems may arise from mobile devices moving in and out of network coverage, which can cause stale or outdated data to accumulate on the device. However, these risks can be mitigated by striking a "manageable compromise" between laissez-faire and authoritarian ideologies (Doargajudhur & Dell, 2020; Doargajudhur & Hosanoo, 2023).

The contrasting views of employers and employees regarding the effects of constant connectivity on them have also been supported by several studies. While some employees claimed that using their personal devices to conduct job-related tasks led to work-life conflict (WLC), others reported that finding the right balance between work and family was simpler. In addition, they purported that the technologies improved their view of their job satisfaction and productivity at work, but numerous other employees claimed that the technologies increased their workload and degree of job-related stress (McDaniel et al., 2021; Arunprasad et al., 2022; Fan & Moen, 2024). The main goal of this chapter is to explain these seemingly contradictory findings in light of these discussions and findings through an in-depth exploration.

Additionally, since the global pandemic has changed how businesses operate globally, it is crucial to shed light on this gap from both a theoretical and practical standpoint. Organizations began to promote teleworking as a response of the Covid 19 emergency lockdown, which increased the relevance of personal technology adoption - especially at a time when organizational tools were scarce. Because it allows workers to conduct business-related tasks and communicate despite social distancing policies and altered working conditions, the BYOD practice eventually became a standard practice.

According to earlier research (Barlette et al., 2021; Mazmanian et al., 2013), the pervasiveness of new devices has changed how employees feel about their jobs by giving them more autonomy in selecting their preferred locations and flexible times to complete organizational duties. Additionally, employees' self-imposed blended work-life behaviors keep them constantly on call during weekends and holidays, increasing stress levels and reducing their ability to disconnect. Hence, the chapter further discusses how exploring the experiences of constantly connected BYOD knowledge workers contributes to a deeper understanding of contemporary work dynamics and provides valuable insights for organizations to thrive in this evolving landscape.

LITERATURE REVIEW

BYOD Trend

The turning point for BYOD started with the advent of smartphones and tablets. Although laptops were certainly more portable, making it possible to commute from home to work, tablets and smartphones made it even simpler to live in two different worlds. According to Moschela et al. (2004), it was a completely new world that kept changing. After its inference by Intel in 2009, this practice of allowing employees to use their personal devices has garnered significant attention from researchers in recent years.

Remote Working Trend

Over time, the definition of "remote working" has evolved. It is also known as teleworking (Bailey & Kurland, 2002), and telecommuting (Torten et al., 2016). It is not clear if the use of different terminologies has significantly altered the definition of remote working. Bailey and Kurland (2002, p.384) define telework as "working remotely and interacting with the traditional workplace through telecommunications or computer-based technology". In the 1970s, the idea of remote working was introduced, but it did not catch on as quickly as anticipated (Torten et al., 2016). After the current COVID-19 epidemic, workers were ordered to report home to work worldwide (Chang et al., 2021). Previously considered an optional work arrangement, remote employment is now required under COVID-19, even for professions deemed essential (Chang et al., 2021).

Remote and hybrid working had been gradually growing before the pandemic. Between January and December 2019, almost 1 in 10 (12%) of the workforce in the UK reported working at least one day from home the week before, and about 1 in 20 (5%) said they did so on a regular basis. This significantly rose throughout the pandemic, reaching a peak in June 2020 when 49% of workers in Great Britain (GB) worked at least one day from home; 11% of the workforce did so, and 38% did so exclusively. These numbers have steadily reduced again once the pandemic restrictions were released, although they are still greater than before the outbreak. In September 2022, almost 1 in 5 (22%) British citizens. These numbers have steadily reduced again once the pandemic restrictions were released, although they are still greater than before the outbreak. Around 1 in 5 (22%) of the workforce in

Great Britain worked from home at least once the previous week, and 1 in 8 (13%) did so exclusively as of September 2022 (Haan, 2023).

REMOTE WORKING ACROSS THE GLOBE

The opportunities for working from home are less in developing nations than in developed ones. Not all employees in a given nation have the same opportunities to go from an office setting to a remote one. Furthermore, it may be more difficult to work from home successfully if there are infrastructure or service access issues. Productivity, job satisfaction, stress levels, and work-life balance can all be impacted by working from home (Viollaz, 2022).

Since the COVID-19 epidemic, remote working has become the norm for millions of individuals worldwide. However, depending on where you live, your rights as a remote worker may be significantly different. According to surveys, many who had to work remotely due to pandemic lockdowns now choose to work from home instead of commuting. According to a 2021 World Economic Forum research, two thirds of workers preferred the flexibility to work remotely (Broom, 2023).

In many countries, it is lawful for employees to ask for flexible or remote work arrangements. For instance, the UK has announced that it is removing a 26-week qualifying period and making it a day-one right when starting a new employment to make it easier to work remotely. In a similar vein, a 2022 McKinsey & Company study indicated that nearly 6 in 10 Americans had been given the option to work from home at least one day per week, underscoring the rising popularity of remote working among both businesses and employees. Over a third of respondents claimed they could currently work from home five days each week. 87% of those who had the option to work flexibly claimed to have done so (Broom, 2023).

More than 2,000 remote workers from around the world participated in a 2022 study by the social media management platform Buffer. The results showed that the experience was extremely favorable; 97% of respondents indicated they would prefer remote working versus being office-based. Some people have taken remote working a step further by moving to foreign countries, particularly those whose jobs, like programming and digital design, can be done from anywhere with a half-decent internet connection (Haan, 2023).

At the most recent count, 49 countries currently issue visas to remote workers, often known as digital nomads, allowing some of them, like Spain, to come to their country with hefty tax incentives. Others provide financial inducements for remote workers to relocate to rural areas in an effort to revitalize local economies. While some nations permit employees to work remotely under specific situations, in others, doing so may require your employer's permission, which isn't always given, particularly when the economy deteriorates (Haan 2023; Broom, 2023).

The COVID-19 pandemic has prompted businesses to rapidly adapt and stress-test their systems like never before. As organizations undergo transformation, new challenges and priorities emerge, including the need for real-time decision-making, online staff training, and risk management for business continuity, with cybersecurity topping the list. According to a recent report, the pandemic has witnessed a surge in cyber threats, with approximately 600 malware attacks, 800,000 spam messages, and 50,000 hits on malicious websites reported since May 2020. Spam emails have seen a staggering 300-fold increase from February to March 2020, accompanied by a 300% rise in malicious URLs (Cook, 2020).

THEORETICAL UNDERPINNING

A thorough review of the literature on work outcomes revealed a number of widely accepted theoretical stances, including the social cognitive theory, the job control model, the cognitive model of stress, the job characteristics model, and the job demands-resources model, among others. Although these lenses have been extensively used in earlier research, it is believed that they also have serious shortcomings. Some of the frameworks, according to Bakker and Demerouti (2014), put a lot of emphasis on work motivation and stress. Additionally, each of these models only pays attention to one of the following outcomes: job stress, work motivation, work-life balance, work-life conflict, job performance. These models are criticized as well for neglecting the foundations of other current frameworks by using a relatively simple approach. These theories are generally acknowledged to be somewhat static, advancing that only a few constructs are recognized to showcase various task settings in a homogenous manner. Furthermore, research often fails to acknowledge how unstable occupations might be in today's rapidly evolving workplaces. These theories in the literature on satisfaction, motivation, performance, work-life balance, stress, and work-life conflict have ignored one of these crucial components (Bakker & Demerouti, 2014). The job-demand resources model is appropriate for the present study, according to the authors, who also argue that additional work-related outcomes like job performance, job autonomy, work pressure and work-life conflict, should be studied concurrently.

The Job Demands-Resources (JD-R) model (Demerouti et al., 2001) is a parsimonious and flexible framework which describes occupational stress as an imbalance between the demands placed on an individual and the resources they have access to at work. Jobs with high demands incur expenses on a physical, organizational, social, and psychological level and involve continual mental or physical effort. Job resources, on the other hand, are characteristics that help employees achieve their goals at work, promote learning and growth in themselves and others, and lessen the pressures of their jobs and the negative effects that come with them.

The JD-R theory also makes the premise that the two types of job features—job resources and job demands, initiate two separate processes, namely motivational impairment and health impairment. Job resources have a motivational effect and result in excellent performance, increased engagement, satisfaction, and low cynicism. Job resources also meet fundamental human needs like relatedness, autonomy, and competence (Deci & Ryan, 2013). The second job category, job demand, on the other hand, is characterized by occupational expectations, which are thought to be the most significant risk factors for psychosomatic health symptoms, repetitive strain injury, tiredness, and a number of other connected health problems. This is primarily caused by the inadequacy of job design or, occasionally, by recurring workplace demands such work overload, intense work pressure, and emotional demands (Demerouti et al., 2001).

The JD-R theory has been used in numerous research to support the dual routes to employee well-being and how they affect crucial organizational outcomes. The JD-R is a useful theoretical framework for this research for a number of reasons. First, this theory conceptualizes health impairment and occupational well-being that result from working conditions but are ignored by the models outlined above. The JD-R framework offers a better overview of workers' work behaviors regarding the usage of their private tools because the primary focus of this research is on work outcomes. Thirdly, this model illustrates how the resources given to employees and the expectations imposed on them interact in ways that the other lenses had missed. Therefore, this study enlightens scholars on how resources like perceived job autonomy are leveraged to lessen demands like perceived workload on the employees. Finally, the dynamic aspect of the JD-R model is appropriate for the evolving nature of work in the current unstable work environment.

METHODOLOGY

We followed a systematic literature review process to identify the effect of constant connectivity while working remotely on outcomes connected to the workplace. Employing an extensive search strategy, we used specific search terms such as "Constant connectivity", "remote working", "teleworking", "BYOD", "Work Outcomes", "Job Performance", "Autonomy", "Covid-19 and Remote Working" in available reputable online databases like Proquest, Google Scholar, Science Direct, Emerald. As a result, a number of articles, reports and conference papers were produced. These were subsequently narrowed down to only those that were most recent and pertinent to our research. The next step was to carefully evaluate each paper's title, keywords, and abstract to see whether it was appropriate for this research focus. Quality assessment measures were applied to enhance the reliability and validity of the selected studies. This involved examining elements such research design, methodological approach, sample size, data collection techniques, and statistical analyses employed in each study. This rigorous methodological approach was adopted to address potential publication biases and consider the publication years, ensuring a balanced representation of the literature.

The extensive literature search process uncovered often applied theoretical lenses like those that are discussed in the section below. Even though some studies widely adopted these perspectives, they also have drawbacks. The discussion that follows delves into each of those work outcomes theories with the goal of analyzing the impact of remote working, through constant connectivity, on employees' performance and well-being. This comprehensive methodology ensures the robustness and reliability of our literature review, subsequent analysis and research findings.

WORK OUTCOMES ASSOCIATED WITH REMOTE WORKING OF BYOD USERS

The variables—perceived job autonomy, perceived job performance, perceived work motivation, organizational commitment, perceived job satisfaction, perceived workload, perceived work pressure, work-life conflict, job stress, and burnout—were extracted from the aforementioned analysis. Figure 1 below illustrates the work outcomes resulting from the impact of remote working of constantly connected knowledge workers. It should be emphasized that the list of work outcomes is not exhaustive, and more researchers may find other consequences related to work that arise from BYOD users' remote working.

Figure 1. Definition of constructs

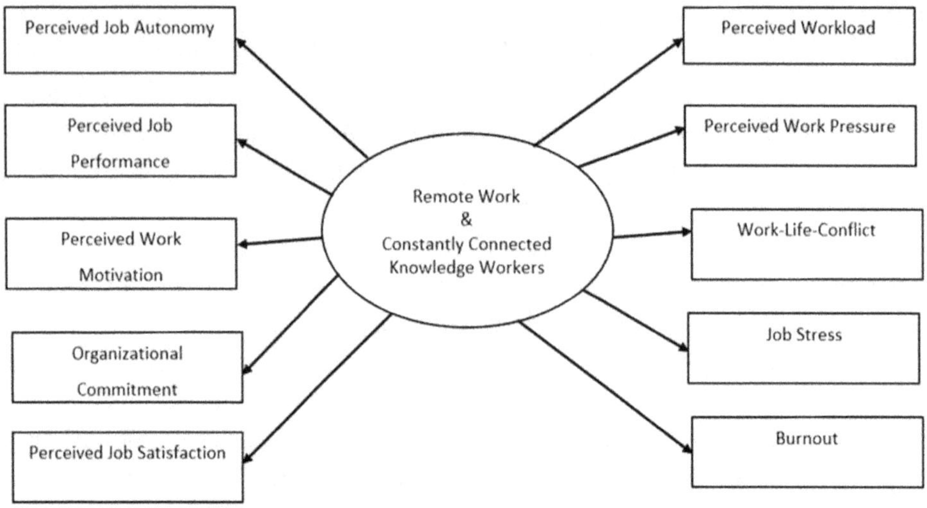

Table 1. 10 derived variables

Bring Your Own Device (BYOD)	The utilization of employees personally owned mobile technologies such as smartphones, tablets, laptops, and other tools, including their software applications namely word processor, video conferencing, spreadsheets, instant messaging software, web browsers, and presentation software to complete work-related activities (Doargajudhur & Dell, 2018, 2020).
Perceived Job Autonomy	Perceived Job autonomy is the degree to which workers perceive that they are independent and can control their decisions regarding when, where and how to fulfill their tasks (DeVaro et al., 2007)
Perceived Job Performance	Job performance refers to an individual's managed behaviors and activities that support the organization's goals. It is an employee's assessment of the general amount, effectiveness, and quality of work performed (Rotundo & Sackett, 2002).
Perceived Work Motivation	Work motivation is the desire to enjoy one's work and focus efforts on things that one is interested in (Ryan & Deci, 2000).
Organizational Commitment	Organizational commitment is the degree to which an individual identifies with and participates in the organization (Mowday et al., 1982).
Perceived Job Satisfaction	According to Smith et al (2018), work satisfaction is the degree of contentment a person has with their employment.
Perceived Workload	Perceived Workload refers to the tension that workers experience due to the amount of tasks that they must undertake. It is often characterized by the feeling of being very busy without any reference to any specific target (Lee & Ashford, 1996).

continues on following page

Table 1 continued

Perceived Work Pressure	Work pressure is characterized as a physiological and psychological condition that results in a person feeling strained or under pressure to complete activities consistently and on time. Depending on how the employee rates the completion of tasks in relation to the workload and goals, the state changes as work pressure levels do (Schieman & Young, 2013).
Work-Life Conflict	Work-life conflict is the degree to which commitments to one area of life—such as one's career or personal life—obstruct commitments to another area of life (Bell et al., 2012).
Job Stress	Stress is described as a relationship between an individual and their surroundings that is deemed to be demanding and dangerous for that individual's health. It is a situation or occurrence that is viewed as demanding, threatening or difficult (Lazarus & Folkman,1984; Harddie et al., 2005).
Burnout	Burnout is a prolonged reaction to ongoing emotional and interpersonal workplace pressures (Maslach et al., 2001).

FINDINGS AND DISCUSSION

Key Variables in Remote Work and Constant Connectivity

In this section, the ten derived variables associated with remote working and constant connectivity of BYOD users are elaborated to demonstrate their relationship with several job outcomes.

Remote Working, Constant Connectivity and Perceived Job Autonomy

The job resource factor of job autonomy may be greatly impacted by BYOD remote workers. Remote working and perceived job autonomy are interconnected since remote work gives workers the freedom to fulfil their tasks as they see fit by embracing their personal devices. It is therefore argued that BYOD-ers who work remotely frequently can afford greater freedom over how they organize their time and handle their jobs. Workers can choose their own work schedule by ensuring their deadlines are met, and they value the independence they acquire when deciding on the ideal working atmosphere and their capacity to prioritize their tasks and make choices surrounding work tasks and timescale (Niehaves et al., 2012, 2013). It seems plausible that workers would cherish having the independence and freedom to utilize their personal technology outside of the office to fulfil their professional obligations. This greater autonomy might, however, present some difficulties, such as the requirement for self-control, efficient time management, and clear communication with coworkers and superiors (Doargajudhur & Hosanoo, 2023).

Remote Working, Constant Connectivity and Perceived Job Performance

The wide range of functions available on employees' personal technologies has made them more productive while utilizing their own equipment. Making the technology work has been reported as a major challenge remote workers face (Flores 2019). Doargajudhur and Hosanoo (2023)'s study on the employees' adoption of BYOD, found that respondents had improved productivity when their technological devices and technologies were appropriate to undertake their work-related tasks, with their privately owned devices. Their findings are consistent with Goodhue and Thompson (1995) who reported workers demonstrate higher productivity when there is a closer fit between the technological features of their devices and the tasks that should to be accomplished. Employee productivity is also increased due to device mobility, especially among tech-savvy workers. This is the case when BYOD employees can complete work-related tasks from home because they think their tools are more suitable for the job and because they can reply to inquiries right away, which improves their productivity. It can also be argued that due to fewer meetings, fewer interruptions from coworkers, and the flexibility to design a personalized workspace, some professionals feel that working remotely increases their productivity. This logical deduction is consistent with a substantial body of prior research (Doargajudhur & Hosanoo, 2023; Huang & Chang, 2016).

Remote Working, Constant Connectivity and Perceived Work Motivation

BYOD employees that work remotely frequently enjoy greater flexibility in terms of their workspace and hours. As employees have the freedom and flexibility to customize their work to fit their requirements and tastes, this enhanced autonomy is likely to raise motivation since they have the ease of working from home. By utilizing their devices to work remotely, it gives workers the freedom to modify their work schedule to accommodate their interests, lifestyle and personal obligations. Hence, it can be contended that employees may perceive an increase in their motivation as a result of being able to balance work and personal obligations. Moreover, BYOD workers avoid the daily commuting time, and hence this contributes to increasing their motivation since it lowers their stress and saves time. It can also be argued that BYOD workers feel more motivated when they have the right tools and technology to accomplish their organizational duties. This logical deduction is consistent with other research (Carlston et al., 2017; DeVaro et al., 2007, Doargajudhur & Hosanoo, 2023), which explains why BYOD workers are motivated about their jobs while serving their business purposes at their own convenience.

Remote Working, Constant Connectivity, and Organizational Commitment

Remote working may also lead to an increase in employees' commitment to their organization. Organizational commitment referring to an attitude relating to an employee's mindset about the organization, is perceived as a psychological connection that people have with their organization, which is often distinguished by substantial identification with the organization and a deep-rooted motivation to advance the fulfillment of organizational objectives (Meyer & Allen 1997). It is therefore contended that remote workers' dedication to the company may increase when they are given the freedom and trust to handle their own responsibilities using their preferred devices. Similarly, one of the main components of organizational commitment is work-life balance, which can be enhanced via remote work. Hence, it can be argued that employee commitment to the company is higher when they have the freedom to better balance their personal and professional life with less pressure from the boss. This argument is consistent with several other studies (Chang et al., 2015; Klein et al., 2012). Moreover, when BYOD employees have the opportunity to work remotely, this can help firms retain top talent since workers feel more emotionally attached to their companies when given this freedom, especially if it suits their personal preferences.

Remote Working and Perceived Job Satisfaction

BYOD employees who work remotely frequently have more flexibility regarding their work schedule and location. Hence, they enjoy more job satisfaction as a result of being able to better manage their personal and professional lives. In addition, working remotely reduces daily travel stress and saves time thereby leading to increased job satisfaction (Grant, 2021). This can be explained by the fact that the employees feel they have the autonomy to make their own decisions and decide how to fulfil their organizational duties and is also supported by the JD-R. The flexible working time and constant connectivity through the diffusion of private technologies are much appreciated by BYOD remote users, and this relationship is supported by the wider community (Doargajudhur & Hosanoo, 2023; Johari et al., 2018).

Remote Working, Constant Connectivity and Perceived Workload

While BYOD remote workers perceive greater freedom, this comes along with the perception of having more work to fulfil. Some workers find it difficult to switch off from their jobs and end up completing their work anywhere and anytime, which in turn results in increased job demands - workload. Those workers find it more difficult to create distinct work-life boundaries in the absence of a traditional office. According to Ferguson et al (2016) and Leclercq-Vandelannoitte (2017), employees perceive more work since they are continuously connected to their devices and are therefore tied to their jobs. In a similar vein, remote work frequently includes less direct supervision as compared to completing work at the office. Hence, employees who are away from their workplace are constantly connected to their device and may take on more work and responsibilities or even work longer hours beyond the normal working time, and these subsequently add to their workload. This can be explained by the fact that the self-imposed or expected 24/7 responsiveness, availability, and always-connected mindset brought about by remote work and BYOD adoption, work is increasingly perceived as being done more at home by employees. This suggests that working using personal devices after regular business hours extends the workplace into the home, which raises perceptions of workload, and is congruent with the results of the studies conducted by (Niehaves et al., 2012, 2013).

Remote Working, Constant Connectivity and Work-Life Conflict

It may be difficult to draw clear lines between business and personal life when you operate remotely. Since there is no physical barrier separating the workplace, it is harder to "leave work at the office." Conflict can arise because of conflicting demands between work and personal time. BYOD remote workers who are constantly connected find it challenging to take time off, which would result in longer workdays and more work-life conflict. The lack of separation can make it alluring to check emails and work after hours. In addition, workers feel emotions of loneliness and isolation when working remotely due to the lack of social interaction with coworkers (Hill et al., 2022). We can argue that conflict between the job and personal life may result from these emotions, which can have a negative effect on the person's overall wellbeing. For instance, people who have caregiving duties, such as looking after children or elderly family members, may find it particularly difficult to work remotely. In this context, work-life conflict can arise from the stress of juggling work and caring obligations.

Remote Working, Constant Connectivity and Stress

It is reported that remote employees are deprived of the support and social interactions that come with working in a physical office, and hence have feelings of stress and loneliness. While working remotely and being constantly connected to their devices, BYOD workers have less social interaction with their peers, and can negatively impact their well-being and lead to more stress since they are expected to be always available around the clock – even on weekends and vacations, which causes them a great deal of stress and makes them feel as though they can never truly unplug. In addition, the lack of face-to-face interactions can affect the effective communication and collaboration among team members (Singh et al., 2022). Constant use of digital communication tools might lead to a heightened sense of isolation, making it more difficult to understand the emotions and intentions of colleagues. Consequently, the rise in misunderstandings and reduced sense of collaboration introduces another layer of stress for remote employees.

Remote Working, Constant Connectivity and Burnout

Burnout, another negative impact of the constant connectivity of teleworkers, through their personal devices. The rising expectation of being constantly available at any time, any place, increases the workload of the remote workers, hence, the employees experience work-related stress, fatigue, and exhaustion. The invasion of home or personal spaces by personal and work technologies are associated to drawback, for instance, employees' constant state of connection. Hyper connectivity of employees is associated to varied psychosocial issues such as stress and mental overshoot, high-tech anxiety, or burnout (Derks & Bakker, 2012). The resulting stress arising from the spillover effect from work to home to home to work is likely to exacerbate work-family conflicts, ultimately resulting in work-related burnout (Hayes et al., 2020). The emotional, quantitative, and mental challenges leading to work stress are also consistent with the identified sources of work-related burnout (Peeters et al., 2005).

Remote Working, Constant Connectivity and Work Pressure

Remote working and voluntary constant connectivity tend to escalate the perception of work pressure due to the permeable boundaries between professional and personal life spheres (Derks & Bakker, 2014), resulting in the double edge impact of increased perception of job autonomy and work pressure simultaneously. Demerouti and Bakker (2023) propose that the job demands such as job pressure exhaust the employees' energy, while job autonomy, which is a job resource promotes the connected employees' coping ability with the job demands. Hence, constant connectivity increases the work pressure of the teleworkers, who are challenged by excessive workload. The spill-over effect between work-related and personal life spheres and time pressure can cause ill health and the cognitive demands are related to burnout (Bakker et al., 2023). The progressively fluid and remote nature of work and the ubiquitous characteristics of ICTs has engendered the voluntary desirability of after-hour connectivity by knowledge workers. While voluntary after-hour work connectivity might facilitate flexibility and autonomy, such autonomy might still be felt as an intrusion of professional into personal lives of the knowledge workers due to the inability to detach from work, hence boosting work pressure (Heissler et al., 2022).

IMPACT OF REMOTE WORKING ON WORKERS ACROSS EUROPE AND GLOBALLY

Research indicates that employees perceive flexible work arrangements to have both benefits and drawbacks. Increased wellbeing, self-reported productivity and work satisfaction, less work-life conflict, new ways to interact, and more inclusive working practices using technology are just a few advantages of remote and hybrid working for employees. Challenges vary from difficulty unplugging from work, health problems, longer hours, fewer social interactions, harder work, distractions, and fewer opportunity for advancement and learning (Eurofound, 2022).

The health and welfare of employees can be impacted by remote and hybrid working in both positive and negative ways. According to ONS data, 47% of people who worked from home in some capacity reported feeling better in February 2022 (Mutebi & Hobbs, 2022). Both positive and negative health effects depend on sociodemographic traits as well as personal variables, like a worker's level of job satisfaction and personal circumstances. One of the most frequent sources of workplace stress during the pandemic has been forced home working; however, due to the pandemic's broader effects on people's mental health and wellbeing, it is challenging to attribute findings on health and wellbeing from data collected during the pandemic to remote and hybrid working (Brans, 2023).

Work-life balance can be impacted by remote and hybrid working in both positive and negative ways. According to ONS data, in February 2022, more than 78% of people who worked from home in some capacity claimed that having the flexibility to do so helped their work-life balance (Mutebi & Hobbs, 2022). The boundaries between work and personal life may become unclear as a result of remote and hybrid working, and there may also be more unpaid overtime hours worked. It may be challenging for employees to switch off when they use information and communication technologies to do tasks linked to their jobs outside of normal working hours.

According to self-reported surveys, almost two-thirds or more of employees who work from home claim to have accomplished at least as much as they did before the pandemic. Employee self-reported productivity varies, with younger employees reporting less productivity and disabled employees reporting more productivity (Mutebi & Hobbs, 2022; Brans, 2023).

Prior to the COVID-19 outbreak, remote workers had a worse chance of getting promoted and having access to training opportunities. There is not enough information to say whether this trend has persisted throughout the pandemic, and it might alter if more individuals start working from home more frequently. There is a "flexibility stigma," or a biased attitude, against remote workers, according to research conducted before and during the lockdowns, albeit there are some signs that the COVID-19 lockdowns have lessened this stigma (Mutebi & Hobbs, 2022).

Various research conducted globally reported positive findings of remote working for the organizations and the employees alike. For instance, diverse organizations in the United States have reported that the incorporation of remote workers within their workforce due to their increased productivity, while countering pertinent transactional cost. It also promotes higher rates of efficiency and productivity (Flores 2019). This shift towards remote work not only yields increased efficiency and productivity but also brings about significant positive implications for organizational dynamics. Beyond individual well-being and work-life balance, remote work contributes to broader positive outcomes for companies, including cost savings and access to a more extensive talent pool. By fostering diverse teams, remote work allows organizations to leverage the unique skills and perspectives of a geographically dispersed workforce, promoting adaptability and innovation (da Silva et al., 2022).

Remote work, with its increasing prevalence, has also spurred discussions on the changing nature of leadership (Krehl & Büttgen, 2022). Traditional leadership models built on physical presence and direct supervision face challenges in a virtual work environment. Leaders are now encouraged to adapt their approaches to foster team cohesion, provide effective communication, and build trust in a digital space. The absence of face-to-face interactions necessitates the development of new leadership skills, such as leveraging technology for team collaboration and ensuring inclusivity across remote teams (Farndale & Vaiman, 2022). Furthermore, the shift to remote work has prompted organizations to reevaluate performance metrics and employee evaluations. Companies are exploring innovative ways to assess productivity, engagement, and overall contributions, moving beyond traditional time-based measurements (Straus et al., 2023). The evolving landscape of remote work requires continuous exploration of leadership dynamics and performance evaluation strategies to ensure sustainable and thriving virtual work environments.

The widespread adoption of remote work not only reshapes the traditional work environment but also offers significant contributions to environmental sustainability (Crawford, 2022). One key consequence is the reduction in the need for daily commuting, as employees can perform their tasks from their homes. This shift results into reduced dependency on transportation, leading to a significant decrease in carbon emissions associated with daily commutes (Orzeł & Wolniak, 2022; Ray et al., 2022). Moreover, the decrease in demand for physical office spaces contributes to a reduction in the environmental footprint linked with the construction, maintenance, and energy consumption of traditional office buildings (Mukherjee & Narang, 2023). As organizations integrate remote work practices, they actively participate in mitigating the environmental impact of the workforce, fostering a more eco-friendly and sustainable approach to modern work dynamics.

While these benefits are promising, ongoing research is needed to delve deeper into the long-term effects of remote work on organizational structures, team collaboration, work outcomes and the overall business environment. Further exploration of these aspects can provide valuable insights for businesses adapting to the evolving nature of work and address pertinent challenges.

STRATEGIES AND RECOMMENDATION FOR REMOTE WORKING FOR BYOD KNOWLEDGE WORKERS

In this changing work environment, especially for those who are always connected through their personal devices (BYOD), it is crucial to use effective strategies to optimize work outcomes. Based on a comprehensive analysis following the systematic literature review, we propose the following best practices and recommendations. These would serve as practical insights and help enhance the overall remote work experience, productivity, and well-being in the context of remote working for BYOD-ers.

Establishing Boundaries and Clarifying Expectations

The analysis of existing literature underscores the significance of establishing clear boundaries between one's professional and personal life (Chen et al., 2020). This involves encouraging employees to explicitly set expectations and communicate their work hours, response times, and availability for work-related matters. Hence, organizations should ensure that constantly connected BYODers are aware of their roles in a remote setting. Such initiative is likely to create a harmonious integration of professional responsibilities and personal life commitments. The flexible work schedule associated with remote working complements this approach by allowing employees to customize their working hours to better align with their peak productivity times and personal commitments (Rudolph et al., 2021).

Offering Training and Professional Growth Opportunities

In a similar vein, organizations can provide resources or training sessions on how to assist BYODers in effectively navigating remote work as well as using digital tools. It is also essential to create awareness on technological boundaries, by emphasizing on the proper utilization of devices. It can be argued that consistent training stands out as a recurring and key strategy, providing employees the means to remain updated on technological advancements and elevate their existing skill sets. This empowers them to smoothly navigate tasks that might have undergone transformative changes due to the paradigm shift toward remote work (Chen et al., 2020). This approach aligns with the findings of Gewald et al (2017), underscoring the positive influence of ongoing training to boost employee competence and performance.

Such initiatives would provide insights into importance of managing their time effectively and maintaining work-life balance in a remote environment by unplugging during non-work hours to alleviate the risk of burnout, work-life conflict and stress linked to constant connectivity (Chen et al., 2020). Studies by Kossek et al (2023) and Huo and Jiang (2023) highlight the significance of these practices in fostering a healthier work-life balance for BYOD workers. The implementation of these strategies aligns with recommendations from Qi (2021), stressing the need for organizations to support employees in managing the challenges posed by constant connectivity.

Supporting BYOD-ers' Well-Being and Mental Health

Prioritizing employee well-being is crucial for maintaining productivity and job satisfaction. For instance, regular breaks proved to be as an effective strategy to prevent burnout and enhance overall well-being (Nair et al., 2020). Studies by Rudolph et al (2021) and Kossek et al (2023) highlight the importance of implementing task prioritization which allows BYOD-ers to focus on critical tasks and manage workloads efficiently stress while maintaining a sense of accomplishment. Fostering open communication channels, as suggested by Chen et al. (2020), promotes a supportive work environment to enhance employee well-being, enabling employees to have appropriate platforms to express concerns, seek assistance and address work-related challenges collaboratively. In addition, organizations should consider providing access to counselling, mindfulness training, and employee assistance programs, among other tools and support services, to aid staff in managing their mental health and well-being.

Büchler et al (2020) sheds light on a significant aspect of workplace dynamics, particularly concerning the interplay between constant connectivity and employee well-being. Their findings underscore the imperative for practitioners to prioritize measures aimed at fostering psychological detachment and controlling the pervasive nature of constant connectivity within the workforce. Furthermore, the authors emphasized that organizational support and policies can play a crucial tole for the promotion of psychological detachment through initiatives such as heightening awareness of its importance, facilitating mindfulness training, and implementing internet-based stress-management interventions.

Promoting Work-Life Balance

Moreover, organizations and managers should lead by honoring BYODers' time off and work towards promoting a work-life balance-oriented culture. In order to do so, they should contemplate establishing technology-free time to provide remote workers with a designated period to disconnect from devices, promoting a healthier work-life balance, improved mental health and overall well-being (Galanti et al., 2021; Huo & Jiang, 2023). This practice aligns with the findings of Qi (2021), highlighting the positive impact on employee well-being when organizations actively support time away from constant connectivity. Finally, offering mental health support is crucial in acknowledging the psychological impact of constant connectivity, and studies by Kossek et al (2023) and Huo and Jiang (2023) emphasize the importance of such initiatives for sustaining employee well-being in the context of BYOD.

Seeking Feedback for Continuous Improvement and Professional Development

Organizations can also seek BYODers and constantly connected workers' opinions on working remotely. They could use the information sought to determine areas requiring improvement such as reviewing and improving policies, rules, procedures and technologies related to remote work on a regular basis and adjust to evolving standards and best practices. Simultaneously, monitoring and adjusting policies contribute significantly to professional development by ensuring that organizational guidelines remain aligned with the evolving needs of both employees and the remote work environment (Hoang et al., 2021). A recent study conducted by Palanisamy et al (2022) support the idea that flexible policies positively impact employee engagement and professional advancement in the context of remote work. By incorporating these strategies, organizations can create an environment that fosters continuous learning and development, thereby enhancing the professional path of constantly connected BYOD knowledge workers. It is also recommended that organizations should trust the employees without imposing micromanagement, and they should give regular feedback to the workers as well as recognize their successes.

Considering Cyber Security and the Remote Workforce

The rapid transition to remote work has not only presented employees with unfamiliar work dynamics but has also significantly altered the cyber-risk profile for enterprises globally. With a heightened awareness of prevalent threats such as phishing attacks, conference 'bombing', and ransomware incidents, businesses must proactively deploy appropriate security measures to safeguard enterprise and employee data. By implementing robust security protocols and leveraging effective security tools, organizations can effectively mitigate these threats and bolster their resilience amid the challenges posed by the pandemic (Curran, 2020; Rakha 2023).

Interrelationship Between Remote Work, Constant Connectivity and Work Outcomes

In a nutshell, it can be concluded that there is a dynamic and multifaced relationship between constant connectivity, remote work, and work-related outcomes. Each component influences the other in different ways as illustrated in the discussion section. It can be deduced that this relationship is characterized by improved flexibility, increased productivity, satisfaction and motivation and greater organizational commitment, as well as challenges related to greater workload, more work pressure, burnout, job stress and work-life conflict. We therefore suggest that organizations focus on finding the right balance to optimize the advantages of remote work and constant connectivity while minimizing the challenges to achieve favorable work outcomes by managing digital technologies wisely, encouraging work-life balance, and upholding open communication.

FUTURE RESEARCH DIRECTIONS

All studies have limitations that researchers must consider when assessing and applying the research findings. Future research may explore additional variables linked to constant connectivity and work outcomes through other theoretical perspectives. In addition, the proposed theoretical model provides the groundwork for future research, where hypotheses can be developed and tested empirically by utilizing a particular quantitative technique thereby concluding the remote working constant connectivity and work outcomes relationships to BYOD organizations' policy makers.

Despite the apparent positive effects associated with BYOD remote workers and constant connectivity, it also poses significant security and privacy concerns. Data protection and security are made more difficult by the employees' devices' lack of security as the devices are frequently connected to multiple networks at once, including Cloud systems, Bluetooth, Wi-Fi, and cellular networks. Since personal devices might not be as secure as company-issued technology, businesses must put strong security measures in place to safeguard critical data. Hence, another research avenue is to examine the security implications associated with remote working for both companies and BYOD-ers.

In addition, future investigations can explore the psychosocial aspects of constant connectivity and remote work, particularly understanding the impact on knowledge workers' mental health, stress levels, and overall well-being in the context of BYOD. Research could investigate the coping mechanisms employees employ to manage the challenges of always being connected, and how organizations can support mental health in a constantly connected remote work environment. Understanding the complex relationship between constant connectivity, work outcomes, employee well-being and other key variables can inform the development of targeted interventions and policies for a healthier and more sustainable work environment. Furthermore, exploring the potential influence of organizational culture and leadership in mitigating the negative consequences of constant connectivity can provide valuable insights into cultivating a positive remote work culture. These research avenues contribute to a comprehensive understanding of the multifaceted implications of constant connectivity in remote work.

CONCLUSION

The BYOD trend does not seem to subside as the culture of work-life integration becomes more widespread. The Covid-19 pandemic has increased the prominence of the remote working trend for Bring Your Own Device (BYOD) users, and it is anticipated that this tendency will only grow in the future. The chapter concludes that there are both positive and negative effects (perceived job autonomy, perceived job performance, perceived work motivation, organizational commitment, perceived job satisfaction, perceived workload, perceived work pressure, work-life conflict, job stress, and burnout) of remote working resulting from constant connectivity, as supported by the dual pathway of the JD-R theory. This chapter sheds light on the advantages cited in the literature namely: job autonomy, work commitment, motivation, job performance and job satisfaction. On the other hand, BYOD practice triggers many disadvantages as well, namely, work pressure, work-life conflict, stress, workload, and burnout. This chapter has unfolded that the multitude of advantages associated with BYOD practice can advance autonomy of workers on how and where to perform organizational duties, hence, boosting their job performance. This chapter contributes valuable insights into strategic approaches for optimizing the remote work experience for knowledge workers. The proposed strategies were formulated to address specific challenges identified in the study and include fostering clear boundaries, implementing flexible schedules, providing training on technological boundaries, encouraging regular breaks, prioritizing tasks, fostering open communication, offering mental health support and ongoing training and policy adjustments for continuous skill enhancement. This remote working practice is the new norm for many employees and organizations and is likely to continue in the foreseeable future.

REFERENCES

Arunprasad, P., Dey, C., Jebli, F., Manimuthu, A., & El Hathat, Z. (2022). Exploring the remote work challenges in the era of COVID-19 pandemic: Review and application model. *Benchmarking*, *29*(10), 3333–3355. doi:10.1108/BIJ-07-2021-0421

Bailey, D., & Kurland, N. (2002). A review of telework research: Findings, new directions, and lessons for the study of modern work. *Journal of Organizational Behavior*, *23*(4), 383–400. doi:10.1002/job.144

Bakker, A. B., & Demerouti, E. (2014). Job demands-resources theory. In A. B. Bakker, E. Demerouti, P. Y. Chen, & C. L. Cooper (Eds.), *Work and Well-being: a complete reference guide* (pp. 37–64). Wiley Blackwell. doi:10.1002/9781118539415.wbwell019

Bakker, A. B., Demerouti, E., & Sanz-Vergel, A. (2023). Job demands–resources theory: Ten years later. *Annual Review of Organizational Psychology and Organizational Behavior*, *10*(1), 25–53. doi:10.1146/annurev-orgpsych-120920-053933

Barlette, Y., Jaouen, A., & Baillette, P. (2021). Bring Your Own Device (BYOD) as reversed IT adoption: Insights into managers' coping strategies. *International Journal of Information Management*, *56*, 102212. doi:10.1016/j.ijinfomgt.2020.102212 PMID:32934432

Bell, A. S., Rajendran, D., & Theiler, S. (2012). Job stress, wellbeing, work-life balance and work-life conflict among Australian academics. *E-Journal of Applied Psychology : Clinical and Social Issues*, *8*(1). Advance online publication. doi:10.7790/ejap.v8i1.320

Brans, P. (2023). *As Covid fades away, what's the future of remote work in Europe?* https://www.computerweekly.com/feature/As-Covid-fades-away-whats-the-future-of-remote-work-in-Europe

Broom, D. (2023). *What does remote working look like around the world?* https://www.weforum.org/agenda/2023/02/remote-working-around-the-world/#:~:text=Many%20nations%20give,had%20taken%20it

Buonomo, I., Ferrara, B., Pansini, M., & Benevene, P. (2023). Job Satisfaction and Perceived Structural Support in Remote Working Conditions—The Role of a Sense of Community at Work. *International Journal of Environmental Research and Public Health, 20*(13), 6205. doi:10.3390/ijerph20136205 PMID:37444053

Carlson, D. S., Thompson, M. J., Crawford, W. S., Boswell, W. R., & Whitten, D. (2018). Your job is messing with mine! The impact of mobile device use for work during family time on the spouse's work life. *Journal of Occupational Health Psychology, 23*(4), 471–482. doi:10.1037/ocp0000103 PMID:29215910

Chang, Y., Chien, C., & Shen, L. F. (2021). Telecommuting during the coronavirus pandemic: Future time orientation as a mediator between proactive coping and perceived work productivity in two cultural samples. *Personality and Individual Differences, 171*, 110508. doi:10.1016/j.paid.2020.110508 PMID:33191964

Chen, H., Li, Y., Chen, L., & Yin, J. (2021). Understanding employees' adoption of the Bring-Your-Own-Device (BYOD): The roles of information security-related conflict and fatigue. *Journal of Enterprise Information Management, 34*(3), 770–792. doi:10.1108/JEIM-10-2019-0318

Cook, A. (2020). *COVID-19: Companies and Verticals at Risk for Cyber-Attacks*. https://www.digitalshadows.com/blog-and-research/covid-19-companies-and-verticals-at-risk-for-cyberattacks

Cousins, K., & Robey, D. (2015). Managing work-life boundaries with mobile technologies: An interpretive study of mobile work practices. *Information Technology & People, 28*(1), 34–71. doi:10.1108/ITP-08-2013-0155

Curran, K. (2020). Cyber security and the remote workforce. *Computer Fraud & Security, 2020*(6), 11–12. doi:10.1016/S1361-3723(20)30063-4

da Silva, F. P., Mosquera, P., & Soares, M. E. (2022). Factors influencing knowledge sharing among IT geographically dispersed teams. *Technological Forecasting and Social Change, 174*, 121299. doi:10.1016/j.techfore.2021.121299

Deci, E. L., & Ryan, R. M. (2013). *Intrinsic motivation and self-determination in human behavior.* Springer.

Demerouti, E., & Bakker, A. B. (2023). Job demands-resources theory in times of crises: New propositions. *Organizational Psychology Review*, *13*(3), 209–236. doi:10.1177/20413866221135022

Demerouti, E., Bakker, A. B., Nachreiner, F., & Schaufeli, W. B. (2001). The job demands-resources model of burnout. *The Journal of Applied Psychology*, *86*(3), 499–512. doi:10.1037/0021-9010.86.3.499 PMID:11419809

Derks, D., & Bakker, A. B. (2014). Smartphone use, work–home interference, and burnout: A diary study on the role of recovery. *Applied Psychology*, *63*(3), 411–440. doi:10.1111/j.1464-0597.2012.00530.x

DeVaro, J., Li, R., & Brookshire, D. (2007). Analysing the job characteristics model: New support from a cross-section of establishments. *International Journal of Human Resource Management*, *18*(6), 986–1003. doi:10.1080/09585190701321211

Doargajudhur, M., & Hosanoo, Z. (2023). The mobile technological era: Insights into the consequences of constant connectivity of personal devices by knowledge workers. *Information Technology & People*, *36*(2), 701–733. doi:10.1108/ITP-08-2021-0593

Doargajudhur, M. S., & Dell, P. (2018). Impact of BYOD on organizational commitment: An empirical investigation. *Information Technology & People*, *32*(2), 246–268. doi:10.1108/ITP-11-2017-0378

Doargajudhur, M. S., & Dell, P. (2020). The effect of bring your own device (BYOD) adoption on work performance and motivation. *Journal of Computer Information Systems*, *60*(6), 518–529. doi:10.1080/08874417.2018.1543001

Eurofound. (2022), *The rise in telework: Impact on working conditions and regulations*, Publications Office of the European Union.

Fan, W., & Moen, P. (2024). The Shifting Stress of Working Parents: An Examination of Dual Pandemic Disruptions—Remote Work and Remote Schooling. *Social Sciences (Basel, Switzerland)*, *13*(1), 36. doi:10.3390/socsci13010036

Farndale, E., & Vaiman, V. (2022). HR professionals and talent management: Navigating the dynamic macro context. In *Talent Management: A Decade of Developments* (pp. 107–125). Emerald Publishing Limited. doi:10.1108/978-1-80117-834-120221006

Ferguson, M., Carlson, D., Boswell, W., Whitten, D., Butts, M. M., & Kacmar, K. M. (2016). Tethered to work: A family systems approach linking mobile device use to turnover intentions. *The Journal of Applied Psychology*, *101*(4), 520–534. doi:10.1037/apl0000075 PMID:26653530

Flores, M. F. (2019). Understanding the challenges of remote working and it's impact to workers. [IJBMM]. *International Journal of Business Marketing and Management*, *4*(11), 40–44.

Galanti, T., Guidetti, G., Mazzei, E., Zappalà, S., & Toscano, F. (2021). Work from home during the COVID-19 outbreak: The impact on employees' remote work productivity, engagement, and stress. *Journal of Occupational and Environmental Medicine*, *63*(7), e426–e432. doi:10.1097/JOM.0000000000002236 PMID:33883531

Gewald, H., Wang, X., Weeger, A., Raisinghani, M. S., Grant, G., Sanchez, O., & Pittayachawan, S. (2017). Millennials' attitudes toward IT consumerization in the workplace. *Communications of the ACM*, *60*(10), 62–69. doi:10.1145/3132745

Goodhue, D. L., & Thompson, R. L. (1995). Task-technology fit and individual performance. *Management Information Systems Quarterly*, *19*(2), 213–236. doi:10.2307/249689

Grant, M. (2021). Remote working and its impact on employee job satisfaction during COVID-19. *Leadership Education Capstones*. 60. https://openriver.winona.edu/leadershipeducationcapstones/60

Haan, K. (2023). *Remote work statistics and trends in 2023*. https://www.forbes.com/advisor/business/remote-work-statistics/

Hardie, E., Kashima, E., & Pridmore, P. (2005). The influence of relational, individual and collective self-aspects on stress, uplifts and health. *Self and Identity*, *4*(1), 1–24. doi:10.1080/13576500444000146

Heissler, C., Kern, M., & Ohly, S. (2022). When thinking about work makes employees reach for their devices: A longitudinal autoregressive diary study. *Journal of Business and Psychology*, *37*(5), 999–1016. doi:10.1007/s10869-021-09781-0

Hill, N. S., Axtell, C., Raghuram, S., & Nurmi, N. (2022). Unpacking virtual work's dual effects on employee well-being: An integrative review and future research agenda. *Journal of Management*.

Hoang, A. T., Nižetić, S., Olcer, A. I., Ong, H. C., Chen, W. H., Chong, C. T., Thomas, S., Bandh, S. A., & Nguyen, X. P. (2021). Impacts of COVID-19 pandemic on the global energy system and the shift progress to renewable energy: Opportunities, challenges, and policy implications. *Energy Policy*, *154*, 112322. doi:10.1016/j.enpol.2021.112322 PMID:34566236

Huang, K. Y., & Chuang, Y. R. (2016). A task–technology fit view of job search website impact on performance effects: An empirical analysis from Taiwan. *Cogent Business & Management*, *3*(1), 1253943. doi:10.1080/23311975.2016.1253943

Huo, M. L., & Jiang, Z. (2023). Work–life conflict and job performance: The mediating role of employee wellbeing and the moderating role of trait extraversion. *Personality and Individual Differences*, *205*, 112109. doi:10.1016/j.paid.2023.112109

Johari, J., Tan, F. Y., & Zulkarnain, Z. I. T. (2018). Autonomy, workload, work-life balance and job performance among teachers. *International Journal of Educational Management*, *32*(1), 107–120. doi:10.1108/IJEM-10-2016-0226

Klein, H. J., Molloy, J. C., & Brinsfield, C. T. (2012). Reconceptualizing workplace commitment to redress a stretched construct: Revisiting assumptions and removing confounds. *Academy of Management Review*, *37*(1), 130–151.

Köffer, S., Junglas, I., Chiperi, C., & Niehaves, B. (2014). Dual use of mobile IT and work-to-life conflict in the context of IT consumerization. *Thirty fifth international conference on Information Systems Completed Research Paper.*

Kossek, E. E., Perrigino, M. B., & Lautsch, B. A. (2023). Work-life flexibility policies from a boundary control and implementation perspective: A review and research framework. *Journal of Management*, *49*(6), 2062–2108. doi:10.1177/01492063221140354

Krehl, E. H., & Büttgen, M. (2022). Uncovering the complexities of remote leadership and the usage of digital tools during the COVID-19 pandemic: A qualitative diary study. *German Journal of Human Resource Management*, *36*(3), 325–352. doi:10.1177/23970022221083697

Lazarus, R. S., & Folkman, S. (1984). *Stress, appraisal, and coping*. Springer Publishing Company Inc.

Leclercq-Vandelannoitte, A. (2017). An ethical perspective on emerging forms of ubiquitous IT-based control. *Journal of Business Ethics*, *142*(1), 139–154. doi:10.1007/s10551-015-2708-z

Lee, R. T., & Ashforth, B. E. (1996). A meta-analytic examination of the correlates of the three dimensions of job burnout. *The Journal of Applied Psychology*, *81*(2), 123–133. doi:10.1037/0021-9010.81.2.123 PMID:8603909

Maslach, C., Schaufeli, W. B., & Leiter, M. P. (2001). Job burnout. *Annual Review of Psychology*, *52*(1), 397–422. doi:10.1146/annurev.psych.52.1.397 PMID:11148311

Mazmanian, M., Orlikowski, W. J., & Yates, J. (2013). The autonomy paradox: The implications of mobile email devices for knowledge professionals. *Organization Science*, *24*(5), 1337–1357. doi:10.1287/orsc.1120.0806

McDaniel, B. T., O'Connor, K., & Drouin, M. (2021). Work-related technoference at home and feelings of work spillover, overload, life satisfaction and job satisfaction. *International Journal of Workplace Health Management*, *14*(5), 526–541. doi:10.1108/IJWHM-11-2020-0197

Meyer, J. P., & Allen, N. J. (1997). *Commitment in the workplace: Theory, research, and application*. Sage publications. doi:10.4135/9781452231556

Moschella, D., Neal, D., Opperman, P., & Taylor, J. (2004). *The Consumerization of Information Technology. El Segundo*. CSC Research White Paper. Available at: https://www.smaele.nl/documents/Taylor-Consumerization-2004.pdf

Mowday, R. T., Steers, R. M., & Porter, L. W. (1979). The measurement of organizational commitment. *Journal of Vocational Behavior*, *14*(2), 224–247. doi:10.1016/0001-8791(79)90072-1

Mukherjee, S., & Narang, D. (2023). Digital economy and work-from-home: The rise of home offices amidst the COVID-19 outbreak in India. *Journal of the Knowledge Economy*, *14*(2), 924–945. doi:10.1007/s13132-022-00896-0

Mutebi, N., & Hobbs, A. (2022). *The impact of remote and hybrid working on workers and organisations*. https://researchbriefings.files.parliament.uk/documents/POST-PB-0049/POST-PB-0049.pdf

Nair, A. V., McGregor, A., & Caputi, P. (2020). The impact of challenge and hindrance demands on burnout, work engagement, and presenteeism. A cross-sectional study using the job demands–resources model. *Journal of Occupational and Environmental Medicine*, *62*(8), e392–e397. doi:10.1097/JOM.0000000000001908 PMID:32404829

Niehaves, B., Köffer, S., & Ortbach, K. (2013). The effect of private IT use on work performance-towards an IT consumerization theory. *11th International Conference on Wirtschaftsinformatik*, 39-53

Niehaves, B., Köffer, S., Ortbach, K., & Katschewitz, S. (2012). Towards an IT consumerization theory: a theory and practice review. *18th Americas Conference on Information Systems,* 1-17.

Orzeł, B., & Wolniak, R. (2022). Digitization in the Design and Construction Industry—Remote Work in the Context of Sustainability: A Study from Poland. *Sustainability (Basel)*, *14*(3), 1332. doi:10.3390/su14031332

Palanisamy, R., Norman, A. A., & Mat Kiah, M. L. (2022). BYOD policy compliance: Risks and strategies in organizations. *Journal of Computer Information Systems*, *62*(1), 61–72. doi:10.1080/08874417.2019.1703225

Peeters, M. C., Montgomery, A. J., Bakker, A. B., & Schaufeli, W. B. (2005). Balancing work and home: How job and home demands are related to burnout. *International Journal of Stress Management*, *12*(1), 43–61. doi:10.1037/1072-5245.12.1.43

Qi, C. (2021). Will BYOD induce work-to-life conflict during COVID-19 pandemic period? Exploring the importance of Work Connectivity Behavior After-hours (WCBA) enabled by BYOD. In PACIS (p. 222). Academic Press.

Rakha, N. A. (2023). Ensuring Cyber-security in Remote Workforce: Legal Implications and International Best Practices. *International Journal of Law and Policy*, *1*(3).

Ray, R. L., Singh, V. P., Singh, S. K., Acharya, B. S., & He, Y. (2022). What is the impact of COVID-19 pandemic on global carbon emissions? *The Science of the Total Environment*, *816*, 151503. doi:10.1016/j.scitotenv.2021.151503 PMID:34752864

Rotundo, M., & Sackett, P. R. (2002). The relative importance of task, citizenship, and counterproductive performance to global ratings of job performance: A policy-capturing approach. *The Journal of Applied Psychology*, *87*(1), 66–80. doi:10.1037/0021-9010.87.1.66 PMID:11916217

Rudolph, C. W., Allan, B., Clark, M., Hertel, G., Hirschi, A., Kunze, F., Shockley, K., Shoss, M., Sonnentag, S., & Zacher, H. (2021). Pandemics: Implications for research and practice in industrial and organizational psychology. *Industrial and Organizational Psychology: Perspectives on Science and Practice*, *14*(1-2), 1–35. doi:10.1017/iop.2020.48

Ryan, R. M., & Deci, E. L. (2000). Self-determination theory and the facilitation of intrinsic motivation, social development, and well-being. *The American Psychologist*, *55*(1), 68–78. doi:10.1037/0003-066X.55.1.68 PMID:11392867

Schieman, S., & Young, M. C. (2013). Are communications about work outside regular working hours associated with work-to-family conflict, psychological distress and sleep problems? *Work and Stress*, *27*(3), 244–261. doi:10.1080/02678373.2013.817090

Singh, P., Bala, H., Dey, B. L., & Filieri, R. (2022). Enforced remote working: The impact of digital platform-induced stress and remote working experience on technology exhaustion and subjective wellbeing. *Journal of Business Research*, *151*, 269–286. doi:10.1016/j.jbusres.2022.07.002 PMID:35847196

Smith, S. A., Patmos, A., & Pitts, M. J. (2018). Communication and teleworking: A study of communication channel satisfaction, personality, and job satisfaction for teleworking employees. *International Journal of Business Communication*, *55*(1), 44–68. doi:10.1177/2329488415589101

Straus, E., Uhlig, L., Kühnel, J., & Korunka, C. (2023). Remote workers' well-being, perceived productivity, and engagement: Which resources should HRM improve during COVID-19? A longitudinal diary study. *International Journal of Human Resource Management*, *34*(15), 2960–2990. doi:10.1080/09585192.2022.2075235

Torten, R., Reaiche, C., & Caraballo, E. L. (2016). Teleworking in the new milleneum. *Journal of Developing Areas*, *50*(5), 317–326. doi:10.1353/jda.2016.0060

Violaz, M. (2022). Does working from home work in developing countries? *IZA World of Labor: Evidence-Based Policy Making*. Advance online publication. doi:10.15185/izawol.504

Wang, B., Liu, Y., Qian, J., & Parker, S. K. (2021). Achieving effective remote working during the COVID-19 pandemic: A work design perspective. *Applied Psychology, 70*(1), 16–59. doi:10.1111/apps.12290 PMID:33230359

Weeger, A., Wang, X., & Gewald, H. (2016). IT consumerization: BYOD-program acceptance and its impact on employer attractiveness. *Journal of Computer Information Systems, 56*(1), 1–10. doi:10.1080/08874417.2015.11645795

Yang, L., Holtz, D., Jaffe, S., Suri, S., Sinha, S., Weston, J., Joyce, C., Shah, N., Sherman, K., Hetch, B., & Teevan, J. (2022). The effects of remote work on collaboration among information workers. *Nature Human Behaviour, 6*(1), 43–54. doi:10.1038/s41562-021-01196-4 PMID:34504299

KEY TERMS AND DEFINITIONS

Bring Your Own Device (BYOD): The utilisation of employees personally owned mobile technologies such as smartphones, tablets, laptops, and other tools, including their software applications namely word processor, video conferencing, spreadsheets, instant messaging software, web browsers, and presentation software to complete work-related activities (Doargajudhur & Dell, 2018, 2020).

JD-R Framework: The Job Demands-Resources (JD-R) model (Demerouti et al., 2001) is a parsimonious and flexible framework which describes occupational stress as an imbalance between the demands placed on an individual and the resources, they have access to at work. Jobs with high demands incur expenses on a physical, organisational, social, and psychological level and involve continual mental or physical effort. Job resources are characteristics that help employees achieve their goals at work, promote learning and growth in themselves and others, and lessen the pressures of their jobs and the negative effects that come with them.

Remote Working: Remote working, also known as telecommuting (Torten et al., 2016), telework (Bailey & Kurland, 2002), virtual work, working from home, flexi work (Hill et al., 2024) is a flexible work arrangement allowing workers to fulfil organizational duties from home or a location outside of the traditional workplace. This practice leverages telecommunications and computer-based technology permitting employees to conduct their work duties from home, co-working spaces, or other remote locations.

Work Outcomes: Work outcomes resulting from the impact of remote working of constantly connected knowledge workers encompass the tangible and intangible results of employees' job performance and contributions within an organization. This includes: 1) Perceived job autonomy - The degree to which workers perceive that they have the freedom, independence and control over their decisions regarding when, where and how to fulfill their tasks (DeVaro et al., 2007); 2) Perceived job performance - An individual's managed behaviours and activities that support the organisation's goals. It is an employee's assessment of the general amount, effectiveness, and quality of work performed (Rotundo & Sackett, 2002); 3) Perceived work motivation - The desire to enjoy one's work and focus efforts on things that one is interested in (Ryan & Deci, 2000); 4) Organisational commitment -The degree to which an individual identifies with and participates in the organisation (Mowday et al., 1982); 5) perceived job satisfaction - The degree of contentment a person has with their employment (Smith et al., 2018); 6) Perceived workload - The pressure that workers experience due to the amount of tasks that they must undertake. It is often characterised by the feeling of being very busy without any reference to any specific target (Lee & Ashford, 1996); 7) Perceived work pressure - Characterised as a physiological and psychological condition that results in a person feeling strained or under pressure to complete activities consistently and on time. Depending on how the employee rates the completion of tasks in relation to the workload and goals, the state changes as work pressure levels do (Schieman & Young, 2013)); 8) Work-life conflict – The degree to which commitments to one area of life—such as one's career or personal life—obstruct commitments to another area of life (Bell et al., 2012); 9) Job stress - A relationship between an individual and their surroundings that is deemed to be demanding and dangerous for that individual's health. It is a situation or occurrence that is viewed as demanding, threatening or difficult (Lazarus & Folkman,1984; Harddie et al., 2005); and 10) Burnout - A prolonged reaction to ongoing emotional and interpersonal workplace pressures (Maslach et al., 2001).

Chapter 7
Arduousness in Technological Aspects of Remote Work

Kritika
https://orcid.org/0000-0002-1186-6032
Independent Researcher, India

ABSTRACT

The evolution of internet and Web 3.0 poses an ever-growing trend of remote working allowing people to work at their ease and raising concern over cyber security issues for secured transaction of work in remote environment. The necessity of implementing a thorough cybersecurity framework that includes encrypted data transmission, secure communication channels, and reliable access control is examined. The chapter also explores the importance of continuous monitoring systems and endpoint security solutions in identifying and addressing possible threats quickly. It highlights the challenges and safeguard measures to be taken to prevent the cyber threats in the near future and eventualities of remote working in the years ahead.

DOI: 10.4018/979-8-3693-1314-5.ch007

INTRODUCTION TO REMOTE WORK

The workspace landscape has undergone tremendous changes more so during the pandemic as witnessed by the world globally, the aftermaths of which are still prevalent in the recent times. The introduction of the concept of work from home deep rooted the lives of people from the time of pandemic, changing and shaping the professional arena around us. Remote working, also known as telecommuting or teleworking is a work arrangement that allows employees to work flexibly from the surroundings of their home or remote location without being physically present at the work station to perform the tasks assigned to them. It has been made possible with the advancements in technology like advent of 5G and 6G networks to stay connected and collaborate with their colleagues and clients through video conferencing, instant messaging, cloud-based technologies and much more. The work arrangement serves many benefits such as boosting flexibility, adapt to global talent with reduced costs, engage remote clients and build a strong communication network overseas ensuring continuous business continuity.

Although it was once thought of as a novel concept, remote working has roots well beyond the age of technology. with occupations, like freelancing, content writing, painting, and consulting, which have always accepted a certain amount of versatility in their work schedules. Nonetheless, in recent decades, there has been a noticeable increase in the acceptance of working from home as a common practice due to feasibility in technological development with the advent of 5G networks and Web 3.0 to allow personnel to communicate from anywhere in the world. The conventional office-centric model has gradually given way to a more decentralized, flexible, and technologically advanced work environment in the twenty-first century with video conferences, teams meet, zoom calls etc.

The impact of the telecommuting revolution on people's work-life balance is one of its most noteworthy side effects with the ability to plan their workday and eliminate commuting time that has often led to a more harmonious union of personal and professional life and has increased adaptability on its own set of difficulties by avoiding burnouts as a result of obscuring the lines between the professional and personal space and making a conscious effort in preserving a positive culture along with balance that supports emotional well-being, and provision of essential support system.

Navigating the complicated waters of remote work circumstances requires effective leadership. A sophisticated strategy is needed when managing distributed, remote teams as opposed to on-site teams. Establishing trust, communicating openly, and putting output ahead of appearance are all skills that leaders need to have.

Leading remote teams is more difficult than just assigning work; it also requires being aware of each team member's particular needs and concerns. Empathy, flexibility, and the capacity to use technology to promote a feeling of community and connection are necessary for virtual leadership. Effective remote leaders understand how critical it is to foster a supportive remote culture that is consistent with the goals and values of the company.

Key components of remote working include:

a) *Technology:* Technology: To enable collaboration and workflow, remote work significantly depends on technology, including computers, internet access, video conferencing devices, management of projects software, and communication platforms.
b) *Flexibility:* Working remotely frequently allows for more freedom in terms of location and work hours.
c) *Communication:* Effective communication is essential in remote work environments.
d) *Productivity tracking:* Employers can track and evaluate remote workers' productivity using a variety of techniques.
e) *Cyber Security:* Cybersecurity: Information security and upholding cybersecurity protocols become crucial when working remotely because it entails accessing company systems and data from several locations.

To sum up, the emergence of remote work is a significant turning point in the development of the contemporary workplace. What started out as a reaction to world events has developed into a revolutionary force that questions established ideas about what work is. The incorporation of technology, in conjunction with evolving perspectives on work-life equilibrium, has transformed our perception of and participation in career pursuits.

Unquestionably, the remote work a revolution is entwined with the future of work. The key to navigating the opportunities and complexities presented by this shift for organizations is to prioritize employee well-being, embrace technological advancements, and cultivate an adaptable culture. Businesses can lead the way in this new era where the lines separating actual and virtual workplaces are blurring by doing this.

EVOLUTION OF REMOTE WORKING

From Antiquity to the Industrial Revolution: The Origins of Distance Work

Though they took on very different forms from the digital environments we live in today, the foundations of remote employment were planted centuries ago. Skilled craftsmen and artisans in ancient civilizations frequently worked from home, resulting in a decentralized model that is strikingly similar to modern remote work. Decentralized work began to fade with the arrival of the Industrial Revolution, which was characterized by its emphasis on mass production and centralized factories. But even in this era, some professions continued to be remote and decentralized, like farming and cottage industries.

Telecommuting (1970s–1980s)

The idea of working remotely originated in the 1970s when computers and other technological advancements made it possible for some workers to work from home that emerged the neo terminology known as "telecommuting".

Internet and email (1990s)

The 1990s saw a tremendous rise in the use of email and the internet, which made remote work much easier. Employees no longer needed to be present in person in workstations to collaborate, communicate, and share information.

Technological Advancements in Mobile (2000s)

The widespread use of smart phones and other handheld devices accelerated the trend of remote working with no longer a necessity to be present in person to access emails, records, etc.

Cloud Computing (2010s)

By offering scalable and easily accessible storage solutions, cloud-based technologies significantly contributed to the development of remote work. Teams were able to work together on documents in real time and access common resources from different places thanks to this.

Video Conferencing (2010s)

The emergence of remote collaboration and increased effectiveness of virtual meetings were facilitated by the development of superior video conferencing tools like the likes of Zoom, Skype, and Microsoft Teams. This was crucial for teams that were dispersed throughout various geographical areas.

Change in Work Culture (2010s)

Realizing that workers could be productive outside of the traditional office setting, businesses began embracing a more flexible work culture. The realization that working remotely could boost job satisfaction and the emphasis on work-life balance were two factors that contributed to this change.

Pandemic Acceleration (2020)

Around the world, the COVID-19 pandemic compelled a dramatic and abrupt shift towards remote work (Wheatley, 2021). In order to maintain business continuity, lockdowns and measures of social distancing forced companies to implement remote work practices. During this time, there was a sharp rise in the use of tools for communication and a review of conventional office layouts.

The Revolution of Remote Work: Acceleration in the Face of Global Events

During the COVID-19 pandemic, the potential for change of remote employment gained unprecedented attention. Lockdowns and restrictions on social interaction were imposed by governments across the globe, compelling companies to quickly transition to remote work in order to maintain operations. What started out as a short-term solution to a crisis demonstrated the robustness of remote work structures and quickened an already-occurring change. Businesses that had been reluctant to accept remote work were forced to reconsider their approaches, which resulted in a radical change in the culture of the workplace.

2020s

Hybrid Work Models: Many organizations adopted hybrid work models, which blend remote and in-person work, as they adapted to telecommuting work during the pandemic. This strategy strikes an equilibrium between the benefits of in-person collaboration and remote work by giving employees greater freedom to choose where and how they work.

Future Prospects for Remote Work: Issues to Be Considered

Looking ahead, it seems that remote work is going to be an ever-present aspect of the contemporary workplace. The concept of hybrid employment is becoming more popular as companies look for ways to offer flexibility without sacrificing the feeling of community (Cappeli P, 2021). It combines the advantages of in-person and remote work. Remote work has the potential to unlock several benefits, including shorter commute times, better access to a worldwide talent pool, and environmental benefits. Nonetheless, issues like digital equity, handling the difficulties of handheld leadership, and managing security and legal ramifications continue to be crucial. The part that institutions, decision-makers, and people play in influencing how remote work develops in a responsible and inclusive way.

In summary, the history of remote work is an intriguing one that encompasses centuries and includes changes in society, technology, and culture. What started out as a decentralized model in antiquity saw ups and downs throughout history before finding fresh significance in the digital era and eventually picking up speed in response to world events. Undoubtedly, the development of remote work is a reflection of the future of work. In order to successfully navigate this rapidly changing landscape, we must be flexible, inclusive, and dedicated to creating an environment for work that meets the needs and goals of both individuals and organizations.

LITERATURE SURVEY

A thorough analysis of the literature on remote work finds an expanding corpus of studies examining the variety of facets of this developing work arrangement. There are historical examples of remote labor, and technology developments as credited with the current upsurge (Nilles, 1976). According to (Hill et al. 1998), future trends point to a move towards blended approaches and the incorporation of cutting-edge technology like augmented and virtual reality into remote work procedures. From a managerial standpoint, successful telecommuting arrangements are facilitated by organizational policies and strong leadership (Hill et al., 1998). Alongside advantages like cost savings and more flexibility are drawbacks like communication breakdowns and possible effects on team cohesiveness (Bloom et al., 2014). According to (Bloom et al. 2014), technology enablers, particularly, digital tools and cloud computing play a critical role in determining the nature of distant work experiences.

The economic and social aspects of remote employment can be examined through theoretical frameworks like Transaction Cost Economics and the theory of social exchange (Allen et al., 2015). Research on productivity emphasizes complex conclusions and stresses the significance of context (Grant & Dallner, 2019). The development and use of remote work regulations are also significantly impacted by legal and ethical issues (Grant & Dallner, 2019).

Recognizing how worldwide catastrophes like the COVID-19 epidemic affect remote work becomes crucial as organizations move towards mixed employment arrangements (Allen et al., 2020). In order to maximize human and organizational outcomes in a changing work environment, this literature review emphasizes the necessity of ongoing investigation, adaptation, and comprehension of remote work dynamics.

The study by (Nelviana M. et al, 2022) reveals that transformational leadership does not significantly impact employee performance, but rather, effective leadership style can lead to increased job satisfaction and enthusiasm. However, if leadership is not well implemented, it may not improve performance. Nurlia's research (2017) highlights the importance of effective leadership in determining subordinate performance. Remote working also has a significant effect on employee performance, but it should be carried out effectively and efficiently. Supervision should be more stringent, including conducting audits, conducting Zoom meetings, and issuing strict sanctions. A friendly remote working system can further increase employee performance by providing a flexible schedule and comfortable workplace. The researchers suggest that companies should continue remote working arrangements even after the pandemic. Further research should explore other variables, such as work environment variables, to further improve employee performance.

The nature and scope of remote working still needs to be defined transparently and essential to limit boundaries for distractions and increased occupational stress due to remote working which influences satisfaction on job front, motivational level on intrinsic and extrinsic level and job performance while highlighting the downtime to re-energize the fear of job insecurity (Prasad K., 2023).

For businesses in all sectors, cybersecurity in the wake of remote work has emerged as a top priority. The typical security barrier is no longer clearly defined as employees use company networks from various, frequently unsafe locations. This changing situation calls for the adoption of strong security protocols to protect sensitive data and vital systems, as well as a thorough awareness of the risks and weaknesses specific to remote work settings.

Through an examination of important ideas, new developments, and the situation in terms of research and practice at the moment, this literature review delves into the complex world of cybersecurity in telecommuting. This review attempts to fill in knowledge gaps that call for more research by synthesizing existing knowledge and offering a basis for comprehending the challenges involved in safeguarding remote work settings.

The review initializes from the introduction of neutralization theory in the wake of failure to comply with the information security system and providing neo perspectives to test the effects of sanctions as outlined by deterrence theory (Siponen, M., & Vance, A. (2010)). The rise of user centric security awareness in preventing cyber threats (Furnell & Clarke, 2012) with effective training programs for remote employees. According to (Vance et al., 2012), psychological factors influence adherence of remote employees to various cyber security protocols and promoting strategies for cyber security conscious remote work culture.

The implications of GDPR compliance in remote work landscape and data protection in distributed work environment was laid down by European Union in 2016. According to (Mirkovic et al., 2019) there exists challenges and best practices persistent to incidence response in remote work by developing effective strategies. (Chen et al., 2019) addressed industry specific compliance requirements in remote work with regulatory standards necessary to be followed for preventing large scale cyber attacks. (Huang et al., 2019) enhanced the role of multifactor authentication (MFA) expected to be utilised for human to everything interactions, finds the missing factor by authenticating the user without disclosing sensitive information with future probability.

Social engineering attacks have become a common means to gain access to confidential data of industries often leveraging sophisticated social engineering attacks creating attack vectors via emails, skype, drop box etc. as emphasised in detailed knowledge worker study by (Krombholz, K. et al, 2019) which plays a critical role in telecommunicating as everyone does not get familiarised with cyber security protocols laid down by the firms and falls prey to such attacks.

The COVID-19 pandemic has led to a significant shift towards remote working, affecting job performance and engagement. A study found that negative work-home interactions, home-work interactions, and loneliness were significant predictors of these outcomes. Social support, established using ICT solutions, recorded benefits but was not a perfect substitute for physical contact. This suggests that remote working practice needs to be further observed for sustainability (Slavkovic et al, 2021).

Institutional framework conditions are crucial for sustainable remote working, as they reduce negative work-home interactions and may reduce the need for social support. Managers and supervisors should monitor and adjust social support to ensure job performance and engagement. Training for managers and supervisors is needed to balance the positive effects of social support with potential negative effects. Remote-working overtime has led to adjustments in employees' behaviour, leading to a self-regulated balance between work and home.

As the trend towards decentralised work settings continues to rise, cybersecurity in working from home is becoming increasingly important. The attack surface has grown as a result of employees accessing corporate systems from different places and devices, creating new difficulties. Phishing and attacks through social engineering are common dangers and can be more successful outside of a protected office network. Organisations prioritise multi-factor authentication, secure access restrictions, VPNs, and encryption to reduce these risks and protect data while it's in transit. Strong security measures are also necessary for endpoints, which are devices that connect remotely to business networks. Continuous efforts to improve cybersecurity procedures are necessary to protect sensitive data and keep a robust digital infrastructure as telecommuting becomes more common.

ADVANTAGES OF REMOTE WORKING

The remote working offers several benefits which includes:

a) *Versatility and Work-Life Balance:* Employees who work remotely can better balance their personal and professional lives by creating custom work schedules that suit their needs.

b) *Cost Savings:* Employers and employees can both save money by working remotely.
c) *Accessibility to a Global Skill Pool:* Hiring remote employees from various regions allows employers to take use of a larger talent pool.
d) *Enhanced Productivity:* A lot of people who work remotely are more productive attributing to shorter commutes, less interruptions at work, and the capacity to customize one's workspace, thus boosting efficiency.
e) *Employee Retention*: Providing remote work opportunities can be a beneficial benefit that raises worker satisfaction and retention rates. Workers frequently value the flexibility that remote work affords, and companies that provide this perk could recruit top talent more readily.
f) *Decreased Stress and Time Associated with Commuting:* Employees who work remotely can avoid the everyday stress and time-consuming traffic, transit, and long commutes.
g) *Environmental Impact:* There may be a decrease in the transportation-related environmental impact, such as lower emissions of carbon dioxide, if fewer workers commute to an actual workplace.
h) *Business Continuity:* The ability of an organization to continue operations in the face of unanticipated circumstances, like natural catastrophes or public health emergencies, can be improved by remote work. This was especially clear during the COVID-19 epidemic, when a lot of businesses were still able to operate even though lockdowns were in place elsewhere.
i) *Personalized Workspaces:* The ability to design a workstation that suits one's tastes might help remote employees feel more comfortable and satisfied with their jobs.
j) *Diverse Viewpoints and Inclusion:* By removing geographic boundaries, remote work can advance diversity and inclusion. Employers who accept remote work might discover it simpler to create a more diverse staff by hiring people from different places and backgrounds.

HOW TO SET UP REMOTE WORK STATION?

Workspace Design

1. Selecting the Ideal Site

The first step to a successful remote workspace is choosing the best location. Examine elements like comfort, little distractions, and natural light. Take into account the ergonomic features of your furnishings to avoid long-term health problems.

2. Ergonomic Equipment and Furnishings

Purchasing ergonomic tools and furnishings is a wager in your health. A pleasant and wholesome work environment can be maintained with the help of these accessories, which range from an adjustable desk to an ergonomic chair.

Mounting Up Digital Infrastructure

1. High-Capacity Internet Access

Remote work requires a reliable, fast internet connection as its foundation. Discover how to evaluate your existing connection, investigate possible upgrades, and put strategies into place to improve the dependability of your internet.

2. Picking the Appropriate Equipment

It's important to select the appropriate gadgets, such as a dependable desktop or laptop computer. Recognize the technical requirements necessary for your job and choose gadgets that meet your industry standards.

3. Crucial Programs and Equipment

Examine a range of applications and gadgets made to facilitate remote work. Find the tools and platforms for communication and project management that best fit your needs for teamwork and workflow.

Scheduling and Productiveness

1. Establishing a Daily Schedule

Creating a daily schedule is necessary to keep things organized. To ensure that you remain productive without compromising your wellbeing, learn how to draw boundaries between your personal and professional lives.

2. Time-Stamping Methods

Learn how to block out time effectively to increase output. Learn how to set priorities for your work, manage your time well, and stay away from typical traps that cause procrastination.

3. Making Use of Task Management Software

Investigate task management tools to maintain the structure and direction of your projects. Sophisticated management of projects tools or basic to-do lists, the correct system can greatly increase productivity.

Transmission and Fraternization

1. Useful Communication Techniques

Learn how to communicate effectively with remote teams. Learn how to promote effective and transparent interactions with your clients, coworkers, and partners through both synchronous and asynchronous communication.

2. Protocol for Virtual Meetings

Learn the proper conduct in online meetings. To improve team dynamics, learn how to facilitate productive video conferences, get involved with participants, while getting maximum of collaborative resources.

Emotional and Mental Well Being

1. Finding a Work-Life Balance

Securing a long-term successful work-life balance is essential. Examine methods for establishing limits, handling stress, and avoiding burnout in the context of remote work.

2. Including Rest Periods and Exercise

Learn how important it is to mix in some exercise and breaks throughout your daily schedule. Easy physical activities and mindfulness techniques can enhance concentration, creativity, and general wellbeing.

Setting up a place to work from home involves more than just setting up the furniture and getting online; it involves taking a comprehensive approach to designing a space that encourages cooperation, productivity, and personal wellbeing. With the help of this all-inclusive guide, you can become an expert at working remotely and prosper in the dynamic workplace environment. Recall that the secret to success is to approach your remote workspace with consideration and purpose.

CHALLENGES TO REMOTE WORKING

The remote working along with positives offers several negatives (Allen T et al, 2015) as well which needs to be closely monitored for the industry to bloom at macro level which are as mentioned below:

a) *Communication Barriers:* Working remotely can present a number of communication problems, such as misreading messages, a dearth of in-person interactions, and trouble expressing context and tone. While messaging apps and video conferencing are useful, they might not be able to accurately capture the subtleties of face-to-face communication.
b) *Isolation and Loneliness:* Employees who work remotely may experience feelings of loneliness and isolation because they are deprived of the companionship and social interactions that are present in traditional office settings. This may have an effect on cohesion, morale, and general job satisfaction.

c) *Maintaining Work-Life Balance:* It can be difficult for remote workers to draw boundaries between their personal and professional lives, which can result in overworking. Conversely, when the lines between the physical worlds of home and work become less distinct, some people might find it difficult to "switch shut down and completely detach from their work.
d) Technical Difficulties: Technical difficulties, like hardware failures, software bugs, or internet outages, can impede productivity and interfere with remote work. To reduce these difficulties, remote workers require dependable technology and a helpful IT infrastructure.
e) Distractions at Home: Household duties, family members, and noise are just a few of the possible distractions that can arise in a home. Without a dedicated, quiet workspace, it can be difficult to stay focused and productive.
f) Collaboration within the Team and Culture Development: When teammates are spread out, it can be difficult to establish and preserve a strong team culture. The emergence of a common corporate culture, team-building exercises, and impromptu interactions can all be hampered by remote work.
g) Security Issues: Working remotely creates new cybersecurity difficulties. Workers who work remotely could use multiple networks and devices, which could raise the possibility of security lapses or data breaches.
h) Monitoring of Performance and Accountability: In a remote work environment, some organizations may find it difficult to keep an eye on worker performance and guarantee accountability. Lack of physical supervision can raise questions about accountability and productivity.
i) Training and Onboarding: In a remote setting, it may be more difficult to onboard new hires and provide continuous training. Assimilation of freshmen into the organization's culture, knowledge transfer, and relationship building may take a while when done via remote access.
j) Inequality and Access: Not everyone may find remote work to be practical or equitable. It's possible that some workers don't have access to the required technology, a suitable home workplace setting, or dependable internet. Differences in opportunities and productivity may result from this.

ROLE OF CYBER SECURITY IN REMOTE WORKING

Because of our rising reliance on digital technology and the increasing popularity of flexible and remote work arrangements, cybersecurity plays a critical role in remote work. The key aspects of cyber security in remote working are as follows:

Data Security

Sensitive data transfer and storage are frequent occurrences in remote employment. To protect data from theft, interception, and unauthorized access, cybersecurity precautions are crucial.

Data loss prevention (DLP) technologies, encryption, and secure transmission of files protocols all aid in safeguarding private data while it moves between networks and devices.

Confidentiality to ensure that all the confidential and critical files and infrastructure is maintained without a compromise in remote working when transmitted over internet.

Security of Endpoints

Cyber-attacks can target endpoints like laptops, tablets, and cellphones. Preventing malware and other cyberattacks requires the implementation of strong endpoint security solutions, such as firewalls, antivirus software, and EDR (endpoint detection and response) systems.

Safe Distance Access

It's critical to offer safe remote access to company networks and resources. Multi-factor authentication (MFA) and virtual private networks (VPNs) are frequently used to make sure that only authorized users can access sensitive information and systems.

Cloud Protection

Cloud services are essential for data storage and collaboration in many distant work environments. It is imperative to guarantee the security of cloud-based systems by appropriate configuration, access limits, and consistent monitoring for any questionable activity.

Awareness and Training of Employees: Best practices for cybersecurity should be taught to remote workers. Employees can learn how to spot phishing attempts, how important it is to have strong passwords, and how to follow safe procedures when working remotely with the aid of training programs.

Reaction to and Monitoring of Incidents

Creating efficient incident response strategies and ongoing monitoring systems facilitates the early detection and mitigation of cybersecurity incidents. This is essential to guarantee a prompt reaction to security problems and to lessen the effect of such breaches.

Regulations and Compliance

Navigating different compliance rules and laws is a common task for remote workers. To stay out of trouble with the law and guarantee that sensitive data is protected in compliance with applicable regulations, cybersecurity measures must be in line with these criteria.

Security of Networks

It's critical to protect the infrastructure of the remote work network. To keep a network environment safe, antivirus programs, intrusion detection/prevention systems, and periodic security audits are helpful.

Device Administration

Ensuring remote devices are secured and up to date can be achieved through the implementation of device management policies. This covers patch management, routine software upgrades, and the capacity to remotely lock or wipe devices in the event of theft or loss.

Security of Collaboration

Tools for secure collaboration and communication are crucial for remote teams. As part of this, make sure that any files or messages shared via collaboration platforms are encrypted to prevent illegal access and eavesdropping.

Regular Updates and Patching

Regularly updating operating systems and software is crucial to patch vulnerabilities that could be exploited by cyber threats in both corporate systems and remote work stations.

TAXONOMY OF CYBER THREATS IN REMOTE WORKING

Particularly with the emergence of technological advances and the continued global trends towards flexible work schedules, remote work has grown in popularity. But in terms of cybersecurity, this change has also brought forth new dangers and difficulties. The following are some typical dangers to cybersecurity remote work:

Phishing Offensives

Phishing emails: Phishing emails are a common tactic used by cybercriminals to fool remote workers into divulging private information, like login credentials or bank account information (Kritika, 2023). Attacks involving phishing can also be delivered via text messages (smishing) or voice calls (vishing).

Risky Networks

Risks associated with Public Wi-Fi: Workers who work remotely can connect to unsecured wireless networks, which are prone to insecurity and man-in-the-middle attacks.

Home Network Security

Unsecured home networks can put remote employees at risk from a range of online attacks.

Security of Endpoints

Unprotected Devices: Employees who work remotely occasionally may use personal devices that are not properly secured, leaving them open to ransomware, malware, and other malicious activity.

Security of Video Conferencing

Zoom-bombing and Eavesdropping: Unauthorized access and listening in on private conversations are possible with vulnerable video conferencing systems.

Inadequate Verification

Weak Passwords credentials of remote workers may be compromised if they frequently use the same weak passwords or ones that are used for several accounts.

Low Level of Security Awareness

Lack of Training of employees who work remotely might not have received enough training to identify and steer clear of cybersecurity risks, which leaves them more vulnerable to social engineering scams.

Dark IT

Usage of Unauthorized Apps and Services that haven't gone through proper security assessments may be used by remote workers, creating more security vulnerabilities. (Abomhara et al, 2015).

Data Breach

Unsecured File Sharing: Data leakage can occur from improper use of services for sharing files or improper handling of sensitive data.

Insider Dangers

Employee negligence occurs when remote workers unintentionally (Kritika, 2023) click on harmful hyperlinks or install infected files, they can endanger the company.

Supply Chain Deployments

Cybercriminals may attack remote workers by taking advantage of weaknesses in the platforms' and tools' supply chains.

Device Loss or Theft

Sensitive data may be exposed if remote workers' devices are stolen or lost.

TOOLS TO SAFEGUARD FROM CYBER THREATS IN REMOTE ENVIRONMENT

With the increasing use of remote working, the nature of work has changed dramatically in recent years. Although there are many benefits to this flexibility, there are also new challenges, especially in the domain of cybersecurity. The increasing sophistication of cyber threats poses an exposure to remote workers to the risks that could jeopardize sensitive data and organizational security. A strong cybersecurity toolkit is the basic foundation to combat these threats while ensuring confidentiality, integrity and availability of the information (Chen X et al, 2019).

Private Network Virtualized (VPN)

Virtual Private Networks (VPNs), a key component of remote cybersecurity, shield data from possible eavesdroppers by encrypting it and creating a safe pathway for internet traffic. VPNs are essential for protecting communication between employers and remote employees. Prominent instances like NordVPN, OpenVPN, and Cisco AnyConnect guarantee the security and confidentiality of data transferred over public networks.

Software for Endpoint Protection

Endpoint protection software is a device's first line of defence against ransomware, malware, and other online threats. Real-time protection is a feature of products like Symantec, McAfee, and Bitdefender, which update their threat databases frequently to recognize and eliminate new threats. This software is especially important for employees working remotely who might use a variety of devices, such as personal computers, for work.

Verification with Multiple Factors (MFA)

Cyberattacks can increasingly target passwords alone. By requiring users to present multiple forms of identification, MFA, or multi-factor authentication, adds an additional layer of security. Tools that improve access security include Microsoft Authenticator, Google Authenticator, and Duo Security. These tools make it much more difficult for unauthorized people to enter, even in the event that a password is compromised.

Tools for Safe File Sharing and Communication

Collaboration and file sharing are critical components of remote work, so having secure tools is essential. Systems such as Dropbox Business, Google Drive, and Microsoft OneDrive provide encrypted file transmission and storage, guaranteeing the security of private data while remote teams collaborate.

Email Protection Options: Email is still a popular delivery system for malware, and phishing attempts and attachments containing malware can be very dangerous. Email security products such as Barracuda Email Security, Proofpoint, and Mimecast use sophisticated threat detection algorithms to weed out malicious content, adding another line of defence against email-borne cyberattacks.

Online Safety Resources

The World Wide Web is a huge and sometimes dangerous place. Web security tools that block access to malicious websites and filter internet traffic for potential threats, like Cisco Umbrella, Webroot, and McAfee Web Gateway, protect remote workers. These resources support the upkeep of a safe online workspace for remote work.

Desktop Resources on the Go: Secure remote access to a computer systems or network is made possible by remote desktop services. Working remotely is made possible by programs like Microsoft Remote Desktop, TeamViewer, and AnyDesk, which enable safe connections to office systems. This keeps sensitive data inside the safe boundaries of the company network while also increasing productivity.

Solutions for Mobile Device Management (MDM)

Solutions for mobile device management (MDM) are essential as the number of people using mobile devices for work continues to rise. Mobile device management can be centralized with the help of VMware Workspace ONE, Microsoft Intune, and MobileIron (Thames et al, 2017). These solutions also enforce security policies and guarantee that only authorized devices can access company resources.

Tool for Encryption

An essential function of encryption tools is to safeguard sensitive data. Data at rest is encrypted by VeraCrypt, BitLocker (Windows), and FileVault (Mac). This way, even if a device is lost or stolen, the data on it will remain unreadable without the right decryption key.

Platforms for Cooperation and Communication

For remote teams, effective communication is essential. When configured properly, platforms such as Microsoft Teams, Slack, and Zoom offer secure channels for team collaboration and communication. To safeguard private conversations and shared documents, these tools provide end-to-end encryption along with additional security features.

Training in Security Awareness

Human behaviour is one of the aspects of cybersecurity that is frequently disregarded. Tools for security awareness training, such as PhishMe, SecurityIQ, and KnowBe4, teach remote employees cybersecurity best practices and assist them in identifying and averting possible threats. A vital line of defence against social engineering and other deceptive methods used by cybercriminals is knowledgeable staff.

Solutions for Identity and Access Management (IAM)

It is essential to manage and secure user identities and system access. By managing user authentication and authorization, IAM solutions like Okta, Microsoft Azure Active Directory, and Ping Identity lower the risk of unauthorized access and guarantee that only authorized users can access critical resources.

Security Analytics and Incident Handling

To quickly identify and address issues, proactive security event monitoring and analysis are crucial. With the ability to monitor and analyze logs and events, tools such as Splunk, IBM QRadar, and SolarWinds Security Event Manager enable businesses to quickly identify and address security threats.

FUTURE PROSPECTS OF REMOTE WORK

The nature of work has changed dramatically for people all over the world in recent years, and one notable and long-lasting trend that has emerged is remote working. Driven by technological breakthroughs, modifications in organizational culture, and the aftermath of extraordinary occurrences such as the COVID-19 pandemic, remote employment has evolved from a stopgap measure to a window into the future of work (Slavkovic et al, 2022). We will explore future-facing facets of remote work in this article, looking at how it might affect people individually, companies, and society as a whole.

Technology Advances Changing Remote Work

The development of technology will always be closely linked to the eventuality of remote work with steady evolution of platforms and tools that enable seamless interaction, cooperation, and productivity in the future. Both augmented reality (AR) and virtual reality (VR) are probably going to be important as they offer immersive experiences that help bridge the gap between real and virtual workspaces. The combination of these technologies has the potential to completely change the nature of remote work, from shared virtual offices to virtual team meetings.

The Development of Mixed Work Models

Even though remote work has shown to be feasible, hybrid work models are probably going to become more popular in the near future where companies will try to find a balance between working remotely and in-person while appreciating the advantages of face-to-face communication but also welcoming the flexibility and productivity that come with working remotely. Because of this change, it will be necessary to create new rules, tools, and office layouts that accommodate both in-person and remote workers' needs.

The Effect on Employee Engagement and Company Culture

In a remote work environment, upholding a strong corporate culture and high employee engagement levels presents a special challenge. Thinking businesses will make investments in initiatives that help remote employees feel connected, a part of a community, and have a purpose. Successful remote work surroundings will include digital communication platforms, mentorship programs, and virtual team-building exercises aimed at improving the employee experience.

Remote Employment and Hiring Talent

In the future, geographical limitations won't be as much of a barrier for remote work. Businesses will have more and more access to a worldwide talent pool, which will give them access to a wide variety of perspectives and skill sets. Due to this change, companies will need to reconsider their hiring procedures, putting more of an emphasis on evaluating applicants' capacity for productive remote work and cross-time zone collaboration.

The Significance of Integrating Work and Life

The distinction between one's personal and professional lives will become increasingly hazy as working from home becomes a commonality. Therefore, reassessing work-life balance will be necessity in the future, with a sole aim on work-life integration with employers placing a high priority on the emotional well-being of their workforce, offering tools and assistance to enable people to juggle work and personal obligations without jeopardizing their mental or physical health.

The Prospects for Distance Learning and Expert Development

The effects of working remotely go beyond conventional office environments. Programs for skill development and educational establishments are also adjusting to the new environment. Online learning environments and remote education will grow in popularity in the future, giving people the freedom to learn new things from the ease of their own homes. In the era of remote work, lifelong learning will prove to be vital for career development.

Looking forward to the future of remote employment, it is clear that this change in paradigm is a transformative force that is changing our perception of and interactions with work, not just a passing fad. Future developments in technology, organizational design, and societal change will shape the nature of remote work. It will take initiative from people, companies, and policymakers to embrace this change. We can jointly navigate the opportunities and challenges that lie ahead by anticipating and adjusting to the changing dynamics of remote work. This will ensure that work is no longer just a place, but rather a dynamic and performing experience in the future.

REFERENCES

Abomhara, M., & Køien, G. M. (2015). Cyber security and the internet of things: vulnerabilities, threats, intruders and attacks. *Journal of Cyber Security and Mobility*, 65-88.

Allen, T. D., Golden, T. D., & Shockley, K. M. (2015). How effective is telecommuting? Assessing the status of our scientific findings. *Psychological Science in the Public Interest*, *16*(2), 40–68. doi:10.1177/1529100615593273 PMID:26403188

Allen, T. D., Merlo, K., Lawrence, R. C., Slutsky, J., & Gray, C. E. (2021). Boundary management and work-nonwork balance while working from home. *Applied Psychology*, *70*(1), 60–84. doi:10.1111/apps.12300

Bloom, N., Garicano, L., Sadun, R., & Van Reenen, J. (2014). The distinct effects of information technology and communication technology on firm organization. *Management Science*, *60*(12), 2859–2885. doi:10.1287/mnsc.2014.2013

Cappelli, P. (2021). *The future of the office: Work from home, remote work, and the hard choices we all face*. Wharton School Press.

Hills, J. W., & Jensen, J. F. (1998). Telepresence technology in medicine: Principles and applications. *Proceedings of the IEEE*, *86*(3), 569–580. doi:10.1109/5.662880

Koh, Y., Soepriyanto, G., Aljuaid, M., & Hasan, F. (2022). The effect of transformational leadership and remote working on employee performance during COVID-19 pandemic. *Frontiers in Psychology*, *13*, 919631. doi:10.3389/fpsyg.2022.919631 PMID:36033029

Kritika. (2023). *Demystifying cyber crimes*. IGI Global. https://www.igi-global.com/chapter/demystifying-cyber-crimes/330260

Krombholz, K., Hobel, H., Huber, M., & Weippl, E. (2015). Advanced social engineering attacks. *Journal of Information Security and Applications, 22*, 113-122.

Li, K. C., Chen, X., & Susilo, W. (Eds.). (2019). *Advances in cyber security: Principles, techniques, and applications*. Springer. doi:10.1007/978-981-13-1483-4

Mirkovic, J., Dietrich, S., Dittrich, D., & Reiher, P. (2004). *Internet denial of service: attack and defense mechanisms (Radia Perlman Computer Networking and Security)*. Prentice Hall PTR.

Nilles, J. M., Carlson, F. R., Gray, P., & Hanneman, G. (1976). Telecommuting-an alternative to urban transportation congestion. *IEEE Transactions on Systems, Man, and Cybernetics*, *SMC-6*(2), 77–84. doi:10.1109/TSMC.1976.5409177

Prasad, K. D. V., Vaidya, R., & Rani, R. (2023). Remote working and occupational stress: Effects on IT-enabled industry employees in Hyderabad Metro, India. *Frontiers in Psychology*, *14*, 998. doi:10.3389/fpsyg.2023.1069402 PMID:37063549

Prasad, R., & Rohokale, V. (2020). *Cyber security: the lifeline of information and communication technology*. Springer International Publishing. doi:10.1007/978-3-030-31703-4

Regulation, P. (2016). Regulation (EU) 2016/679 of the European Parliament and of the Council. *Regulation*, *679*, 2016.

Siponen, M., & Vance, A. (2010). Neutralization: New insights into the problem of employee information systems security policy violations. *Management Information Systems Quarterly*, *34*(3), 487–502. doi:10.2307/25750688

Slavković, M., Sretenović, S., & Bugarčić, M. (2022). Remote working for sustainability of organization during the covid-19 pandemic: The mediator-moderator role of social support. *Sustainability (Basel)*, *14*(1), 70. doi:10.3390/su14010070

Thames, L., & Schaefer, D. (2017). *Cybersecurity for industry 4.0*. Springer. doi:10.1007/978-3-319-50660-9

Vance, A., Siponen, M., & Pahnila, S. (2012). Motivating IS security compliance: Insights from habit and protection motivation theory. *Information & Management*, *49*(3-4), 190–198. doi:10.1016/j.im.2012.04.002

Wang, J., Hong, Y., Wang, J., Xu, J., Tang, Y., Han, Q. L., & Kurths, J. (2022). Cooperative and competitive multi-agent systems: From optimization to games. *IEEE/CAA Journal of Automatica Sinica, 9*(5), 763-783.

Wheatley, D., Hardill, I., & Buglass, S. (Eds.). (2021). *Handbook of research on remote work and worker well-being in the post-COVID-19 era*. IGI Global. doi:10.4018/978-1-7998-6754-8

KEY TERMS AND DEFINITIONS

Cybersecurity: Cybersecurity is the protection of systems, networks, and data from digital threats and unauthorized access which involves threat prevention, risk management, access control, encryption, incident response, compliance, and awareness and training using key components such as firewalls, antivirus software, intrusion detection systems, risk management, access control, encryption, incident response, compliance, and employee training.

Endpoint Security Solutions: A set of tools and technologies designed to protect individual devices from cyber threats that includes antivirus and antimalware protection, firewall control, encryption, access control, endpoint detection and response (EDR), device management and policy enforcement, and patch and vulnerability management. These solutions are crucial in protecting the network perimeter, as endpoints are often entry points for cyber threats.

Remote Working: Remote working, also known as telecommuting or working from home, allows employees to perform their duties from their own home or a location of their choosing with key characteristics such as location independence, flexibility, technology-enabled tools, asynchronous communication, and trust-based relationships.

Web 3.0: Also known as the Semantic Web, is the next gen internet evolution aimed at making online data more meaningful, interconnected, and machine-readable which focuses on semantic meaning, decentralization, AI and ML, interoperability, and user-centric ownership. along with emphasizing the ability for seamless information exchange between systems, applications, and data sources.

Chapter 8
Workplace Transformation:
Navigating the Transition From Remote to Hybrid Work

Ridhima Goel
https://orcid.org/0000-0002-0505-9218
Institute of Management Studies and Research, Maharshi Dayanand University, Rohtak, India

Jagdeep Singla
https://orcid.org/0000-0001-8628-9244
HPKV Business School, India

ABSTRACT

Recent years have witnessed a dramatic transformation in the workplace when it comes to flexibility, mostly as a result of advances in technology and changes in societal perceptions regarding the nature of work. This chapter covers the primary influences that lead to this change, the significant effects it has on businesses and the employees they hire, as well as the prospects and obstacles that may follow. In addition, this chapter discusses implications for the path of work, taking into consideration the profound and complex impacts of this fundamental shift on businesses, workers, and society at large. Based on these findings, the chapter provides recommendations to organizations striving to make their way amid this evolving environment so they may establish a more adaptable, long-lasting, favorable work environment.

DOI: 10.4018/979-8-3693-1314-5.ch008

1. INTRODUCTION

In the 21st century, the realm of work has changed significantly due to changes in labor expectations, demographic upheavals, and technological improvements. The transition to a flexible work culture, which includes a variety of work arrangements that stray from the conventional 9–5 office-based paradigm in favor of real-time flexibility, has been one of the most prominent changes (Chen et al., 2019; Jiang et al., 2023). The future of work is being shaped by the emergence of hybrid work models, which have replaced traditional office-based labor alongside working from home. It is a journey that has seen the rise of working from home as the predominant method of conduct (Yang et al., 2021; Gibbs et al., 2021) and, more recently, the arrival of hybrid work models. The framework beneath this evolution is rooted in multiple interlinked forces and variables that have redefined the essence of work.

Technological advances in communication, collaboration, and cloud computing have enabled workers to stay productive and productive virtually from anywhere (Brynjolfsson & McAfee, 2014). Additionally, demographic shifts, specifically the influx of younger folks into workplaces, have ushered in new standards and principles regarding work-life balance and flexibility (Kowske et al., 2010). According to *Bartik et al.* (2020), the COVID-19 pandemic served as a catalyst in this context, accelerating the adoption of remote working and showcasing its feasibility. This transition additionally originates from an urge to adapt to the shifting demands and preferences of the contemporary workforce (Bosua et al., 2019). Employees gradually seek a balance between their private and work lives, aiming at minimizing commute timings as well as expenditures while optimizing their productivity in environments of their choice. Conversely, employers are aware of the prospective benefits of a flexible work environment, including better employee retention, wider access to talent, and lower real estate expenses (Shifrin and Michel, 2022). In recent investigations, it has been claimed that hybrid and remote work structures may transform urban areas and the way individuals operate (Choudhury et al., 2022; Marcus, 2023), akin to previous projections made at the beginning of the 21st century when computers and the internet gained prominence (Glaeser, 1998). Since the global workforce has grown accustomed to a new normal, organizations have been obligated to re-evaluate their work Arrangements and adjust to an improved, flexible, and accommodating workplace environment.

In the following sections, this chapter will present an overview of the substantial change incorporated in the arena of flexible work culture, with a particular emphasis on the transition from the conventional "work from home" (WFH) model to the hybrid work paradigm. This chapter explores the trajectory of flexible work culture, from the onset of remote work to the emergence of hybrid work models.

The historical development of work culture is a fascinating journey designated by significant developments shaped by both technological advancements and emerging societal values. In the pre-industrial era, work was deeply associated with agrarian and craft-based communities, where workers were intricately connected to the land and subject to seasonal rhythms. People were able to retain a personal connection with their occupation as a consequence (Smith, 1776). The Industrial Revolution, which spanned the late 18th and 19th centuries, prompted an abrupt change in work culture. Technological marvels, such as the invention of the steam engine and automated manufacturing, led to the growth of huge factories and formal organizations. Labor became extremely regulated, with workers in centralized industries adhering to predetermined hours (Marx, 1867). This ancient era established the foundation for the conventional 9–5 working schedule that continues to be prevalent today.

Figure 1. Evolution of work culture over time

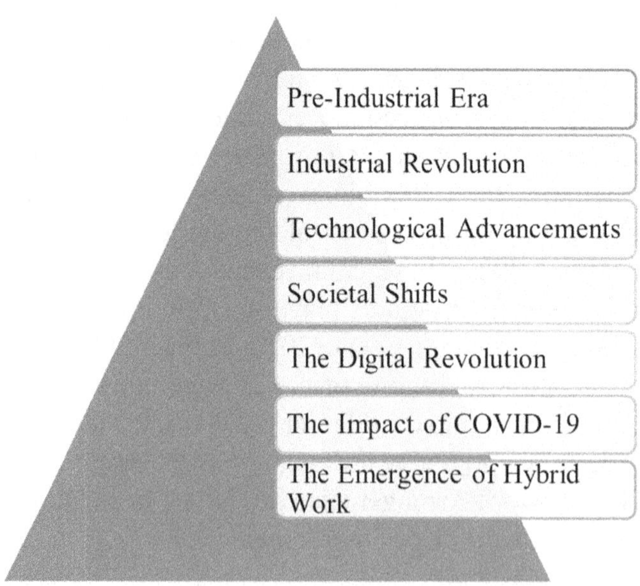

A slew of technological innovations, including the invention of electricity, the telephone, and eventually the computer, were made in the 20th century. These advances revolutionized work culture, promoting more substantial connectivity and communication, which subsequently enabled organizations to expand their scope and collaborate throughout vast distances (Chandler, 1977). Societal attitudes toward work have also evolved over the years. The post-World War II era valued job security and longevity of conventional employment. However, in the second half of the 20th century, values began to shift with an emphasis on individual fulfillment and work-life balance, especially for younger generations joining the workforce (Maslow, 1943).

The 21st century has heralded the digital revolution, a pivotal alteration in work culture. Innovations in cloud computing, connectivity, and teamwork have rendered it feasible for professionals to work from virtually anywhere with an internet connection. This transition defined the rise of working from home, liberating individuals from the constraints of conventional work environments (Brynjolfsson & McAfee, 2014). The COVID-19 pandemic, a worldwide health emergency, served as a catalyst for transition, accelerating the emergence of remote work. Organizations worldwide had to promptly adjust to new realities, emphasizing not only the viability of remote work but also the necessity for higher flexibility in work culture (Bartik et al., 2020). Hybrid work models have emerged as the most recent development in the culture of work. Organizations are currently investigating strategies to integrate the perks of working remotely with in-person collaboration, offering a balanced approach that incorporates both employee preferences and the need for face-to-face interaction (Laumer et al., 2017).

2. RISE OF REMOTE WORK

2.1. Historical Context

The historical context of flexible work culture provides a foundational understanding of how remote work, as a key component of this culture, has evolved over time. This section delves into the emergence of remote work, its early adoption by specific groups, and the pivotal role played by technological advancements in enabling remote work. Flexible work culture, as authors understand it today, has roots tracing back to the mid-20th century, but its prominence has grown significantly in recent decades. Initially, the conventional 9-to-5 office-based model was the norm for most employees. However, this rigid framework began to shift as societal and technological changes created opportunities for new ways of working. Early adopters of remote work primarily consisted of freelancers, consultants, and knowledge workers. These individuals possessed a unique set of skills and responsibilities that allowed them to experiment with flexible work arrangements. For instance, freelancers and consultants often engaged in project-based work, allowing them to break free from the constraints of traditional office spaces. Knowledge workers, who rely heavily on information and communication technologies (ICT), were early pioneers in telecommuting. They recognized that technology could liberate them from the physical confines of an office, allowing them to contribute to their organizations from remote locations (Davenport et al., 1998).

Technological advancements played a pivotal role in making remote work a viable option. The development of personal computers, the internet, and communication tools such as email and video conferencing opened up new possibilities for collaboration across geographical boundaries (Mokhtarian & Salomon, 2001). Notably, the widespread availability of high-speed internet connections in the late 20th century further facilitated remote work, making real-time communication and data sharing feasible. For instance, the advent of the Internet in the 1990s enabled employees to access corporate resources remotely, laying the groundwork for telecommuting. Moreover, the proliferation of mobile devices and wireless technology in the 21st century has untethered work from physical locations, empowering employees to work from virtually anywhere. Notably, during the COVID-19 pandemic, remote work saw an unprecedented surge, showcasing the extent to which technology has transformed work practices. Early adopters, including freelancers, consultants, and knowledge workers, paved the way for this transition. (Davenport et al., 1998; Powell, 2017). However, it is the continuous march of technological progress that has truly enabled remote work to become a mainstream and adaptable component of the modern workforce.

2.2. Remote Work Trends

The landscape of remote work has continued to evolve, influenced by a dynamic interplay of contemporary factors and backed by recent studies that shed light on the shifting paradigms of work. Remote work, or WFH (Working From Home), is a flexible work setup where employees perform their core job tasks from the comfort of their homes (Goel et al., 2024), facilitated by information and communication technology (ICT) relying upon a joint commitment from both employers and employees to uphold business continuity and job retention (International Labour Organisation, 2020). This section delves into these modern trends, elucidating the forces propelling remote work into the forefront of today's work culture.

2.2.1 Factors Contributing to the Growth of Remote Work

The proliferation of remote work can be attributed to a multitude of interrelated factors:

1. **Technological Advancements**: The most fundamental driver has been the continuous evolution of technology. High-speed internet, cloud computing, and a plethora of collaboration tools have made it possible for employees to work from virtually anywhere. The advent of mobile devices has further accelerated this trend (Davenport, 1998; Brynjolfsson & McAfee, 2014). The ubiquitous nature of high-speed internet, cloud-based applications, and robust video conferencing platforms has catalyzed the rise of remote work. Recent studies, such as a survey by Slack (2021), reveal that 78% of knowledge workers believe technology has made remote collaboration more accessible and effective.
2. **Changing Workforce preferences:** The modern workforce increasingly values flexibility and work-life balance. Employees are seeking arrangements that allow them to better integrate their personal and professional lives. Remote work addresses this demand by offering flexibility in terms of when and where work is conducted (PwC, 2021). Today's workforce increasingly values flexibility and autonomy. Research from *Mitchell* (2022) shows that 54% of employees prefer remote or hybrid work arrangements. This trend underscores a shift in employee expectations toward more adaptable work models.
3. **Globalization and Access to Talent:** Employers have recognized the benefits of accessing a global talent pool. Remote work allows organizations to tap into a diverse range of skills and expertise, regardless of geographical location (Katz & Krueger, 2019). Companies are capitalizing on remote work to tap into global talent pools. A study by *Upwork* (2021) reports that 40.7 million Americans freelanced in

2020, contributing to the growing trend of hiring remote freelancers for specialized skills.

4. **Cost Savings:** Remote work can yield significant cost savings for employers. Reduced expenditures on office space, utilities, and other facilities-related expenses contribute to the appeal of remote work arrangements (Bloom et al., 2015).
5. **Pandemic-Induced Paradigm Shift:** The COVID-19 pandemic accelerated remote work adoption. A study by *McKinsey* (2021) found that 20-25% of the workforce in advanced economies could work remotely three to five days a week, opening up new possibilities for work arrangements.

2.2.2 Advantages for Employees and Employers

Remote work brings forth several advantages for employees and employers:

For Employees

1. **Improved Work-Life Balance:** Remote work empowers employees to better manage their time and responsibilities. It enables parents to spend more time with their children, reduces commute-related stress, and allows for a more customized work schedule (Golden et al., 2008).
2. **Enhanced Work-Life Integration:** Recent research by *Buffer* (2022) indicates that 97% of remote workers would like to continue working remotely, citing better work-life balance as a primary reason. This aligns with the growing emphasis on holistic well-being in the modern workforce.
3. **Reduced Commuting Stress:** Eliminating daily commutes translates into tangible benefits such as reduced fuel costs, less wear and tear on vehicles, and a decrease in greenhouse gas emissions. Commuting also consumes a significant portion of an individual's time, which can be redirected toward work or leisure activities (Cohen et al., 2016). Remote work provides a respite from daily commuting. *The U.S. Census Bureau* (2021) reported that the average American spent 225 hours commuting annually. Remote work significantly reduces this time, offering a healthier lifestyle and environmental benefits.

For Employers

From an employer's perspective, remote work offers various advantages:

1. **Access to Diverse Talent Pool:** Remote work enables organizations to recruit and retain top talent regardless of geographical boundaries. This widens the talent pool and enhances workforce diversity, fostering innovation and competitiveness (Al-Youbi et al., 2020). Recent data from LinkedIn (2022) reveals that the majority of job postings now offer remote work options. This widening talent pool is conducive to diversity and inclusion, fostering innovation and creativity (Deloitte, 2021).
2. **Cost Efficiency and Sustainability**: By reducing the need for physical office spaces and associated overhead costs, organizations can achieve substantial savings. These resources can be reinvested in strategic initiatives or distributed as additional benefits to employees (Bloom et al., 2015).

Organizations continue to reap cost savings through remote work. A study by Global Workplace Analytics (2021) suggests that employers can save $11,000 per year per remote worker. This not only enhances profitability but also aligns with sustainability goals by reducing the carbon footprint associated with office space.

2.2.3 Challenges to Remote Working

The modern trends in remote work are driven by advancements in technology, shifting workforce preferences, the enduring appeal of remote work, and the pandemic-induced paradigm shift. These trends bring advantages such as work-life integration and cost efficiency, but organizations must also address communication and well-being challenges to fully harness the potential of remote work in contemporary work environments. While remote work presents numerous benefits, it is not without its challenges:

1. **Communication and collaboration**: Effective communication can be more challenging in remote work settings. The absence of face-to-face interaction can lead to misunderstandings and hinder collaboration. Organizations must invest in communication tools and strategies to mitigate these challenges (Fuchs & Reichel, 2023).

The latest insights from Harvard Business Review (2021) emphasize the importance of adapting communication strategies for remote work. Asynchronous communication tools and clear communication guidelines have become essential to mitigate collaboration challenges.

2. **Loneliness, Isolation, and Mental Well-being:** Remote workers may experience feelings of isolation and loneliness, especially if they lack opportunities for social interaction. This can impact their mental well-being and job satisfaction. Employers need to prioritize initiatives that foster a sense of belonging and connection among remote employees (Grant et al., 2013). Studies published in journals like "The Lancet" highlight the mental health challenges associated with prolonged remote work. Employers are increasingly focusing on initiatives to combat isolation, such as virtual social activities and mental health support (Agarwal et al., 2023).

3. INTRODUCTION TO HYBRID WORKING

3.1. Definition of Hybrid Work

The concept of hybrid work represents a pivotal evolution in contemporary work culture, characterized by a flexible and adaptive approach to where and how work is performed. This section provides a comprehensive explanation of the hybrid work model, emphasizing its core elements, the seamless blend of remote and in-office work, and the emphasis on customizable schedules and flexibility.

3.1.1. Hybrid Work Model

Hybrid work, often referred to as a "hybrid work model" or "hybrid work arrangement," fundamentally redefines the traditional work environment. It centers on a dynamic framework that allows employees to divide their work time between physical office spaces and remote locations, such as their homes or other remote hubs (Detjen & Webber, 2023). Unlike the all-or-nothing approach of fully remote or fully in-office work, hybrid work strikes a balance that integrates the benefits of both worlds.

Blending Remote and In-Office Work

A hallmark of the hybrid work model is the harmonious blending of remote and in-office work, enabling employees to leverage the strengths of each setting. In practice, this means that employees have the flexibility to choose where they work on any given day based on their specific tasks, preferences, and workstyles (Iqbal et al., 2021). For instance, collaborative brainstorming sessions and team meetings may be conducted in the office, whereas focused individual tasks can be accomplished remotely. This hybrid approach recognizes that not all work requires physical presence, and by allowing employees to determine their work environment, organizations can optimize productivity, creativity, and engagement (Gallup, 2023).

Customizable Schedules and Flexibility

Hybrid work is synonymous with schedule customization and flexibility. It dismantles the rigid 9-to-5 workday structure, granting employees the autonomy to craft schedules that align with their personal lives and productivity peaks (Accenture, 2022). This flexibility is grounded in the belief that individuals vary in their chronotype, energy levels, and preferences, and a one-size-fits-all schedule is no longer the most effective approach. Moreover, the ability to customize schedules empowers employees to better integrate work and personal life commitments, contributing to improved work-life balance (Gallup, 2023). For instance, parents can adapt their schedules to accommodate childcare needs, while others may opt for non-traditional work hours that suit their lifestyle.

In conclusion, the emergence of the hybrid work model signifies a transformative shift in how the authors conceptualize work. It transcends the dichotomy of remote versus in-office work by offering a nuanced approach that melds the best of both worlds. Central to this model are the concepts of flexibility, customization, and the seamless blending of physical and remote workspaces, all of which contribute to enhanced job satisfaction, productivity, and employee well-being in the modern workplace.

3.2. Drivers of Hybrid Work

The ascent of the hybrid work model is propelled by a confluence of influential drivers that have gained prominence in the wake of the COVID-19 pandemic. This section elucidates these drivers, emphasizing post-pandemic considerations, the compelling weight of employee preferences for flexibility, and the imperative to enhance organizational resilience.

1. **Post-Pandemic Considerations:** The COVID-19 pandemic, a seismic event that reshaped the global work landscape, stands as a paramount catalyst for the adoption of hybrid work models (Baker et al., 2020). The pandemic forced organizations worldwide to implement remote work practices as a survival strategy during lockdowns. However, as organizations transitioned through various stages of the pandemic, it became evident that remote work was not merely a crisis response but a transformative and sustainable way of working. The pandemic underscored the vulnerability of traditional work arrangements, prompting organizations to reevaluate their long-term strategies. As a result, the concept of hybrid work emerged as a strategic response that integrates lessons learned during the pandemic (Murgia et al., 2020).
2. **Employee Preferences for Flexibility:** The desires and demands of the modern workforce for increased flexibility have grown in prominence and have become a formidable driver of the hybrid work model (PwC, 2021). Employees, particularly millennials and Generation Z, prioritize work arrangements that align with their lifestyles, values, and personal responsibilities. Research consistently shows that flexibility is a highly sought-after attribute when considering job opportunities (Gallup, 2023). Employees' expectations for flexibility extend beyond the choice of where they work. They seek adaptable schedules that permit them to balance work commitments with personal life responsibilities, contributing to enhanced job satisfaction and well-being (Andrzejczyn, 2023). The desire for flexibility is now viewed as a key factor in attracting and retaining top talent, making it a pivotal driver in shaping the future of work.
3. **Enhancing Organizational Resilience:** The quest for enhanced organizational resilience and adaptability is another pivotal driver underpinning the adoption of the hybrid work model (Deloitte, 2021). Organizations recognize that the ability to navigate unexpected disruptions, whether caused by pandemics, natural disasters, or other unforeseen events, is contingent on their capacity to maintain operations while accommodating remote work. By embracing a hybrid work model, organizations fortify their ability to respond swiftly to crises, minimize operational risks, and ensure business continuity (Al-Youbi et al., 2020). This resilience is not only advantageous in times of crisis but also in

a rapidly changing business landscape characterized by technological advancements and evolving customer expectations.

In summary, the ascendancy of the hybrid work model is underpinned by post-pandemic considerations, the burgeoning preference for flexibility among employees, and the strategic imperative for organizations to enhance their resilience in the face of uncertainty. These drivers collectively point to a paradigm shift in the way work is conceptualized and practiced, highlighting the imperative for organizations to adapt to this evolving work ecosystem.

3.3. Advantages and Challenges of Hybrid Work

The adoption of hybrid work models ushers in a range of advantages and challenges that warrant careful consideration. This section delves into the benefits for organizations, including heightened productivity and talent retention, as well as the attendant challenges, such as effective communication and preserving company culture. Additionally, it addresses the imperative of addressing the digital divide in the context of hybrid work.

3.3.1 Benefits for Organizations

1. **Productivity:** A prominent advantage of the hybrid work model is its potential to boost organizational productivity. Research by Bloom et al. (2015) suggests that remote work, when implemented effectively, can enhance productivity by eliminating workplace distractions, reducing commute times, and allowing employees to work during their peak productive hours.
2. **Talent Retention:** The allure of hybrid work contributes to talent retention efforts. A study conducted by PwC (2021) highlights that 83% of employees view the option to work remotely as a significant factor in their decision to stay with their current employer. The capacity to attract and retain skilled employees fosters organizational continuity and competitive advantage.

3.3.2 Challenges of Hybrid Working

1. **Communication:** Effective communication emerges as a critical challenge in hybrid work models. Maintaining clear and consistent

communication becomes more complex in dispersed work environments (Lerman et al., 2020). Without the spontaneous face-to-face interactions that occur in traditional office settings, organizations must rely on digital tools and strategies to bridge the communication gap.

2. **Maintenance of On-site Workplace Culture:** The preservation of on-site work culture can be disoriented due to the implementation of hybrid work settings (Deloitte, 2021). However, virtual team-building activities can help organizations proactively cultivate and reinforce their culture by means of digital platforms and reinforcing leadership practices that uphold the organizational values (Benedic, 2023).
3. **Addressing the Digital Divide:** The digital divide, characterized by disparities in access to technology and reliable internet connectivity, poses an ethical and practical challenge in the context of hybrid work (ITC Next, 2021). Organizations need to consider the inclusivity of remote work arrangements and ensure that all employees have equitable access to the necessary tools and resources.
4. **Application Sprawl and Security Concerns:** In the wake of the global pandemic, companies embraced various technologies, allowing employees to have the freedom to use the finest tools most suitable for their specific duties (Forbes, 2022). This faster adoption rate led to a broad distribution of apps in the workplace, which caused data fragmentation and increased security risks. Organizations are strongly inclined to simplify the usage of devices and applications to improve monitoring and security concerns. A study by Microsoft indicated that more than 75% of employees preferred the adoption of complete solutions to facilitate cooperation across departments or applications.
5. **Identity and Access Management (IAM) Integration Issues**: Organisations have long grappled with finding an effective method to traverse their cloud transformation path, establish a comprehensive Identity and Access Management (IAM) strategy, and maintain the process over the long term (Devlekar and Ramteke, 2022). The prevalent obstacles that organizations encounter throughout their IAM transition include concerns over the handling of data in the cloud with regard to privacy while ensuring data encryption and security throughout data transmission and storage in the cloud environment (Tao et al., 2024). In the Hybrid work model, it is highly challenging to enforce identity verification measures such as multi-factor authentication, which, in turn, may lead to breaches of major security issues (Deloitte, 2023).

6. **Weakened Organizational Ethos:** Studies emphasize that in hybrid workplaces, a lack of continuous, uninterrupted closeness undermines natural ties, making it harder to establish associations and organizational culture (Hirsch, 2021). The dynamism of interaction may also be disrupted since the communication is typically constrained to official networks. Remote workers may experience exclusion from joint decision-making processes and exchanges with their in-office counterparts due to the additional effort necessary for coordination (Benedic, 2023).

In conclusion, hybrid work models offer compelling advantages, including heightened productivity and talent retention. However, they are not without their challenges, particularly in areas related to communication, preserving company culture, and addressing the digital divide. As organizations make strides in adjusting to the hybrid work environment, reverting to the conventional work system is regarded negatively as a setback. Organizations tend to become more obligated to establish novel means of safeguarding the Internet, where they must navigate these complexities to leverage the benefits of hybrid work while mitigating its challenges effectively.

4. OVERVIEW OF THE KEY DISTINCTIONS BETWEEN REMOTE WORK AND HYBRID WORK

Comparative table between remote work and hybrid work, highlighting key differences between the two work arrangements:

5. IMPACT OF REMOTE AND HYBRID WORK ON ORGANIZATIONS AND EMPLOYEES

The impacts of remote and hybrid work on organizational culture and employees' well-being are deep-rooted. This section will explore the impact further in detail.

5.1. Organizational Culture

Organizational culture plays a pivotal role in shaping the work environment and the experiences of both employees and the organization as a whole. The impacts of remote and hybrid work on organizational culture include the shift from an office-centric culture to one centered on trust and results, the fostering of inclusion and diversity, and the adoption of tools and practices for effective remote collaboration, which is explained further in detail.

Table 1. Distinction between remote work and hybrid work

Aspect	Remote Work	Hybrid Work
Work Location	Primarily off-site (home or remote hub).	Combination of on-site (office) and off-site (remote or office).
Frequency at Office	Infrequent or optional.	Regular or scheduled, varies by organization.
Flexibility	High flexibility in work location and schedule.	High flexibility in work location and some schedule customization.
Productivity	May vary; depends on individual work habits and home environment.	Potential for increased productivity due to on-site collaboration.
Employee Engagement	May face challenges in team bonding and company culture integration.	Opportunity for in-person team interactions and culture reinforcement.
Communication	Relies heavily on digital communication tools; may require extra effort to maintain effective communication.	Blend of digital and in-person communication; potential for more spontaneous interactions.
Talent Attraction and Retention	Attracts talent seeking flexibility; may contribute to talent retention.	Appeals to a broader talent pool; can enhance talent retention.
Digital Divide Consideration	Less concern, but organizations should ensure remote employees have adequate technology access.	Critical consideration due to varying access to remote and in-office technology.
Cost Savings for Employers	Potential for reduced office space and overhead costs.	Some cost savings through flexible office space usage, but not as significant as fully remote.
Organizational Resilience	Enhanced resilience in crisis situations.	Improved crisis preparedness with options for on-site operations.

5.1.1 Shifting From Office-Centric to a Culture of Trust and Results

The traditional office-centric culture, characterized by fixed working hours and physical presence in the workplace, has undergone a significant transformation with the advent of remote and hybrid work models. Organizations have shifted towards a culture of trust and results (Davenport et al., 2010). In this culture, the focus shifts from measuring employee productivity based on hours spent in the office to evaluating outcomes and results achieved. This shift is accompanied by a heightened emphasis on autonomy and accountability. Employees are entrusted with greater flexibility in managing their work schedules and environments (Bloom et al., 2015). It implies a higher degree of confidence in employees' ability to deliver results, which can contribute to increased job satisfaction and motivation (Gajendran & Harrison, 2007). In the traditional office-centric culture, work was often defined by physical presence, with an emphasis on hours spent at the office (Davenport et al., 2010). However, remote and hybrid work arrangements focus on outcomes and performance rather than mere attendance. This cultural shift is characterized by a higher level of trust in employees' ability to manage their own work and deliver results independently (Bloom et al., 2015). Organizations grant employees greater autonomy in determining when and where they work, fostering a sense of responsibility and accountability (Gajendran & Harrison, 2007). This shift promotes job satisfaction by affording employees more control over their work environment and schedule.

5.1.2 Fostering Inclusion and Diversity

Remote and hybrid work models can play a pivotal role in fostering inclusion and diversity within organizations (Harvard Business Review, 2021). By enabling employees to work from diverse geographical locations and accommodating different workstyles, these models can facilitate access to a broader talent pool (Al-Youbi et al., 2020). This not only enhances workforce diversity but also fosters a culture of inclusion and equity. Organizations are increasingly prioritizing diversity and inclusion initiatives to create a sense of belonging for all employees, regardless of their location or background (Deloitte, 2021). To do so effectively, they implement policies, practices, and training programs that promote diversity awareness and inclusivity (Choudhury, 2022). Remote and hybrid work models offer unique opportunities to foster inclusion and diversity within organizations. Inclusion becomes a core component of the culture, emphasizing equal opportunities for all employees, regardless of their location or background (Harvard Business Review, 2021). To nurture a culture of diversity and inclusion, organizations implement policies and practices aimed at creating a sense of belonging (Deloitte, 2021). Inclusion is an ethical imperative and a driver of innovation and organizational success.

5.1.3. Tools and Practices for Remote Collaboration

The success of remote and hybrid work models hinges on the effective use of tools and practices for remote collaboration (McKinsey, 2021). Organizations invest in digital technologies such as video conferencing, project management software, and collaborative document-sharing platforms to facilitate seamless communication and collaboration among remote teams (Accenture, 2022). Moreover, the adoption of best practices for remote collaboration becomes paramount, including clear communication protocols, regular check-ins, setting expectations, and defining team goals and roles. Organizations that excel in these areas maintain high levels of productivity and nurture a culture of teamwork and cooperation across dispersed teams. Effective remote collaboration is essential for the success of remote and hybrid work arrangements (McKinsey, 2021). Organizations invest in digital technologies and practices to facilitate seamless communication and collaboration among remote teams (Accenture, 2022). This includes video conferencing platforms, project management software, and collaborative document-sharing tools. Moreover, organizations must establish best practices for remote collaboration (Jung et al., 2017). Organizations that excel in these areas maintain high levels of productivity and teamwork, even in a dispersed work environment.

5.2. Employee Well-Being

The impact of remote and hybrid work on employee well-being is multifaceted. They highlight the importance of mental health considerations, work-life balance, and strategies for employee engagement to ensure the well-being and success of employees in a remote or hybrid work environment, explained further in detail,

5.2.1. Mental Health Considerations

The adoption of remote and hybrid work models underscores the importance of mental health considerations for employees. Remote work can blur the boundaries between work and personal life, potentially leading to burnout and increased stress levels (Eurofound, 2020). Organizations need to prioritize mental health initiatives to support their employees (Gibbs et al., 2021). This may involve offering access to mental health resources, promoting stress management techniques, and creating a culture that encourages employees to take breaks and seek support when needed. By addressing mental health, organizations can enhance employee well-being and job satisfaction.

5.2.2. Balancing Work and Personal Life

One of the key advantages of remote and hybrid work is the potential for improved work-life balance. However, achieving this balance requires deliberate effort on the part of both employees and organizations. Organizations can implement policies that encourage employees to set clear boundaries between work and personal life (Greenhaus et al., 2012). Employees, on the other hand, must proactively manage their time and prioritize self-care.

5.2.3. Strategies for Employee Engagement

Engaging remote and hybrid employees presents unique challenges. Organizations must adopt strategies that foster connection and engagement among dispersed teams (Cohen & Bailey, 1997). This may include virtual team-building activities, regular communication, and opportunities for social interaction. Additionally, organizations should provide opportunities for skill development and career growth, ensuring that remote and hybrid employees have the same opportunities for advancement as their in-office counterparts (Eurofound, 2020).

6. TRENDS IN REMOTE AND HYBRID WORK

The future of flexible work culture is shaped by a dynamic interplay of trends in remote and hybrid work, reflecting the evolving needs of organizations and employees. This section explores these trends, focusing on the long-term sustainability of remote work, technological advancements, including virtual reality and AI, and the regulatory and legal considerations that are emerging in this transformative landscape.

1. ***Long-Term Sustainability of Remote Work:*** The sustainability of remote work as a long-term strategy has emerged as a prominent trend in the wake of the COVID-19 pandemic. Initially, remote work was primarily adopted as a crisis response, but it has since evolved into a strategic and sustainable practice (Katz & Krueger, 2019). Organizations that have

experienced the benefits of remote work, including cost savings and access to a broader talent pool, are increasingly inclined to make it a permanent feature of their work culture (PwC, 2021).

2. ***Technology Trends (Augmented Reality, AI, etc.):*** Technological advancements are poised to redefine the landscape of remote and hybrid work in the future. Emerging technologies such as Augmented reality (AR) and artificial intelligence (AI) hold the potential to transform the way remote work is conducted (Accenture, 2022). This trend is underscored by the recognition that remote work can offer a competitive advantage in attracting and retaining top talent (Al-Youbi et al., 2020). To sustain remote work practices, organizations are investing in technologies, policies, and practices that support remote teams and foster collaboration, creativity, and engagement (Deloitte, 2021).
3. ***Virtual Reality (VR):*** VR technologies are gaining traction as tools for remote collaboration and immersive experiences. VR can facilitate virtual meetings, conferences, and training sessions, offering a more interactive and engaging alternative to traditional video conferencing (Contreras-Cruz et al., 2023). As VR hardware becomes more accessible and affordable, it is likely to play a more significant role in remote work.
4. ***Artificial Intelligence (AI):*** AI-driven tools and platforms are becoming increasingly essential for remote work. AI can automate repetitive tasks, analyze large datasets, and enhance decision-making (Frey & Osborne, 2017). Chatbots, virtual assistants, and AI-powered analytics tools are already being used to streamline remote work processes and improve productivity (McKinsey, 2021).
5. ***Regulatory and Legal Considerations:*** As remote and hybrid work arrangements become more prevalent, regulatory and legal considerations are emerging as vital trends that organizations must navigate. Different regions and jurisdictions may have varying regulations regarding remote work, employment contracts, tax implications, and data security (Baker et al., 2020). These considerations require organizations to stay informed and compliant to avoid legal issues and liabilities. Moreover, issues related to employee rights, such as data privacy, overtime, and workplace safety, are coming to the forefront (Iqbal et al., 2021). Organizations must establish clear policies and practices that address these concerns while respecting the rights and well-being of their remote and hybrid workforce.

7. RECOMMENDATIONS FOR BUSINESSES

In the era of hybrid work, businesses face the challenge of adapting to new work models while maintaining productivity, employee well-being, and organizational culture. Businesses can successfully navigate the challenges and opportunities presented by hybrid work by implementing clear policies, investing in technology and training, and fostering open communication and collaboration. By prioritizing employee well-being, organizations can create a flexible and inclusive work culture that supports the evolving needs of their workforce in the hybrid work era. This section provides detailed recommendations for businesses to successfully implement hybrid work, including strategies for implementation, the importance of providing training and support and maintaining communication and teamwork in a dispersed workforce.

7.1 Strategies for Successful Implementation of Hybrid Work

1. ***Technology Infrastructure*:** Invest in robust technology infrastructure to support remote and hybrid work. Ensure that employees have access to the necessary tools, software, and hardware for seamless collaboration and productivity (McKinsey, 2021).
2. ***Flexible Work Schedules*:** Offer flexibility in work schedules that allow employees to balance their personal and professional lives effectively. Consider flexible hours or alternative workdays to accommodate diverse workstyles and needs (Deloitte, 2021).
3. ***Performance Metrics*:** Shift from measuring employee productivity based on time spent in the office to outcome-based performance metrics. Focus on results, milestones, and key performance indicators (KPIs) that align with organizational goals (PwC, 2021).
4. ***Well-being Initiatives:*** Prioritize employee well-being by offering wellness programs and mental health resources and promoting a culture of work-life balance. Address potential issues related to isolation and burnout proactively (Eurofound, 2020).

7.2 Providing Training and Support for Remote and Hybrid Work

1. ***Training Programs*:** Organisations should offer comprehensive training programs from professionals and experts that may equip employees with the skills and knowledge of the latest tools and software and how to operationalize them for effective remote work experience. This should include training on secure integration of the IAM hybrid cloud, managing time effectively, and dealing with the concerns of cybersecurity breaches (Benedic, 2023; Accenture, 2022).
2. ***Technical Support:*** Organizations should provide the most effective and highly efficient tools to their remote workers along with their technical help support team, which can enhance the productivity of the output of remote employees while addressing all technical glitches or issues. Availability of the latest tools and software like Microsoft Teams, Zoom Meetings, and Any Desk, Slack may lead to better connectivity and performance (Ilag, 2021; Slack, 2021). Employers need to ensure that help desks are easily accessible to remote employees for troubleshooting and assistance. (Harvard Business Review, 2021).
3. ***Mentorship and Coaching*:** Organizations should Establish high-end mentorship and coaching programs for remote employees where seniors and experienced remote or hybrid employees can guide newcomers. This may lead to bridging knowledge gaps and promoting a sense of belongingness among the company personnel (Deloitte, 2021).

7.3 Ways to Overcome the Challenges Faced by Remote and Hybrid Working

1. **Effective Communication Tools:** Organizations must invest their time in one-to-one communication with all employees and promote the use of effective communication tools. Encourage regular check-ins, video conferences, and team meetings to maintain open lines of communication (Katz & Krueger, 2019).
2. **Virtual Team-Building:** Organize virtual team-building activities and social events to foster connections and maintain a sense of camaraderie among remote and hybrid teams (Cohen & Bailey, 1997).
3. **Collaborative Platforms:** Implement collaborative platforms and project management tools that facilitate teamwork and project coordination.

Ensure that documents and project resources are easily accessible (McKinsey, 2021).

4. **Leadership and Inclusivity:** Encourage leadership practices that prioritize inclusivity and ensure that remote and hybrid employees have equal opportunities for growth and advancement (Choudhury, 2022).
5. **Establish Well-defined Policies and Guidelines:** Establish clear policies and guidelines that define the expectations and parameters of hybrid work. These policies should address issues such as work hours, communication protocols, and performance evaluation criteria (Bloom et al., 2015).

8. CONCLUSION

The transformation from the conventional office-centric work culture to the dynamic landscape of flexible work culture, encompassing both work-from-home and hybrid work arrangements, stands as a profound reflection of the adaptive nature of the modern workplace. Throughout this review paper, we have meticulously examined the multifaceted dimensions of this transformation, delving into its historical context, the driving forces behind it, and the manifold advantages and challenges it presents. In this concluding section, we synthesize the key insights, emphasizing the enduring significance of embracing flexible work culture in the ever-evolving realm of work. The historical context of flexible work culture reveals a gradual evolution from the early adoption by freelancers, consultants, and knowledge workers to the transformative impact of technological advancements. It underscores how flexible work practices have progressively integrated into the fabric of contemporary work, driven by the democratization of technology and changing workforce expectations (Harvard Business Review, 2021).

The growth of remote and hybrid work has been driven by a convergence of factors. The COVID-19 pandemic played an unprecedented role in accelerating this transformation, showcasing the agility of organizations in adapting to crises (McKinsey, 2021). Employee preferences for flexibility have become a defining feature of the modern workforce, and organizations are harnessing this as a strategic advantage (PwC, 2021). Additionally, the imperative to enhance organizational resilience in the face of uncertainty has driven the adoption of flexible work models (Deloitte, 2021). The advantages of a flexible work culture are myriad. For employees, it offers enhanced work-life integration and reduced commuting stress, aligning with contemporary well-being priorities (Buffer, 2022; U.S. Census Bureau, 2021). Employers benefit from access to a diverse talent pool and cost efficiency (LinkedIn, 2022; Global Workplace Analytics, 2021). However, challenges related to communication and isolation underscore the need for deliberate strategies to mitigate these issues (Harvard Business Review, 2021; Agarwal et al., 2023).

9. EMBRACING THE FUTURE

As organizations continue to evolve in response to changing circumstances, understanding and embracing flexible work culture becomes essential for achieving success in the future of work. This transformation is not merely a response to a specific moment in time but a reflection of an ongoing paradigm shift. It calls for continuous adaptation, innovation, and a commitment to employee well-being and organizational resilience (Al-Youbi et al., 2020; Eurofound, 2020). In conclusion, the metamorphosis in the dimension of flexible work culture signifies an enduring shift in how work is perceived, organized, and practiced. It is a testament to the capacity of organizations and employees to adapt to change and harness new possibilities. As we stand at the cusp of the future of work, embracing flexible work culture stands as an imperative for navigating the ever-evolving landscape of the modern workplace.

REFERENCES:

Accenture. (2022). *The Future of Work 2022*. Available at: https://www.accenture.com/th-en/insights/consulting/future-work

Agarwal, A., Gupta, P. K., Singh, A., & Kar, S. K. (2023). Conceptual model of Remotely Accessed Mobile Application for Psychiatric disorder Screening (RoAdMAPS): A self-screening and referral aid for mental health in India. *The Lancet Regional Health. Southeast Asia*, *8*, 100100. doi:10.1016/j.lansea.2022.100100 PMID:37384138

Al-Youbi, A. O., Al-Hayani, A., Rizwan, A., & Choudhry, H. (2020). Implications of COVID-19 on the Labor Market of Saudi Arabia: The Role of Universities for a Sustainable Workforce. *Sustainability (Basel)*, *12*(17), 7090. doi:10.3390/su12177090

Andrzejczyn, D. K. (2023). *Effects of Flexible Work Arrangements and Gender on Employee Job Satisfaction* [Doctoral dissertation]. Capella University.

Baker, S., Bloom, N., Davis, S., & Terry, S. (2020). *COVID-Induced Economic Uncertainty*. doi:10.3386/w26983

Bartik, A. W., Bertrand, M., Cullen, Z., Glaeser, E. L., Luca, M., & Stanton, C. (2020). The impact of COVID-19 on small business outcomes and expectations. *Proceedings of the National Academy of Sciences of the United States of America*, *117*(30), 17656–17666. doi:10.1073/pnas.2006991117 PMID:32651281

Benedic, M. (2023). Hybrid work for small businesses: Strategies for reaping benefits. *The Journal of Business Strategy*, *45*(1), 41–47. doi:10.1108/JBS-09-2022-0163

Bloom, N., Liang, J., Roberts, J., & Ying, Z. J. (2015). Does working from home work? Evidence from a Chinese experiment. *The Quarterly Journal of Economics*, *130*(1), 165–218. doi:10.1093/qje/qju032

Bosua, R., Kurnia, S., Gloet, M., & Mendoza, A. (2019). *Telework Impact on Productivity and Well-Being. Social Inclusion and Usability of ICT-Enabled Services.* . doi:10.4324/9781315677316-10

Brynjolfsson, E., & McAfee, A. (2014). *The Second Machine Age: Work, Progress, and Prosperity in a Time of Brilliant Technologies*. W. W. Norton & Company.

Buffer. (2022). *2022 State of Remote Work*. Available at: https://buffer.com/state-of-remote-work/2022

Chandler, D. (1997). *An Introduction to Genre Theory*. Academic Press.

Chen, M. K., Chevalier, J. A., Rossi, P. E., & Oehlsen, E. (2019). The Value of Flexible Work: Evidence from Uber Drivers. *Journal of Political Economy*, *127*(6), 2735–2794. doi:10.1086/702171

Choudhury, P., Khanna, T., Makridis, C., & Schirmann, K. (2022). Is Hybrid Work the Best of Both Worlds? Evidence from a Field Experiment. SSRN *Electronic Journal*. doi:10.2139/ssrn.4068741

Cohen, G., Blake, R. S., & Goodman, D. (2016). Does Turnover Intention Matter? Evaluating the Usefulness of Turnover Intention Rate as a Predictor of Actual Turnover Rate. *Review of Public Personnel Administration*, *36*(3), 240–263. doi:10.1177/0734371X15581850

Cohen, S. G., & Bailey, D. E. (1997). What Makes Teams Work: Group Effectiveness Research from the Shop Floor to the Executive Suite. *Journal of Management*, *23*(3), 239–290. doi:10.1177/014920639702300303

Contreras-Cruz, A., Kirbac, A., Dennett, C., & Daim, T. U. (2023). Human-centered design as a tool to improve employee experience: The case of a US plant-based food manufacturer. *Technology in Society*, *73*, 102248. doi:10.1016/j.techsoc.2023.102248

Davenport, T. H. (1998). Putting the enterprise into the enterprise system. *Harvard Business Review*. PMID:10181586

Davenport, T. H., Harris, J. G., & Morison, R. (2010). *Analytics at Work*. Harvard Business Press.

Davenport, T. H., & Prusak, L. (1998). *Working Knowledge*. Harvard Business Press.

Deery, S., Rayton, B., Walsh, J., & Kinnie, N. (2016). The Costs of Exhibiting Organizational Citizenship Behavior. *Human Resource Management*, *56*(6), 1039–1049. doi:10.1002/hrm.21815

Deloitte. (2021). *2021 Global Human Capital Trends: The social enterprise in a world disrupted*. Deloitte Insights.

Deloitte. (2023). *Addressing Identity and Access Management challenges in Today's hybrid IT environment*. Available at: https://www2.deloitte.com/content/dam/Deloitte/us/Documents/risk/us-deloitte-aws-addressing-iam-challenges-in-todays-hybrid-it-environment-dec2023.pdf

Detjen, J., & Webber, S. S. (2023). Leading hybrid teams in a transition to the future knowledge workplace. *Strategy and Leadership*, *51*(6), 16–21. doi:10.1108/SL-06-2023-0065

Devlekar, S., & Ramteke, V. (2022). Identity and Access Management: High-level Conceptual Framework. *Cardiometry*, (24), 393–399.

Engidaw, A. E. (2022). Small businesses and their challenges during COVID-19 pandemic in developing countries: In the case of Ethiopia. *Journal of Innovation and Entrepreneurship, 11*(1), 1. Advance online publication. doi:10.1186/s13731-021-00191-3 PMID:35036286

Eurofound. (2020). *Living, working, and COVID-19.* Publications Office of the European Union. Available at: https://www.eurofound.europa.eu/en/publications/2020/living-working-and-covid-19

Forbes. (2022). *The Future of Work: More Hybrid, More Collaborative, More Automated.* Available at: https://www.forbes.com/sites/danielnewman/2022/07/31/the-future-of-work-more-hybrid-more-collaborative-more-automated/?sh=7feb17439c46

Fuchs, C., & Reichel, A. (2023). Effective communication for relational coordination in remote work: How job characteristics and HR practices shape user–technology interactions. *Human Resource Management, 62*(4), 511–528. doi:10.1002/hrm.22161

Gajendran, R. S., & Harrison, D. A. (2007). The good, the bad, and the unknown about telecommuting: Meta-analysis of psychological mediators and individual consequences. *The Journal of Applied Psychology, 92*(6), 1524–1541. doi:10.1037/0021-9010.92.6.1524 PMID:18020794

Gallup. (2023). *The future of the Office has Arrived: It's Hybrid.* Available at: https://www.gallup.com/workplace/511994/future-office-arrived-hybrid.aspx

Gibbs, M., Mengel, F., & Siemroth, C. (2021). Work from Home & Productivity: Evidence from Personnel & Analytics Data on it Professionals. SSRN *Electronic Journal.* doi:10.2139/ssrn.3841457

Glaeser, E. L. (1998). Are Cities Dying? *The Journal of Economic Perspectives, 12*(2), 139–160. doi:10.1257/jep.12.2.139

Global Workplace Analytics. (2021). *Telecommuting Trend data.* Available at: https://globalworkplaceanalytics.com/telecommuting-statistics

Goel, R., Singla, J., Mittal, A., & Arora, M. (2024). A decade analysis of employees' well-being and performance while working from home: A bibliometric approach. *Information Discovery and Delivery.* Advance online publication. doi:10.1108/IDD-03-2023-0030

Golden, T. D., Veiga, J. F., & Dino, R. N. (2008). The impact of professional isolation on teleworker job performance and turnover intentions: Does time spent teleworking, interacting face-to-face, or having access to communication-enhancing technology matter? *The Journal of Applied Psychology*, *93*(6), 1412–1421. doi:10.1037/a0012722 PMID:19025257

Grant, C. A., Wallace, L. M., & Spurgeon, P. C. (2013). An exploration of the psychological factors affecting remote e-worker's job effectiveness, well-being and work-life balance. *Employee Relations*, *35*(5), 527–546. doi:10.1108/ER-08-2012-0059

Greenhaus, J. H., Ziegert, J. C., & Allen, T. D. (2012). When family-supportive supervision matters: Relations between multiple sources of support and work–family balance. *Journal of Vocational Behavior*, *80*(2), 266–275. doi:10.1016/j.jvb.2011.10.008

Harvard Business Review. (2021a). *Remote Work Should Be (Mostly) Asynchronous*. Available at: https://hbr.org/2021/12/remote-work-should-be-mostly-asynchronous

Harvard Business Review. (2021b). The New Reality of Remote Work. *Harvard Business Review*.

Hirsch, P. B. (2021). Sustaining corporate culture in a world of hybrid work. *The Journal of Business Strategy*, *42*(5), 358–361. doi:10.1108/JBS-06-2021-0100

Ilag, B. N. (2021). Tools and Technology for Effective Remote Work. *International Journal of Computer Applications*, *174*(21), 13–16. doi:10.5120/ijca2021921109

International Labour Organization. (2020). *An Employers' Guide on Working from Home in Response to the Outbreak of COVID-19*. Retrieved from: https://www.ilo.org/wcmsp5/groups/public/---ed_dialogue/---act_emp/documents/publication/wcms_745024.pdf

Iqbal, K. M., Khalid, F., & Barykin, S. Y. (2021). Hybrid Workplace: The Future of Work. In B. Khan, M. Kuofie, & S. Suman (Eds.), *Handbook of Research on Future Opportunities for Technology Management Education* (pp. 28–48). IGI Global. doi:10.4018/978-1-7998-8327-2.ch003

Jiang, L., Pan, Z., Luo, Y., Guo, Z., & Kou, D. (2023). More flexible and more innovative: The impact of flexible work arrangements on the innovation behavior of knowledge employees. *Frontiers in Psychology*, *14*, 1053242. Advance online publication. doi:10.3389/fpsyg.2023.1053242 PMID:37179861

Jung, Y.-W., Lim, Y., & Kim, M. (2017). Possibilities and limitations of online document tools for design collaboration: The case of Google Docs. *Proceedings of the 2017 ACM Conference on Computer Supported Cooperative Work and Social Computing.*

Katz & Krueger. (2019). Understanding Trends in Alternative Work Arrangements in the United States. *RSF: The Russell Sage Foundation Journal of the Social Sciences,* *5*(5), 132. doi:10.7758/rsf.2019.5.5.07

Kowske, B. J., Rasch, R., & Wiley, J. (2010). Millennials' (Lack of) Attitude Problem: An Empirical Examination of Generational Effects on Work Attitudes. *Journal of Business and Psychology*, *25*(2), 265–279. doi:10.1007/s10869-010-9171-8

Laumer, S., Maier, C., & Weitzel, T. (2017). Information quality, user satisfaction, and the manifestation of workarounds: A qualitative and quantitative study of enterprise content management system users. *European Journal of Information Systems*, *26*(4), 333–360. doi:10.1057/s41303-016-0029-7

Lerman, D. C., O'Brien, M. J., Neely, L., Call, N. A., Tsami, L., Schieltz, K. M., Berg, W. K., Graber, J., Huang, P., Kopelman, T., & Cooper-Brown, L. J. (2020). Remote Coaching of Caregivers via Telehealth: Challenges and Potential Solutions. *Journal of Behavioral Education*, *29*(2), 195–221. doi:10.1007/s10864-020-09378-2 PMID:36093285

LinkedIn. (2022). *Has remote work Peaked?* Available at: https://www.linkedin.com/pulse/has-remote-work-peaked-get-hired-by-linkedin-news/

Marcus, J. S. (2023). COVID-19 and the Shift to Remote Work. In Beyond the Pandemic? Exploring the Impact of COVID-19 on Telecommunications and the Internet. Emerald Publishing Limited. doi:10.1108/978-1-80262-049-820231003

Marx, K. (1867). Capital: A critique of political economy. Volume 1, Part 1: The process of capitalist production. New York, NY: Cosimo.

Maslow, A. H. (1943). A theory of human motivation. *Psychological Review*, *50*(4), 370–396. doi:10.1037/h0054346

McKinsey & Company. (2021). *What executives are saying about the future of hybrid work*. McKinsey & Company. Available at: https://www.mckinsey.com/capabilities/people-and-organizational-performance/our-insights/what-executives-are-saying-about-the-future-of-hybrid-work

Mitchell, T. (2022). *COVID-19 Pandemic Continues To Reshape Work in America*. Pew Research Center. Retrieved from: https://www.pewresearch.org/social-trends/2022/02/16/covid-19-pandemic-continues-to-reshape-work-in-america/

Mokhtarian, P. L., & Salomon, I. (2001). How derived is the demand for travel? Some conceptual and measurement considerations. *Transportation Research Part A, Policy and Practice*, *35*(8), 695–719. doi:10.1016/S0965-8564(00)00013-6

Murgia, A., Bozzon, R., Digennaro, P., Mezihorak, P., Mondon-Navazo, M., & Borghi, P. (2020). Hybrid Areas of Work Between Employment and Self-Employment: Emerging Challenges and Future Research Directions. *Frontiers in Sociology*, *4*, 86. Advance online publication. doi:10.3389/fsoc.2019.00086 PMID:33869406

Next, I. T. C. (2021). *ITC NEXT: Reimagining Businesses for a Redefined Future*. Retrieved from- https://www.itcportal.com/about-itc/ChairmanSpeakContent.aspx?id=2394&type=B&news=2021

Poleg, D. (2021). What is the Future of Offices When Workers Have a Choice? *The New York Times*. https://www.nytimes.com/2021/01/04/upshot/work-office-from-home.html

Powell, R. E., Henstenburg, J. M., Cooper, G., Hollander, J. E., & Rising, K. L. (2017). Patient Perceptions of Telehealth Primary Care Video Visits. *Annals of Family Medicine*, *15*(3), 225–229. doi:10.1370/afm.2095 PMID:28483887

PwC. (2021). *Global Crisis Survey 2021: India Insights*. Available at: https://www.pwc.in/assets/pdfs/services/crisis-management/pwcs-global-crisis-survey-2021.pdf

Shifrin, N. V., & Michel, J. S. (2022). Flexible work arrangements and employee health: A meta-analytic review. *Work and Stress*, *36*(1), 60–85. doi:10.1080/02678373.2021.1936287

Slack. (2021). *The great executive-employee disconnect.* Retrieved from: https://slack.com/intl/en-in/blog/news/the-great-executive-employee-disconnect

Smith, A. (1776). *An inquiry into the nature and causes of the wealth of nations: Volume One*. Printed for W. Strahan and T. Cadell.

Tao, Y., You, S., Zhu, J., & You, F. (2024). Energy, climate, and environmental sustainability of trend toward occupational-dependent hybrid work: Overview, research challenges, and outlook. *Journal of Cleaner Production, 442*, 141083. doi:10.1016/j.jclepro.2024.141083

Upwork. (2021). *Future Workforce Report 2021: How Remote Work is Changing Businesses Forever.* Retrieved from: https://www.upwork.com/research/future-workforce-report

U.S. Census Bureau. (2021). *Census Bureau Estimates Show Average One-Way Travel Time to Work Rises to All-Time High.* Available at: Census.gov. https://www.census.gov/newsroom/press-releases/2021/one-way-travel-time-to-work-rises.html

World Bank. (2021). *From Remote Work to Smart Work in the Post-COVID Era: An Illustrative Case for Leapfrogging into Smart Government.* Retrieved From- https://www.worldbank.org/en/events/2021/02/02/from-remote-work-to-smart-work-in-the-post-covid-era-an-illustrative-case-for-leapfrogging-into-smart-government

Yang, L., Holtz, D., Jaffe, S., Suri, S., Sinha, S., Weston, J., Joyce, C., Shah, N., Sherman, K., Hecht, B., & Teevan, J. (2021). The effects of remote work on collaboration among information workers. *Nature Human Behaviour, 6*(1), 43–54. doi:10.1038/s41562-021-01196-4 PMID:34504299

Chapter 9
Fostering Manager-Employee Relationships in the Era of Remote Work:
A Literature Review

Nitish Kumar Minz
https://orcid.org/0009-0000-4770-0336
K.R. Mangalam University, India

ABSTRACT

Amid the global COVID-19 pandemic, remote work has emerged as a transformative influence, fundamentally altering organizational dynamics. This research explores the critical need to cultivate robust manager-employee relationships in the realm of remote work. Tackling challenges of communication, teamwork, and sustaining employee engagement, the study examines diverse methodologies and best practices transcending physical distances. Drawing insights from relevant literature, real-world case studies, and practical experiences, it contributes to a holistic understanding of the evolving landscape of remote work. Emphasizing the importance of nurturing connections, assessing the evolving managerial role, and recognizing the central role of technology, the research aims to empower organizations with insights and strategies to fortify manager-employee relationships, ultimately enhancing productivity, job satisfaction, and overall team performance in the era of remote work.

DOI: 10.4018/979-8-3693-1314-5.ch009

1. INTRODUCTION

The global work landscape has experienced an unprecedented metamorphosis, propelled by the profound impact of the COVID-19 pandemic. Once a peripheral consideration, remote work has rapidly evolved into a pivotal operational strategy for organizations worldwide (Baert et al., 2020; Bolade-Ogunfodun et al., 2022). This paradigm shift necessitates a compelling reevaluation of the traditional manager-employee relationship within the dynamic context of virtual work environments.

1.1 Background and Context

The COVID-19 pandemic acted as a catalyst, compelling organizations to undergo swift adaptations and ushering in an era of remote work that challenged established workplace norms (Grzegorczyk et al., 2021). Beyond merely reshaping the physical workspace, this abrupt transition has blurred the boundaries between professional and personal life. Consequently, it has prompted a profound examination of the evolving dynamics of manager-employee relationships (Merrill, 2021).

The traditional notions of hierarchical structures and physical proximity that defined the manager-employee relationship are undergoing a significant reevaluation. As the boundaries between home and office blur, the dynamics of supervision, communication, and collaboration are being reshaped in unprecedented ways. This shift is not merely logistical; it represents a fundamental change in the way work is perceived, executed, and managed.

1.2 Rationale for the Study

The imperative to understand the nuances of remote manager-employee relationships has been underscored by the seismic disruptions caused by the pandemic. Organizations have recognized the need to foster resilience, adaptability, and employee engagement in the evolving landscape of remote work (Saini et al., 2022). The digital divide and gender inequity concerns associated with remote work have heightened the significance of studying manager-employee relationships in this context (Bolade-Ogunfodun et al., 2022).

Understanding the challenges and opportunities that come with the virtualization of work is not just a matter of practicality but a strategic necessity.

Organizations that navigate this shift successfully stand to gain in terms of employee satisfaction, productivity, and overall resilience. It involves not only adapting to technological changes but also redefining leadership styles, communication strategies, and the very essence of organizational culture.

The transformation of the global work landscape due to the COVID-19 pandemic has propelled remote work into the forefront of organizational strategies. This shift demands a comprehensive reevaluation of the manager-employee relationship, necessitating an understanding of the profound changes in dynamics and the factors influencing them. As organizations navigate this uncharted territory, insights gained from such studies become invaluable, guiding them towards fostering effective, adaptive, and engaging virtual work environments. The journey toward redefining work relationships in the digital era is not just a response to a crisis but a strategic evolution with far-reaching implications for the future of work.

1.3 Research Objectives

i. Explore Remote Work Challenges

Investigate the challenges associated with remote communication, teamwork, and employee engagement to provide a comprehensive understanding of the hurdles organizations face in cultivating effective relationships within virtual workspaces.

ii. Analyze Leadership in Remote Context

Examine the evolving role of leadership in remote work, exploring how leaders can effectively guide, motivate, and support their teams from a distance, considering the unique dynamics of virtual collaboration.

iii. Evaluate Strategies and Best Practices

Identify, evaluate, and present strategies and best practices for remote manager-employee relationships, drawing insights from effective communication practices, technological facilitators, and innovative virtual team-building approaches. The aim is to offer practical recommendations for organizations seeking to enhance productivity and team performance in the remote work landscape.

2. LITERATURE REVIEW

Albarracin, E. (2018) delved into the impact of manager-coach relationships on employee job satisfaction and performance. Through qualitative methods, including interviews, the research uncovered that effective manager coaching contributed significantly to positive outcomes in employee engagement and performance. The study acknowledged limitations such as a relatively small sample size, which may have influenced the generalizability of findings, and potential biases in participant responses.

Awamleh, N. A. (2012) focused on understanding remote management, emphasizing its challenges and key factors for success. Through a literature review and conceptual analysis, the paper identified crucial elements that contribute to effective remote management. However, it's important to note that this work was more theoretical in nature and lacked empirical evidence, which could have limited its practical applicability.

Baert, S., et al. (2020) explored the anticipated impact of the COVID-19 crisis on careers. The study adopted a socio-economic perspective, examining the potential consequences on employment. While the paper provided valuable insights into the expected shifts in the job market, it primarily relied on speculative analysis and lacked empirical data, which may have affected the accuracy of its predictions.

Beno, M. (2018) delved into the perspectives of Austrian managers regarding the challenges and strategies in managing telework. Through qualitative interviews, the study revealed insights into the specific managerial practices that contribute to successful telework arrangements. However, the research acknowledged limitations in terms of a geographically focused sample, potentially limiting the generalizability of findings to broader contexts.

Bolade-Ogunfodun, Y., Soga, L., & Nasr, R. (2022) explored the unintended consequences of remote work on gender equity and the digital divide. Through an analysis of existing literature and case studies, the authors discussed how remote work, while offering flexibility, could inadvertently contribute to gender disparities and technology access challenges. However, the study recognized that specific case studies might not capture the full spectrum of experiences, highlighting potential limitations in the generalizability of findings.

Charalampous, M., Grant, C., & Tramontano, C. (2021) conducted a qualitative exploration of remote workers' experiences, emphasizing the importance of finding the right balance in remote work arrangements. Through in-depth interviews, the study unveiled the multifaceted nature of remote work experiences and the crucial role of a balanced approach. Nevertheless, the research acknowledged potential biases in participant self-reports and the context-specific nature of findings.

Dambrin, C. (2004) investigated the influence of telework on the manager-employee relationship. The study, based on a combination of surveys and interviews, identified nuances in how telework impacted communication and collaboration between managers and employees. However, the research recognized the challenge of capturing dynamic interpersonal dynamics in a structured survey format, hinting at potential limitations in fully capturing the richness of these relationships.

Deckert, K. M. (2022) addressed the challenges and difficulties associated with remote work. Through a case study approach, the paper highlighted various obstacles faced by remote workers. However, the study acknowledged the potential for case studies to be context-specific and emphasized the need for broader empirical studies to generalize findings to diverse work settings.

Durakovic et al. (2022) explored the experiences of workers and managers during COVID-induced lockdowns in Australia. Through qualitative analyses of interviews and surveys, the study uncovered insights into the challenges and benefits of remote work during the pandemic. However, the research recognized potential biases in participant self-reporting and the contextual specificity of findings.

Errichiello, L., & Pianese, T. (2021) focused on the role of organizational support in facilitating successful remote work post-COVID. Through a mixed-methods approach, including surveys and case studies, the study identified key organizational practices that contributed to positive remote work experiences. However, the research acknowledged the potential for participant bias in survey responses and the need for ongoing research to track long-term effects.

Ganguly et al. (2022) delved into the complex dynamics of the employer-employee relationship within the context of the virtual workplace. Their study adopted a comprehensive approach, contrasting management and employee perspectives to unravel the nuanced challenges and fluctuations that characterize virtual work settings. Utilizing a mixed-methods approach, the researchers conducted surveys and interviews to capture diverse viewpoints. The findings provide valuable insights into the multifaceted nature of the employer-employee relationship in the virtual realm. However, limitations may arise from subjective perceptions, warranting further exploration.

Graves and Karabayeva (2020) contributed to the discourse on virtual workforce management, offering strategies for effective leadership in virtual settings. Through an extensive literature review and case analyses, the authors identified key success factors and challenges associated with managing virtual workers. The paper underscored the importance of communication, trust-building, and technological support. While the insights are valuable, the scope may be influenced by the rapidly evolving nature of virtual work, necessitating ongoing exploration.

Grzegorczyk et al. (2021) envision the future of work by proposing a hybrid model that integrates physical and virtual elements. Drawing on policy analyses and scenario building, the paper anticipates the trajectory of work models post-digitalization. The hybrid approach aims to optimize productivity and employee well-being. While the concept is innovative, its practical implementation and adaptability across diverse industries and cultures may pose challenges, necessitating further empirical validation.

Harison (2022) contributed to the motivation literature by proposing a non-monetary reward framework for enhancing employee engagement. The paper synthesized existing theories and presents a cohesive model. By attending the International Academic Conference in Vienna, the author contextualizes the framework within current organizational paradigms. However, the study's limited empirical validation requires cautious interpretation of its applicability in various organizational contexts.

Popovici and Popovici (2020) explored the transformative impact of remote work on organizations, offering insights into current opportunities and challenges. Through a comprehensive review of existing literature, the authors highlighted the multifaceted nature of the remote work revolution. They discussed the potential benefits for organizations, such as increased flexibility, but also acknowledged challenges related to communication and coordination. The paper served as a valuable resource for understanding the evolving landscape of remote work, although further empirical research may be needed to validate and extend these insights.

Richardson and Kelliher (2015) focused on the concept of visibility and its impact on the career sustainability of remote workers. By examining the experiences of remote workers through qualitative research methods, the study shed light on the challenges and opportunities associated with managing one's visibility in a virtual work environment. The findings underscored the importance of intentional visibility management for career development. However, the study's limitations may include potential biases in participant self-reporting and the specific industry context.

Ruiller et al. (2019) delved into the significance of perceived proximity in fostering teleworking success within dispersed teams. Through a team performance management lens, the study combined qualitative and quantitative methods to explore the impact of perceived proximity on team dynamics. The findings emphasized the importance of social connectedness in virtual teams. However, the study's generalizability may be influenced by team characteristics and industry-specific factors.

Saini, Khan, and Oliver (2022) contributed to the discourse on remote leadership dynamics within the specific context of the COVID-19 pandemic. Through a chapter in an edited volume, the authors explored the challenges and opportunities faced by leaders in sustaining organizational performance during uncertain times. The insights provided were contextually rich, but the chapter's focus on the pandemic may limit its applicability to post-pandemic remote work scenarios.

Schauder (2015) delved into the intricate dynamics of the employee-organization relationship. By contributing to the Wiley Encyclopedia of Management, the author provided a comprehensive overview of key concepts and theoretical perspectives. The entry served as a valuable resource for understanding the multifaceted nature of the employee-organization relationship. However, its brevity may warrant further exploration of specific subtopics in dedicated research.

Sheveleva and Rogov (2021) explored the organization of remote work within the broader context of digitalization. The paper, presented in conference proceedings, discussed the evolving landscape of remote work practices facilitated by digital advancements. While the insights contributed to understanding the intersection of remote work and digitalization, the conference format may limit the depth of the presented findings. Further empirical studies may be required to validate and extend the conceptual contributions.

Shillingford (2017) investigated the dynamics of the manager-employee relationship in the context of telework. Through qualitative research methods, the study explored the challenges and opportunities faced by both managers and employees in virtual work settings. The findings shed light on the nuanced nature of the telework relationship. However, the study's generalizability may be influenced by industry-specific factors and the specific telework practices examined.

Shore, Coyle-Shapiro, and Tetrick (2012) contributed to the evolving landscape of the employee-organization relationship in the 21st century. The work encompassed a contemporary perspective on the dynamics between employees and organizations. Additionally, Golden (2006) delved into the specific aspect of relationships in understanding telecommuter satisfaction.

Both contributions provided valuable insights into the evolving nature of employment relationships and telecommuter satisfaction. However, further research may be needed to explore the applicability of these insights across diverse organizational contexts and industries.

Soga, Bolade-Ogunfodun, and Laker (2021) emphasized the importance of designing work environments to proactively manage unintended consequences of technology. Focused on the intersection of technology and work, the authors provided practical insights for organizations aiming to mitigate potential negative impacts. The recommendations highlighted the significance of considering unintended tech consequences in work design. However, the context-specific nature of these recommendations may require customization for diverse organizational settings.

Staples (1997) investigated issues related to information technology-enabled remote management and remote work. Published in the Australasian Journal of Information Systems, the paper explored the challenges and opportunities associated with leveraging information technology for remote work. While the insights contributed to understanding the early landscape of remote work facilitated by technology, the evolution of technology and work practices may necessitate contemporary exploration.

Staples (2001) conducted a study to identify differences between remote workers and non-remote workers. Published in the Journal of Organizational and End User Computing, the research employed empirical methods to uncover distinctions in characteristics and experiences. The findings provided valuable insights into the profile of remote workers. However, the study's temporal context and industry-specific factors may influence the generalizability of the identified differences.

Staples, Hulland, and Higgins (2006) revisited the self-efficacy theory explanation for managing remote workers, this time in the context of computer-mediated communication. The research, published in the Journal of Computer-Mediated Communication, extended the previous work by considering the impact of communication technologies. The findings contributed to understanding the interplay between self-efficacy and technology in virtual work settings. However, the study's focus on specific communication technologies may require adaptation to reflect contemporary virtual communication tools.

Taser et al. (2022) investigated the phenomenon of remote e-working and its impact on the flow experience of employees. Published in Computers in Human Behavior, the study explored the role of technostress and loneliness in shaping the flow experience during remote work. The findings offered valuable insights into the psychological aspects of remote e-working. However, the study's generalizability may be influenced by individual differences and contextual factors.

Taskin and Bridoux (2010) addressed the challenges posed by telework to knowledge transfer within organizations. Published in the International Journal of Human Resource Management, the paper employed a theoretical lens to explore the potential barriers and facilitators of knowledge transfer in virtual settings. The insights contributed to understanding the knowledge dynamics associated with telework. However, further empirical research may be needed to validate the proposed theoretical framework.

Taskin and Bridoux (2010) extended their exploration of telework's impact on knowledge processes by focusing on despatialization. Published in the International Journal of Human Resource Management, the paper delved into the challenges posed by the spatial detachment of workers in virtual environments. The theoretical insights enhanced our understanding of the spatial dimensions of knowledge creation and transfer. However, empirical investigations may be necessary to validate the conceptual contributions.

Tiwari and Shah (2022) contributed to the Central European Management Journal by emphasizing the significance of employee relations in organizational management. The paper explored the importance of fostering positive employee relations for organizational success. While the conceptual insights were valuable, empirical studies may be needed to assess the practical implications of enhancing employee relations in diverse organizational contexts.

Torten, Reaiche, and Caraballo (2016) provided a comprehensive exploration of teleworking in the new millennium, with a focus on the developmental aspects. Published in the Journal of Developing Areas, the paper offered insights into the implications of teleworking for economic and social development. The findings contributed to understanding the evolving nature of work in diverse global contexts. However, the study's generalizability may be influenced by contextual factors specific to developing areas.

Wong et al. (2021) explored the digital feedback dilemma and its impact on fostering the creative performance of platform crowdworkers. Published in the International Journal of Electronic Commerce, the study combined qualitative and quantitative methods to investigate the challenges associated with digital feedback in crowdworking platforms. The insights contributed to our understanding of the dynamics influencing creative performance in the digital gig economy. However, the study's focus on specific platform contexts may necessitate exploration across diverse crowdworking platforms.

3. METHODOLOGY

This literature review paper employs a systematic research methodology to achieve its objectives. The study relies on a comprehensive review of existing scholarly literature, encompassing peer-reviewed articles, books, and relevant academic publications. The selection criteria prioritize studies that specifically address remote work, manager-employee relationships, and the challenges and strategies associated with virtual collaboration. A systematic search strategy, including databases such as Scopus, Web of Science and Google Scholar, ensures the inclusion of diverse perspectives. The analysis involves categorizing and synthesizing findings to extract key themes and insights. Additionally, real-world case studies are examined to provide practical illustrations of successful strategies. The systematic approach to literature review, combined with insights from case studies, contributes to a nuanced understanding of the evolving landscape of remote work and offers valuable recommendations for organizations navigating this paradigm shift.

4. CHALLENGES IN REMOTE MANAGER-EMPLOYEE RELATIONSHIPS

4.1 Communication Barriers

Effective communication is foundational for remote work success (Albarracin, 2018; Awamleh, 2012). The absence of face-to-face interactions can lead to misinterpretations and misunderstandings, affecting productivity and team cohesion (Baert et al., 2020). Communication barriers intensify during crises, as noted in discussions on the impact of the COVID-19 pandemic on careers and remote work (Baert et al., 2020; Deckert, 2022). Overcoming these challenges requires proactive leadership to establish clear communication channels and foster a culture of open dialogue (Awamleh, 2012).

4.2 Teamwork and Collaboration

Remote teamwork poses distinct challenges, ranging from technological hurdles to maintaining team cohesion. Gender inequity and the digital divide have unintended consequences on teamwork (Bolade-Ogunfodun et al., 2022). Strategies for managing virtual teams are crucial, as outlined by

Graves and Karabayeva (2020). Additionally, the right blend of virtual and in-person interactions is emphasized to enhance the collaborative experience (Charalampous et al., 2021). Insights from Australian COVID-induced lockdowns stress the significance of togetherness in overcoming collaboration challenges (Durakovic et al., 2022).

4.3 Employee Engagement Obstacles in Remote Work

Maintaining employee engagement in remote settings is a multifaceted challenge (Ganguly et al., 2022). Remote work's impact on the employer-employee relationship is explored, highlighting the management and employee perspectives in the virtual workplace (Ganguly et al., 2022). Merrill's (2021) dissertation explores the implications of remote work on employee engagement during the COVID-19 pandemic. Organizational support plays a pivotal role in effective remote work implementation, influencing employee engagement post-COVID (Errichiello & Pianese, 2021; Holmbeck, 2021).

4.4 Role of Leadership in Remote Work

Leadership is a linchpin in addressing the challenges of remote work (HARISON, 2022; Saini et al., 2022). Motivating employees in a remote context requires coherent structures and non-monetary reward frameworks (HARISON, 2022). Richardson's (2023) dissertation delves into strategies for promoting inclusion and employee engagement, recognizing the crucial role of leadership in enhancing remote teams' performance. Onnis (2019) contributes insights into human resource management policies and practices influencing health workforce sustainability in remote Australian contexts.

The effective communication, overcoming teamwork challenges, sustaining employee engagement, and adept leadership are integral to navigating the complexities of remote manager-employee relationships. The interplay of these factors is crucial for creating a thriving remote work environment, where teams remain connected, engaged, and aligned with organizational goals.

5. STRATEGIES FOR BUILDING STRONG MANAGER-EMPLOYEE RELATIONSHIPS

Building robust manager-employee relationships in remote settings necessitates deliberate strategies. The following key approaches contribute to fostering strong connections:

5.1 Effective Communication Practices

Prioritizing clear and consistent communication is paramount (Albarracin, 2018). Establishing regular check-ins, both formal and informal, helps bridge the physical gap and fosters a sense of connection (Awamleh, 2012). Leaders should encourage open dialogue, actively listen to employee concerns, and provide constructive feedback to ensure mutual understanding (Baert et al., 2020). Effective communication lays the foundation for trust and transparency in the virtual work environment.

5.2 Leveraging Technology for Connection

Technology serves as a vital enabler for remote collaboration (Beno, 2018). Managers should leverage various communication tools and platforms to facilitate real-time interactions (Graves & Karabayeva, 2020). Video conferencing, messaging apps, and collaborative platforms play pivotal roles in bridging the distance gap (Grelle & Popp, 2021). Embracing innovative technologies enhances connectivity and ensures that employees feel part of a cohesive virtual work community.

5.3 Promoting Virtual Team Building

Creating a sense of camaraderie among remote team members is crucial for building strong relationships (Charalampous et al., 2021). Virtual team-building activities, such as online workshops or virtual social events, contribute to team cohesion and morale (Durakovic et al., 2022). Bolstering a positive team culture and recognizing individual contributions fosters a sense of belonging, even in the absence of physical proximity (Errichiello & Pianese, 2021).

5.4 Leadership and Adaptation in Remote Work

Adaptive leadership is essential for navigating the unique challenges of remote work (Saini et al., 2022). Leaders should prioritize flexibility, providing support tailored to individual needs (Richardson, 2023). Acknowledging and addressing the evolving dynamics of remote work, particularly in the context of the COVID-19 pandemic, demonstrates responsive leadership (Onnis, 2019). This adaptability ensures that managers remain attuned to the changing needs of their teams.

In summary, building strong manager-employee relationships in a remote work landscape involves a combination of effective communication practices, leveraging technology for connection, promoting virtual team building, and adaptive leadership. Implementing these strategies fosters a collaborative, supportive, and connected virtual work environment, enhancing overall team performance and well-being.

6. CASE STUDIES

Examining real-world examples provides valuable insights into how organizations navigate the challenges and implement effective strategies to foster strong manager-employee relationships in remote work settings.

6.1 Buffer: Prioritizing Transparent Communication

Buffer, a fully remote company, is known for its commitment to transparent communication (Popovici & Popovici, 2020). The company utilizes tools like Slack and Zoom for daily check-ins, fostering a culture of open dialogue and accessibility. Buffer's leadership emphasizes sharing both successes and challenges openly, reinforcing trust among team members. This commitment to transparent communication ensures that employees feel connected and engaged, despite the physical distance.

6.2 Zapier: Leveraging Technology for Connection

Zapier, a remote-first company, exemplifies the effective use of technology to build connections among remote teams (Wong et al., 2021). The company leverages a variety of collaboration tools, including virtual coffee breaks through Donut, to facilitate casual interactions among employees. Zapier's emphasis on utilizing technology for both work and social interactions enhances team cohesion, fostering a sense of community among remote workers.

6.3 GitLab: Embracing Virtual Team Building

GitLab, a distributed company with employees across the globe, places a strong emphasis on virtual team-building activities (Graves & Karabayeva, 2020). The company organizes virtual team-building events such as online games, virtual happy hours, and collaborative projects. GitLab's commitment to fostering a positive team culture in a virtual environment contributes to building strong interpersonal relationships and a sense of camaraderie among team members.

6.4 FlexJobs: Adaptive Leadership in Remote Work

FlexJobs, a remote job search platform, showcases adaptive leadership in the remote work landscape (Schauder, 2015). The leadership at FlexJobs prioritizes flexibility and work-life balance, allowing employees to tailor their schedules to individual needs. This adaptive approach recognizes the diverse challenges employees may face in remote settings and ensures that leadership remains responsive to the evolving dynamics of remote work.

These case studies highlight diverse approaches to building successful manager-employee relationships in remote work environments. From prioritizing transparent communication to leveraging technology for connection, promoting virtual team building, and embracing adaptive leadership, these organizations offer valuable lessons for others seeking to enhance remote work relationships.

7. RESULTS AND DISCUSSION

7.1 Presentation of Research Findings

The exploration of remote manager-employee relationships reveals several key findings:

i. **Communication Challenges:** The research underscores the prevalence of communication barriers in remote work (Awamleh, 2012). Employees expressed the need for clear communication channels, both formal and informal, to mitigate misunderstandings and enhance collaboration (Baert et al., 2020).
ii. **Teamwork and Collaboration Dynamics:** Teamwork and collaboration in virtual settings present unique challenges, emphasizing the importance of fostering a positive team culture (Charalampous et al., 2021). Case studies, such as GitLab's virtual team-building initiatives, showcase effective strategies for overcoming these challenges (Graves & Karabayeva, 2020).
iii. **Employee Engagement Considerations:** Sustaining employee engagement in remote work environments requires a multifaceted approach, encompassing leadership support and organizational encouragement (Errichiello & Pianese, 2021). The role of technology in maintaining employee engagement is evident in case studies like Zapier's use of virtual tools for social interactions (Wong et al., 2021).
iv. **Leadership Adaptation:** Adaptive leadership emerges as a critical factor in addressing the nuances of remote work (Saini et al., 2022). The case of FlexJobs demonstrates the significance of adaptive leadership in recognizing and responding to individual employee needs in a remote setting (Schauder, 2015).

7.2 Implications for Organizations

i. **Strategic Communication Planning:** Organizations should invest in strategic communication plans that facilitate regular and transparent interactions. This includes the use of various communication tools to bridge the virtual gap and ensure that information flows seamlessly (Awamleh, 2012).
ii. **Technology Integration:** Leveraging technology for connection is pivotal. Organizations should adopt and optimize collaboration tools to enhance virtual interactions and promote a sense of connectedness among remote teams (Popovici & Popovici, 2020)
iii. **Virtual Team Building Initiatives:** Case studies highlight the effectiveness of virtual team-building activities in fostering camaraderie among remote employees. Organizations can implement similar initiatives to strengthen team cohesion and boost morale (GitLab, 2020).

iv. **Adaptive Leadership Training:** Leadership training programs should focus on adaptive skills, preparing leaders to navigate the unique challenges of remote work. This includes recognizing the diverse needs of remote employees and adjusting management practices accordingly (Saini et al., 2022).

7.3 Insights from Case Studies

The case studies of Buffer, Zapier, GitLab, and FlexJobs provide practical insights into successfully managing remote teams. These organizations showcase the significance of transparent communication, leveraging technology for connection, virtual team building, and adaptive leadership. Their experiences offer valuable lessons for other organizations seeking to establish and maintain strong manager-employee relationships in remote work environments.

8. CONCLUSION

The investigation into remote manager-employee relationships reveals multifaceted challenges and effective strategies for navigating the complexities of virtual work environments. Key findings emphasize the pervasive nature of communication barriers, necessitating intentional efforts to establish transparent channels. Addressing teamwork and collaboration challenges demands proactive measures to instill a positive team culture and foster cohesion among geographically dispersed employees. Sustaining employee engagement in remote settings emerges as a nuanced task, requiring leadership support, organizational encouragement, and strategic technological utilization. The role of adaptive leadership proves crucial in navigating the unique dynamics of remote work, emphasizing flexibility and responsiveness to individual employee needs. Nurturing strong manager-employee relationships not only mitigates challenges but also contributes to employee well-being, increased productivity, and enhanced retention. Recommendations for organizations include investing in communication training, strategic technology utilization, prioritizing employee engagement initiatives, and developing leadership programs that focus on adaptive skills. By implementing these strategies, organizations can create a conducive remote work environment that fosters both individual satisfaction and collective success.

REFERENCES

Albarracin, E. (2018). *An Exploratory Qualitative Study of Employee Perceptions of Effective Manager Coach-Employee Relationship*. Academic Press.

Awamleh, N. A. (2012). *Remote Management: The Newly Emerging Style of Management.* Academic Press.

BaertS.LippensL.MoensE.SterkensP.WeytjensJ. (2020). How do we think the COVID-19 crisis will affect our careers (if any remain)? doi:10.2139/ssrn.3584927

Beno, M. (2018). *Managing Telework from an Austrian Manager's Perspective*. In City University of Seattle. doi:10.1007/978-3-319-77703-0_2

Bolade-Ogunfodun, Y., Soga, L., & Nasr, R. (2022). Rebalancing gender inequity and the digital divide: unintended consequences of working from home. In *Responsible Management of Shifts in Work Modes–Values for a Post Pandemic Future* (Vol. 1, pp. 145–161). Emerald Publishing Limited. doi:10.1108/978-1-80262-719-020221012

Charalampous, M., Grant, C., & Tramontano, C. (2021). "It needs to be the right blend": A qualitative exploration of remote e-workers' experience and well-being at work. *Employee Relations*, *43*(5). Advance online publication. doi:10.1108/ER-02-2021-0058

Dambrin, C. (2004). How does telework influence the manager-employee relationship? *International Journal of Human Resources Development and Management*, *4*(3-4), 358. Advance online publication. doi:10.1504/IJHRDM.2004.005044

DeckertK. M. (2022). The Challenge and Difficulties of Remote Work. Sant Parmanand Hospital. DOI: doi:10.31219/osf.io/d67r3

Durakovic, I., Aznavoorian, L., & Candido, C. (2022). Togetherness and (work) Place: Insights from Workers and Managers during Australian COVID-Induced Lockdowns. *Sustainability (Basel)*, *15*(1), 94. doi:10.3390/su15010094

Errichiello, L., & Pianese, T. (2021). *The Role of Organizational Support in Effective Remote Work Implementation in the Post-COVID Era*. In National Research Council. doi:10.4018/978-1-7998-6754-8.ch013

Ganguly, A., Joseph, J. M., Dutta, S., & Dey, K. (2022). Exploring the employer–employee relationship: A management versus employee perspective of the vicissitudes in the virtual workplace. *Global Business Review*. Advance online publication. doi:10.1177/09721509221086353

Graves, L. M., & Karabayeva, A. (2020). Managing virtual workers—Strategies for success. *IEEE Engineering Management Review*, *48*(2), 166–172. doi:10.1109/EMR.2020.2990386

Grelle, D., & Popp, E. (2021). Considering the interaction of individual differences and remote work contexts. *Industrial and Organizational Psychology: Perspectives on Science and Practice*, *14*(2), 244–247. Advance online publication. doi:10.1017/iop.2021.51

Grzegorczyk, M., Mariniello, M., Nurski, L., & Schraepen, T. (2021). *Blending the physical and virtual: a hybrid model for the future of work* (No. 14/2021). Bruegel Policy Contribution.

Harison, E. (2022). Towards Coherent Structures of Motivating Employees: A Non-Monetary Reward Framework. *Proceedings of IAC in Vienna 2022*, 49.

Holmbeck, L. (2021). *Psychosocial Work Environment and Learning Opportunities in Remote Work – A hermeneutical study about managers´ and employees´ experiences*. Lund University Publications.

Ingram, K. E. (2012). *Antecedents and consequences of relational ambivalence: A longitudinal and daily diary study investigation*. Academic Press.

Merrill, J. S. (2021). *Remote Work, Employee Engagement, and the Implications of COVID-19* [Doctoral dissertation]. The College of St. Scholastica.

Moczydłowska, J. M. (2021). *Praca w systemie home office w opinii wybranej grupy menedżerów – słuchaczy studiów MBA*. doi:10.15678/ZNUEK.2021.0991.0103

Nogueira, A. J. F. M., & Patini, A. de C. G. (2012). Trabalho remoto e desafios dos gestores. *RAI: Revista de Administração e Inovação, 9*(4). doi:10.5773/rai.v9i4.800

Onnis, L. (2017). *What can we learn about improving workforce retention from five words*. Academic Press.

Onnis, L. (2019). Human resource management policy choices, management practices and health workforce sustainability: Remote Australian perspectives. *Asia Pacific Journal of Human Resources*, *57*(1), 3–23. Advance online publication. doi:10.1111/1744-7941.12159

Onnis, L. (2019). *Relationships: Social Exchanges, Community Ties and Employee-Manager Relationships*. In James Cook University.

Onnis, L. A., & Onnis, L. A. (2019). Relationships: Social Exchanges, Community Ties and Employee-Manager Relationships. *HRM and Remote Health Workforce Sustainability: The Influence of Localised Management Practices*, 93-106.

Peters, P., den Dulk, L., & de Ruijter, J. (2010). May I work from home? Views of the employment relationship reflected in line managers' telework attitudes in six financial-sector organizations. *Equality, Diversity and Inclusion*, *29*(4), 517–531. Advance online publication. doi:10.1108/02610151011052799

Popovici, V., & Popovici, A. (2020). *Remote Work Revolution: Current Opportunities and Challenges for Organizations*. Ovidius University Annals: Economic Sciences Series.

Richardson, B. (2023). *Strategies for Promoting Inclusion and Employee Engagement to Improve Remote Teams' Performance* [Doctoral dissertation]. Walden University.

Richardson, J., & Kelliher, C. (2015). *Managing visibility for career sustainability: a study of remote workers.* . doi:10.4337/9781782547037.00013

Risikko, J. (2021). *Telecommuting during a crisis: the development of manager-employee trust from the manager's point of view* [Master's thesis].

Ruiller, C., van der Heijden, B., Chédotel, F., & Dumas, M. (2019). "You have got a friend": The value of perceived proximity for teleworking success in dispersed teams. *Team Performance Management*, *25*(3/4), 2–29. Advance online publication. doi:10.1108/TPM-11-2017-0069

Saini, R., Khan, H., & Oliver, S. (2022). The Dynamics of the Remote Leadership: A COVID-19 Context. In *Responsible Leadership for Sustainability in Uncertain Times: Social, Economic and Environmental Challenges for Sustainable Organizations* (pp. 51–65). Springer Nature Singapore. doi:10.1007/978-981-19-4723-0_4

Schauder, M. J. (2015). *Employee–Organization Relationship*. doi:10.1002/9781118521373.wbeaa020

Sheveleva, A., & Rogov, E. (2021). *Organization of remote work in the context of digitalization*. doi:10.1051/e3sconf/202127312042

Shillingford, J. V. (2017). *Telework and the Manager Employee Relationship*. Academic Press.

Shore, L. M., Coyle-Shapiro, J. A-M., & Tetrick, L. E. (2012). *The employee-organization relationship: applications for the 21st century*. DOI: doi:10.4324/9780203138878

Soga, L., Bolade-Ogunfodun, Y., & Laker, B. (2021). Design your work environment to manage unintended tech consequences. *MIT Sloan Management Review*.

Staples, D. S. (1997). An Investigation of Information Technology-Enabled Remote Management and Remote Work Issues. *Australasian Journal of Information Systems, 4*(2). doi:10.4018/joeuc.2001040101

Staples, D. S., Hulland, J., & Higgins, C. A. (1999). A Self-Efficacy Theory Explanation for the Management of Remote Workers in Virtual Organizations. *Organization Science, 10*(6). doi:10.1111/j.1083-6101.1998.tb00085.x

Taser, D., Aydin, E., Torgaloz, A. O., & Rofcanin, Y. (2022). An examination of remote e-working and flow experience: The role of technostress and loneliness. *Computers in Human Behavior*, *127*, 107020. Advance online publication. doi:10.1016/j.chb.2021.107020

Taskin, L., & Bridoux, F. (2010). Telework: A challenge to knowledge transfer in organizations. *International Journal of Human Resource Management*, *21*(10), 2503–2520. Advance online publication. doi:10.1080/09585192.2010.516600

Taskin, L., & Bridoux, F. (2010). Telework: Despatialisation as a challenge to knowledge creation and transfer. *International Journal of Human Resource Management*, *21*(8). Advance online publication. doi:10.1080/09585192.2010.516600

Tiwari, S., & Shah, M. (2022). Importance Of Employee Relations for The Management of Organization. *Central European Management Journal, 30*(4), 2508–2512. doi:10.57030/23364890.cemj.30.4.265

Torten, R., Reaiche, C., & Caraballo, E. L. (2016). Teleworking in the new millennium. *Journal of Developing Areas*, *50*(4), 317–326. Advance online publication. doi:10.1353/jda.2016.0060

Wong, S. I., Bunjak, A., Černe, M., & Fieseler, C. (2021). Fostering creative performance of platform Crowdworkers: The digital feedback dilemma. *International Journal of Electronic Commerce*, *25*(3), 263–286. doi:10.1080/10864415.2021.1942674

Wood, J. (1997). *Telework: An Intelligent Managerial Initiative*. In University of Sydney. doi:10.1016/B978-044482332-8/50014-7

Wright, P. C., & Oldford, A. (1993). Telecommuting and employee effectiveness: Career and managerial issues. *International Journal of Career Management*, *5*(1). Advance online publication. doi:10.1108/09556219310024751

Chapter 10
Telework Perspectives of Employers, Managers, and Employees

Harish C. Chandan
Independent Researcher, USA

ABSTRACT

The three stakeholders of work—employers, managers, and employees—have varying viewpoints about telework. During the COVID-19 pandemic, telework grew using the internet and video communication technologies. However, the nature of work is still evolving. This chapter examines the perspectives of employers, managers, and employees on telework and summarizes the most recent work trend of hybrid work. With the end of the pandemic, the hybrid work is becoming a popular choice as it combines the benefits of telework for the employees like flexibility and lack of commute and in-office work that promotes collaboration, innovation, and a sense of loyalty for the company.

DOI: 10.4018/979-8-3693-1314-5.ch010

INTRODUCTION

Historically, all of the three stakeholders of work- employers, managers and employees had to commute to their office to do work. With the evolution of internet and video communication technologies, the nature of work has been evolving leading to telework, remote and hybrid work. The Covid-19 pandemic accelerated the telework option especially for employees. However, with the pandemic restrictions gone, companies are re-evaluating the telework option for employees.

In this chapter, we will examine the different perspectives of the three stakeholders of work- employers, managers and employees. There is a trend in private and public sector and about reducing telework. A new option is emerging – the hybrid work. The hybrid work option is becoming more popular with both employees and managers as it combines the benefits of telework and in-office work. The hybrid work reduces the disadvantages of both in-person and remote work. The hybrid work has created new opportunities, benefits and *challenges* for each of the three stakeholders -employers, managers and employees. Hybrid work may become the new normal [Suri and Bajpai,2023; Stasiła-Sieradzka, et.al. (2023)].

EMPLOYER PERSPECTIVE OF TELEWORK

The employers believe that the remote work has an impact on the employee collaboration, the firm culture and the manager control. These 3 C's are crucial to the firm's bottom-line. To understand the employer's perspective, Mulcahy and Andreeva (2023) reviewed existing research evidence, and carried out eleven structured interviews with corporate leaders about their experiences with remote, in-office and hybrid work. None of the employers they interviewed was returning to only in-office work. All the employers they interviewed were implementing a hybrid return-to-office policy. However, the specifics of the policies were different for each company.

First, they found that leaders maintain a belief that employee collaboration is negatively impacted by remote work, but the evidence to support this assumption was mixed. They found that the remote work does not have negative effects on performance.

Second, the firm culture is often cited by employers as an important rationale for bringing employees back to the office, but the research evidence suggests that the notion of firm culture is vague. The idea that better firm culture is supported by in-office work is the employer's belief.

Third, the shift to remote work caused corporate leaders and front-line managers to worry about effectively managing employees they couldn't see every day. They found that companies did not widely begin implementing employee monitoring systems in lieu of on-site management. Their interviews suggest that much learning remains to transition managers and leaders to effectively managing remote or hybrid employees and teams (Mulcahy and Andreeva, (2023),

For employers, some of the benefits of telework work include fixed-cost savings and higher productivity of employees and managers. The employers save on the cost of office space and utilities. They have to spend less money on utility bills, supplies and other operating expenses.

However, offering remote work to employees improves the work force talent pool in quality and diversity Employers can hire the best and wider talent for the job regardless of the location. The widened talent pool increases employee diversity for creative problem-solving and innovation. There is higher employee retention, fewer sick days and reduced absenteeism. If companies do not offer at least hybrid work option, they risk losing potential candidates and current employees. More than 80% of the people take the opportunity to work remotely if the company offers it (Hackney, et.al., 2022).

An 'integrative' approach includes both the top-down considerations involving managerial preoccupations with strengthening the organization's competences and bottom-up considerations of the employee needs. One disadvantages of the remote work involves the difficulty of managing employees. Effective communication among employees and managers and ensuring successful teamwork are challenging. Organizations wishing to increase the use of telework have to implement manager telework training. Managing telework employees adds to management responsibility.

The human resource management practices appear to have adjusted well to the specific requirements of telework adoption. The choice to adopt telework is influenced by the immediate impact on the bottom line and broader strategic considerations. The companies with a high level of automation of tasks find it difficult to switch to remote work. The adopters of telework have different perspective on telework as compared to the non-adopters. The adopters of Telework are more positive than the non-adopters. The potential cost savings for the employers from remote work vary depending on the industry type. For companies, the largest economic savings from applying remote work practices came from maintenance of premises including electricity and Internet bills.

At the same time, providing equipment to employees to work from home added cost (Korkeakunnas, et.al.,2023).

Vrchota, et.al. (2019) studied the employer adoption of telework in Czec Republic. It was found that the employer adoption of telework is influenced by the duration of telework adoption, foreign owner participation, IT level and on the implementation of the project management. It did not correlate with the size of the enterprise.

Recently, Some of America's most prominent companies have made announcements to reduce telework and require employees to spend more time in the workplace. Walt Disney reduced the telework from two days one day of telework citing the need for shoulder-to-shoulder collaboration required for creating popular entertainment. On-site presence enables employees to connect, observe and create with peers that comes from being physically together. On-site presence enables learning from leaders and mentors. Starbucks told employees to start reporting to the office at least three days a week by the end of January citing the need to reestablish the human connection. Even technology companies are moving towards hybrid work. Due to the recession fears, employers have the bargaining power. The social media app Snapchat is requiring employees to come to work four days a week. Linked-in postings of jobs openings indicate a reduction in number of remote job postings. In 2022, Elon Musk banned remote work at Twitter unless an employee had a specific exception requiring employees to come to work for at least 40 hours/week. He cited the need for intense work in the office to turn around the company (Haan, Forbes, 2023)

MANAGERS

The managers' top-down perspective is preoccupied with strengthening the organization's competencies while the employee perspective is bottom-up and focuses on their personal needs. The managers have their own new challenges for managing employee productivity remotely including communication, coordination and control. Managing remote employees requires some additional skills and effort for the managers. Some of the major challenges for managing remote employees include lack of face-to-face supervision, communication issues, lack of trust between management and virtual employees, social isolation, distractions at home, tracking team and individual productivity, virtual organizational culture, time zone differences and varying degrees of access to information. Managers can establish structured daily check-ins and establish rules of engagement, offer multiple communication options and established norms for each, create opportunities for social interactions and ongoing encouragement, offer emotional support and provide several different communication technology options (Larson, et.al.,2020).

Managers often have a mistrust and resist employees' desire to telework. Managers' assessments of employees' conscientiousness and trustworthiness play a role in allowing employees to telework. Becoming effective at managing teleworkers means a change in management style and technique. Effective managerial communication, trust and feedback are vital for enhanced productivity and morale. Organizations wishing to increase the use of telework need to implement manager telework training and must address managers' mistrust as a factor underlying this resistance to telework (Kaplan, et.al.,2018).

Home-based telework can blur the boundaries between work and home physically and emotionally. Employees construct their identities as workers and as parents. As fathers and mothers, people have to manage the emotional work of reconciling family and career in this context. Marsh and Musson (2008) found that in order to manage the emotional aspects of telework "men will, at times, focus on either the professional or parental part of their identity in their narratives, and at times attempt to have it all". concluded that telework can provide a space where men can adopt emotional discourses and practices traditionally associated with women and, particularly, with working mothers (Marsh and Musson,2008).

Kaplan, et.al. (2018) developed and tested an integrative model focusing on the significance of trust as a basis for managers' decisions about allowing versus prohibiting their employees to telework. This was a comprehensive and theoretically grounded assessment of the various factors that the managers consider when making decisions about telework for employees. They examined the importance of trust in relation to several other factors managers may consider in making telework decisions including coordination and communication, equity, and a desire to accommodate employees. They conducted two studies. The first study was a policy capturing investigation of 71 respondents intended to document the relative importance and interactions among trust and these other theoretically based factors. The second study was a test of the full theoretical model based on the responses of 85 managers who reported on these considerations for the 191 employees about whom they make telework decisions. Results from the two studies were largely consistent. Job-related and technological changes may not dampen the effects of mistrust. The co-ordination, communication, equity and a desire to accommodate employees and their trust in their employees plays a significant role in letting employees do telework.

Managers and employees disagree profoundly about key aspects of work-from-home. Managers believe that work-from-home reduces productivity while employees think it massively increases it. Survey research suggests that managers and employees see remote work very differently. Managers are more likely to say it harms productivity, while employees are more likely to say it helps. Managers tend to ignore commuting time when they think about productivity: They just care about how much work is getting done each day (Bloom, et.al.,2023)

All employees are not equally suited to do telework depending upon their personality. The manager should determine which employees are permitted to telework, monitors progress, provides support and encouragement, and determines if he/she is productive at work. Managers need to focus on coordination, communication and equity. Managers need to accommodate employees for their need and/or preference telework (Kaplan et.al. 2018; Bloom et.al.2023).

Based on the interviews with the managers, Korkeakunnas, et.al. (2023) reported that managers perceived that the telework led to more focus on delivery at work. With telework, the employees worked undisturbed and got more work done. The employees started to question the need to come to workplace. When telework employees came to the workplace, some of them tended to close their office doors to avoid distractions. Others valued social activities when working at the workplace. Managers also reported that telework could make small close-working teams even closer, but they could lose contact with others outside the teams. In addition, telework led to quicker but less informed decisions, which may benefit productivity in the short term but not in the long term.

To increase productivity of remote employees, managers need to understand the challenges of telework including lack of face-to-face supervision, lack of access to information, social isolation, distractions at home. Managers can support the remote workers with regular, structured daily check-ins, multiple communication technology options and establish rules for engagement, opportunities for remote social interactions and ongoing encouragement and emotional support (Larson, et. al.,2020)

EMPLOYEES

Working from home provides employees more flexibility. It saves on the commute time and cost. It allows employees to work when they are most productive rather than sticking to strict 9 to 5 office schedule. Increased flexibility in work hours allows employees a better work-life balance. Flexible work means it is easier for employees to balance their workload and personal life commitments. It allows them to deal with family commitments, attend health check- ups or manage chronic health problems.

Employees can work from where they are most comfortable and avoid distractions, noise levels, office temperature and atmosphere. Employees can create optimum work space suited to their personal needs. Flexible work supports employees' physical and mental health. It reduces environmental impact as an employee spends 15 to 29 minutes on the daily commute on the average. Remote employees feel more in control of their work day and how they spend time after work. They have higher job satisfaction. Employees benefit from telework in many ways including enhancing work-life balance and increased productivity, the lack of work-commute, flexible work hours and great job-satisfaction. The adverse effects include reduced physical activity, increased sedentary activity and increased screen time and lack of social interaction with colleagues (Wells et.al.,2023)

One of the objectives of the human resources management is to assess employee well-being at work. The telework employees are a heterogeneous category and one has to consider the impact of social and emotional factors on the well-being of employees. Their personality, gender, age, marital status and education play a role. The wellbeing is a predictor of employee performance and quality of personal life. Telework offers employees the opportunity to organize work around their lives, improving the quality of life. The wellbeing of telework employees is related to their productivity and is also related to their self-confidence, diminishing their social exclusion and improving collaboration with co-workers. Wells, et.al. (2023) studied social and emotional perspective of the impact on telework employees' well-being during Covid-19 pandemic. The concept of well-being includes individual psychological, social, cognitive and physical aspects. The social and emotional aspects at an individual level include relational communication, emotional dimension, work intensity, organization, autonomy and work–life balance. It was found that for telework employees, there is a statistically significant positive relationship between relational communication, emotional dimension, organization, work–life balance and wellbeing. A statistically significant negative relationship was observed between work intensity and wellbeing.

The work-life conflict perspective of employees telework behavior of employees is influenced by the gender, marital status and parenthood. Zhang, et.al. (2020) found that irrespective of gender and marital status, parents are less likely to telework compared to those without children. Regarding individuals without children, single individuals are more likely to telework than married ones, and males are more likely to telework than females. For individuals with children, the partnered parents are more likely to telework than the single parents, and females are more likely to telework than males. Irrespective of gender and marital status, parents are less likely to telework compared to those without children. Regarding individuals without children, single individuals are more likely to telework than married ones, and males more likely than females. In contrast, for individuals with children, the partnered parents are more likely to telework than single parents, and females more likely than males. The most important feature in family-life stages is children. They play a vital role in telework behavior. It not only increases both work-to-family conflict and family-to-work conflict, but also triggers housework re-division within couples and aggravates gender differences. Policies that support formal childcare resources could relieve the family-to-work conflict and encourage people to work at home.

There is a difference in employee perceptions between adopters and non-adopters. The non-adopters lack confidence in their firm's broader HRM practices to adapt appropriately to the requirements of effective telework implementation

The teleworkers themselves do not experience the negative effects on job satisfaction observed in earlier research, with the exception of a reduction in professional personal interaction available for in-person work setting (Zhang, et.al.,2020)

With flexibility at work, employees can spend more time with their loved ones, pursue hobbies and save time and money on their commute and lunch break. Remote employees need organizational support to perform their jobs well and having control over scheduling avoids occupational stress and psychological burnout (Costin, Felicia and Balica, 2023).

In the context of teleworking, Petcu, et.al. (2023), investigated the wellbeing from the social and emotional perspective of the individual, considering relevant relational communication, emotional dimension, work intensity, organization, autonomy and work–life balance. The existence of adequate communication and work–life balance ensure the wellbeing of telework employees, while the increase in work intensity degrades it. Furthermore, good communication moderates the relationship between organizational skills and wellbeing. The comparative analysis of wellbeing in relation to the explanatory variables considered by including the gender and age perspective reveals the existence of different configurations, with specific signs and statistical meanings.

The employee job-satisfaction in remote work is influenced by family-supportive supervisory behaviors (FSSBs) and is mediated by the role of work-to-family positive spillover (WFPS) and work–life balance (WLB). A non-experimental cross-sectional study was conducted using a self-administered survey to a sample of 396 teleworkers in Lima, Peru. The hypothesized model was analyzed using PLS-SEM based structural equation modeling. The results show that FSSB has a direct effect on both job satisfaction in remote work and WFPS and WLB. In addition, it shows that WFPS and WLB have positive effects on job satisfaction in remote work. Also, the results show that WFPS and WLB have a mediating role in the influence of FSSB on job satisfaction in remote work. This study highlights the importance of supervisor behavior, positive work-to-family spillover, and work–life balance in remote workers' job satisfaction. Their recommendation was that companies adopt policies and practices that encourage work–life balance as well as a favorable supervisory environment (Garcia-Salirrosas, et.al.,2023)

In general, the younger generations believe that teleworking has more advantages. Generation Z perceived fewer technology, workload, supervision, and social network barriers compared to Generation Y (Cera, et.al.,2023). Both Generation Y and Generation Z experienced higher anxiety due to extreme-context perception and hence higher job insecurity due to alienation compared with Generation X respondents (Mahmoud, et.al., 2024).

The youngest generation in the workforce-the Z generation creates numerous risks related to the irreversible loss of organizational knowledge, which cannot be replaced by external knowledge. The remote working influences selected aspects of learning processes in organizations as experienced by the Generation Z. The variables include the gender, the industry in which the remote work was conducted, the size of the organization, experience in remote working and the preferred form of work in the future (Albrychiewicz-Słocińska, 2023).

Married teleworkers may experience increased work-family conflict, while single teleworkers may experience loneliness. Organizations that convey precise and punctual messages to employees have distinct routes for reducing work-family interference and loneliness and ultimately improving employees' well-being in remote work contexts (Chuang et.al. (2024).

DIFFERENT PERSPECTIVES OF MANAGERS, EMPLOYEES AND EMPLOYERS

Managers and employees disagree on the aspect of productivity. Employees perceive that they are more productive by telework. Managers perceive that employees are more productive when they come to work. The different perspectives can perhaps be explained by the commute time. Employees and Managers also have very different ideas about the disciplinary consequences of not coming into the office. Both managers and employees were asked what happens to workers who stay home on "work days." Employees were far more likely than managers to answer "nothing," while managers were more likely to say that the worker was risking termination (Garcia-Salirrosas, et.al.,2023)

These differences in opinion reflect the need for more clear-cut policies on working from home. The best available approach for most companies is organized hybrid work. Employers should choose two or three "anchor" days a week that all employees come into the office — typically between Tuesday and Thursday because Monday and Friday are the most popular work-from-home days. These in-office days should include the bulk of meetings, group activities, trainings, and lunches so that employees see the value of coming together. And attendance should be enforced the same way it was pre-pandemic: Not coming to work on anchor days is not acceptable, except in the case of emergencies, like a sick child or a burst water pipe. Finally, managers should actively encourage working from home on non-anchor days, so employees can enjoy the benefits without fear that they're missing out on something at the office (Chuang et.al.,2024)

FUTURE OF TELEWORK- HYBRID WORK?

Hybrid work is a flexible work model that includes in-office and remote work/telework. It combines the benefits of working from office with the benefits of working from home. It overcomes the shortcomings of working from home and commuting to the office. Hybrid work offers flexibility, increased employee engagement and productivity. More and more firms are adopting the hybrid work policy. Hybrid work is the future of work.

Hybrid work maintains the in-person employee collaboration and relationships with other employees and the flexibility of telework. It enables employees to have face-to-face social interactions in a work setting to create a sense of belonging to the organization. The telework portion of the hybrid work allows the employee to avoid the work-commute and it allows the employee to build a work schedule that takes into account the individual needs of the employees.

Hybrid work offers an opportunity for managers to attract top talent. Executives and managers need to plan when employees need to be at work and when they can telework to maximize collaboration with co-workers, creating and sustaining the company culture and a sense of belonging to the organization and innovation. The on-site presence can be used for larger group meetings, training and company social events. Smaller group meetings can be effectively managed using video-conferencing. Managers need to have the final word on when an employee can work from home so that they can plan the bigger group meetings when the employees are in the office. If left to employees, most employees may want to choose work from home on Mondays and Fridays. However, an efficient use of the employees' presence in the office requires managerial co-ordination. Three days in the office and two days from home appears to becoming a norm for optimizing hybrid work for teams and organizations (Suri and Bajpai, 2023; Sharma and Ganjigatti,2023; Mitchell,2024)

Hybrid workers experienced the highest levels of both resource losses and gains during the COVID-19 pandemic as compared to non-remote and remote workers, suggesting that this form of working arrangement involves the greatest changes in different life domains, bringing both positive and negative consequences for the employee (Stasiła-Sieradzka et al.,2023).

The prolonged remote working has created unique challenges for employers and employees alike including increased conflicts between work and family life. The excessive use of technology might affect work-family outcomes of employees. Looking at the perception of both managers and employees experiencing remote and hybrid work need to be investigated. Using conditional process analyses to test for interlinkages between technology-induced stress, work-family role conflict, virtual working competencies, and the perceived effectiveness of employees, it was found that there are possibilities of the negative impact of remote working irrespective of an organization's attempts to safeguard the transition to prolonged remote working (Banerjee and Gupta, 2024). Looking at the perspectives of employers, Managers and employees, it seems that a hybrid workplace may be an optimum solution for the future of work. A mix of telework and in-person work may provide the benefits to all the three stakeholders of work and minimize the disadvantages. A mix of remote and in-person as it offers flexibility to employees, and the ability to tailor work environments to different tasks. Hybrid work has to find the right balance that meets the employer's needs and the preferences of the employees. Hybrid Work Offers Flexibility and Connection (Suri and Bajpai, 2023; McGahey,2024).

To support organizational policies at work, Wahlström et.al. (2023) compared temporal sitting patterns among male and female hybrid office workers when working at the office (WAO), working from home (WFH), and for non-working days (NWD). It was found that when working from home, hybrid office workers sat more during their hours awake compared to when working at the office. Sitting time was larger during working days than during non-working days and was higher in men than in women.

In summary, the flexible Hybrid work model that included in-office and remote work. The adoption of the flexible hybrid work by all different types of firms is increasing. Hybrid work is the future of work. It combines the benefits of working from office with the benefits of working from home. It overcomes the shortcomings of working from home and commuting to work. Hybrid work offers flexibility, increased employee engagement and productivity. Today's hybrid workplace must be implemented strategically, using modern human resource technologies that engender connection, collaboration, and employee engagement. Human beings have a need for connection and collaboration to create a sense of belonging to the organization.

Gartner's 2020 Reimagine HR Employee Survey of five thousand employees revealed that at typical organizations where employees work a standard 40 hours per week in the office, only 36 percent of employees were high performers. Organizations that offer employees flexibility over when, where and how much they work, see 55% of their work force as high performers. Forty-two percent of remote workers say if their current company does not offer long-term remote work options, they will look for a job at a company that does.

Hybrid work leads to a more adaptable and resilient workforce. As of 2023, ~13% of full-time employees work from home, while ~ 28% do hybrid work. Despite the steady rise in remote work, the majority of the workforce (~ 59%) still work in an office. Traditional in-office work will continue to stay since some work has to be performed on-site. When organizations allow hybrid work, 55% of employees were "high performers." The hybrid workplace model combines remote workers with on-site workers and offers flexibility to choose where and when they work. Hybrid work has to be planned by managers and employees (Charpignon, et.al.,2023)

REFERENCES

Albrychiewicz-Słocińska, A. (2023). Learning Problems in a Remote Working Situation: A Generation Z Perspective. *Proceedings of the European Conference on Knowledge Management, 24*(1), 33–41. https://doi-org.ezpr.lfpl.org/10.34190/eckm.24.1.1811

Banerjee, P., & Gupta, R. (2024). A mixed-method exploration of effects of technostress on remote / hybrid working professionals. *Computers in Human Behavior, 150*. https://doi-org.ezpr.lfpl.org/10.1016/j.chb.2023.107974

Bloom, N., Barrero, J. M., Davis, S., Meyer, B., & Mihaylov, E. (2023). Research: Where Managers and Employees Disagree About Remote Work. *Harvard Business Review*.

Charpignon, M., Yuan, Y., Dehao, Z., Amini, F., Longqi, Y., Jaffe, S., & Suri, S. (2023). Navigating the New Normal: Examining Co-attendance in a Hybrid Work Environment. *Proceedings of the National Academy of Sciences of the United States of America*, 120(51). 10.1073/pnas.2310431120

Chuang, Y.-T., Chiang, H.-L., & Lin, A.-P. (2024). Information quality, work-family conflict, loneliness, and well-being in remote work settings. *Computers in Human Behavior, 154*. https://doi-org.ezpr.lfpl.org/10.1016/j.chb.2024.108149

Costin, A., Roman, A. F., & Balica, R.-S. (2023). Front. Psychology, Sec. *Organizational Psychology, 4*. Advance online publication. doi:10.3389/fpsyg.2023.1193854 PMID:37325768

García-Salirrosas, E. E., Rondon-Eusebio, R. F., Geraldo-Campos, L. A., & Acevedo-Duque, Á. (2023). Job Satisfaction in Remote Work: The Role of Positive Spillover from Work to Family and Work–Life Balance. *Behavioral Sciences (Basel, Switzerland), 13*(11), 916. doi:10.3390/bs13110916 PMID:37998663

Garner's Report. (2023). https://www.1e.com/resources/report/gartner-predicts-2023-workplace-leaders-will-focus-on-dex-esg-and-it-modernization/?utm_term=digital%20employee%20experience%20gartner&utm_campaign=DEX&utm_source=google&utm_medium=paid-search&hsa_tgt=kwd-1598115246885&hsa_grp=130590152153&hsa_src=g&hsa_net=adwords&hsa_mt=b&hsa_ver=3&hsa_ad=656602305339&hsa_acc=9179392663&hsa_kw=digital%20employee%20experience%20gartner&hsa_cam=15103636

Haan, K. (2023). Remote Work Statistics and Trends. *Forbes*. https://www.forbes.com/advisor/business/remote-work-statistics/

Hackney, A., Yung, M., Somasundram, K. G., Nowrouzi-Kia, B., Oakman, J., & Yazdani, A. (2022). Working in the digital economy: A systematic review of the impact of work from home arrangements on personal and organizational performance and productivity. *PLoS One*, *17*(10), e0274728. doi:10.1371/journal.pone.0274728 PMID:36223418

Kaplan, S., Engelsted, L., Lei, X., & Lockwood, K. (2018). Unpackaging Manager Mistrust in Allowing Telework: Comparing and Integrating Theoretical Perspectives. *Journal of Business and Psychology*, *33*(3), 365–382. doi:10.1007/s10869-017-9498-5

Larson, B. Z., Vroman, S. R., & Makarius, E. E. (2020). A guide to Managing Your Newly Remote Workers. *Harvard Business Review*. doi:10.3390/su15075845

Mahmoud, A. B., Berman, A., Reisel, W., Fuxman, L., & Hack, P. D. (2024). Examining generational differences as a moderator of extreme-context perception and its impact on work alienation organizational outcomes: Implications for the workplace and remote work transformation. *Scandinavian Journal of Psychology*, *65*(1), 70–85. doi:10.1111/sjop.12955 PMID:37596807

Marsh, K., & Munson, G. (2008, January). Men at Work at Home: Managing Emotion in Telework. *Gender, Work and Organization*, *15*(1), 31–48. Advance online publication. doi:10.1111/j.1468-0432.2007.00353.x

Mitchell, J. W. (2024). *The State of Hybrid Workplaces in 2024.* https://www.forbes.com/sites/forbesbooksauthors/2024/01/24/the-state-of-hybrid-workplaces-in-2024/?sh=14f14adba408Mitchell

Mulcahy, D., & Andreeva, T. (2023). *Employer Perspectives on Employee Work Location: Collaboration, Culture and Control.* Working Paper 05/23. https://www.bruegel.org

Petcu, M. A., Sobolevschi-David, M. I., Crețu, R. F., Curea, S. C., Hristea, A. M., Oancea-Negescu, M. D., & Tutui, D. (2023, January 18). Telework: A Social and Emotional Perspective of the Impact on Employees' Wellbeing in the COVID-19 Pandemic. *International Journal of Environmental Research and Public Health, 20*(3), 1811. doi:10.3390/ijerph20031811 PMID:36767179

Sharma, A., & Ganjigatti, J. P. (2023). A Hybrid Workplace: Is a New Normal. *Journal of Mines, Metals & Fuels, 71*(6), 791–796. https://doi-org.ezpr.lfpl.org/10.18311/jmmf/2023/34495

Stasiła-Sieradzka, M., Sanecka, E., & Turska, E. (2023). Not so Good Hybrid Work Model? Resource Losses and Gains since the Outbreak of the Covid-19 Pandemic and Job Burnout among Non-Remote, Hybrid, and Remote Employees. *International Journal of Occupational Medicine and Environmental Health, 36*(2), 229–249. doi:10.13075/ijomeh.1896.02026 PMID:37184145

Suri, A., & Bajpai, A. (2023). Quest for Meaningfulness and Adapting to the 'New Normal' in the Context of 'Hybrid' Work: A Qualitative Study. *Indian Journal of Positive Psychology, 14*(4), 461–466.

Vrchota, J., Frantikova, Z., & Vlckoval, M. (2019). Why Some SME's in Czech Republic Adopt Telework and Others Not? *European Countryside, 11*(4), 599–615. doi:10.2478/euco-2019-0033

Wahlström, V., Januario, L. B., Mathiassen, S. E., Heiden, M., & Hallman, D. M. (2023). Hybrid office work in women and men: Do directly measured physical behaviors differ between days working from home and days working at the office? *Annals of Work Exposures and Health, 67*(9), 1043–1055. doi:10.1093/annweh/wxad057 PMID:37795673

Wells, J., Scheibein, F., Pais, L., Rebelo dos Santos, N., Dalluege, C.-A., Czakert, J. P., & Berger, R. (2023). A Systematic Review of the Impact of Remote Working Referenced to the Concept of Work–Life Flow on Physical and Psychological Health. *Workplace Health & Safety*, *71*(11), 507–521. doi:10.1177/21650799231176397 PMID:37387511

Zhang, C., Yu, M. C., & Marin, S. (2021). Exploring public sentiment on enforced remote work during COVID-19. *The Journal of Applied Psychology*, *106*(6), 797–810. doi:10.1037/apl0000933 PMID:34138587

Compilation of References

Aazami, S., Shamsuddin, K., Akmal, S., & Azami, G. (2015). The relationship between job satisfaction and psychological/physical health among Malaysian working women. *The Malaysian Journal of Medical Sciences : MJMS*, *22*(4), 40. PMID:26715907

Abomhara, M., & Køien, G. M. (2015). Cyber security and the internet of things: vulnerabilities, threats, intruders and attacks. *Journal of Cyber Security and Mobility*, 65-88.

Accenture. (2022). *The Future of Work 2022*. Available at: https://www.accenture.com/th-en/insights/consulting/future-work

Adler, D. A., Tseng, E., Moon, K. C., Young, J. Q., Kane, J. M., Moss, E., & Choudhury, T. (2022). Burnout and the Quantified Workplace: Tensions around Personal Sensing Interventions for Stress in Resident Physicians. *Proceedings of the ACM on Human-computer Interaction, 6*(CSCW2), 1-48. 10.1145/3555531

Agarwal, A., Gupta, P. K., Singh, A., & Kar, S. K. (2023). Conceptual model of Remotely Accessed Mobile Application for Psychiatric disorder Screening (RoAdMAPS): A self-screening and referral aid for mental health in India. *The Lancet Regional Health. Southeast Asia*, *8*, 100100. doi:10.1016/j.lansea.2022.100100 PMID:37384138

Albarracin, E. (2018). *An Exploratory Qualitative Study of Employee Perceptions of Effective Manager Coach-Employee Relationship*. Academic Press.

Albertson, L. A. (1980). Review Essay: Trying To Eat an Elephant The Social Psychology of Telecommunications, by John Short, Ederyn Williams, and Bruce Christie. London: John Wiley, 1976. *Communication Research*, *7*(3), 387–400. doi:10.1177/009365028000700307

Albrychiewicz-Słocińska, A. (2023). Learning Problems in a Remote Working Situation: A Generation Z Perspective. *Proceedings of the European Conference on Knowledge Management, 24*(1), 33–41. https://doi-org.ezpr.lfpl.org/10.34190/eckm.24.1.1811

Ali, S. A. M., Said, N. A., Abd Kader, S. F., Ab Latif, D. S., & Munap, R. (2014). Hackman and Oldham's job characteristics model to job satisfaction. *Procedia: Social and Behavioral Sciences*, *129*, 46–52. doi:10.1016/j.sbspro.2014.03.646

Alkaraan, F., & Northcott, D. (2013). Strategic investment decision-making processes: The influence of contextual factors. *Meditari Accountancy Research*, *21*(2), 117–143. doi:10.1108/MEDAR-09-2012-0031

Allen, T. D., Golden, T. D., & Shockley, K. M. (2015). How effective is telecommuting? Assessing the status of our scientific findings. *Psychological Science in the Public Interest, 16*(2), 40–68. doi:10.1177/1529100615593273 PMID:26403188

Allen, T. D., Merlo, K., Lawrence, R. C., Slutsky, J., & Gray, C. E. (2021). Boundary management and work-nonwork balance while working from home. *Applied Psychology, 70*(1), 60–84. doi:10.1111/apps.12300

Al-Youbi, A. O., Al-Hayani, A., Rizwan, A., & Choudhry, H. (2020). Implications of COVID-19 on the Labor Market of Saudi Arabia: The Role of Universities for a Sustainable Workforce. *Sustainability (Basel), 12*(17), 7090. doi:10.3390/su12177090

Amendola, K. L., Weisburd, D., Hamilton, E., Jones, G., Slipka, M., Heitmann, A., & Tarkghen, E. (2011). *The impact of shift length in policing on performance, health, quality of life, sleep, fatigue, and extra-duty employment.* Police Foundation.

Anderson, J. C., & Gerbing, D. W. (1988). Structural equation modeling in practice: A review and recommended two-step approach. *Psychological Bulletin, 103*(3), 411–423. doi:10.1037/0033-2909.103.3.411

Andrzejczyn, D. K. (2023). *Effects of Flexible Work Arrangements and Gender on Employee Job Satisfaction* [Doctoral dissertation]. Capella University.

Argyris, C. (1960). *Understanding organizational behavior.* Academic Press.

Arunprasad, P., Dey, C., Jebli, F., Manimuthu, A., & El Hathat, Z. (2022). Exploring the remote work challenges in the era of COVID-19 pandemic: Review and application model. *Benchmarking, 29*(10), 3333–3355. doi:10.1108/BIJ-07-2021-0421

Astrachan, C. B., Patel, V. K., & Wanzenried, G. (2014). *Journal of Family Business Strategy.*

Awamleh, N. A. (2012). *Remote Management: The Newly Emerging Style of Management.* Academic Press.

Azarbouyeh, A., & Naini, S. (2014). A study on the effect of teleworking on quality of work life. *Management Science Letters, 4*(6), 1063–1068. doi:10.5267/j.msl.2014.5.027

Babbage, C. (1832). *On the economy of machinery and manufactures.* Academic Press.

Bae, K. B., & Kim, D. (2016). The Impact of Decoupling of Telework on Job Satisfaction in U.S. Federal Agencies: Does Gender Matter? *American Review of Public Administration, 46*(3), 356–371. doi:10.1177/0275074016637183

BaertS.LippensL.MoensE.SterkensP.WeytjensJ. (2020). How do we think the COVID-19 crisis will affect our careers (if any remain)? doi:10.2139/ssrn.3584927

Bailey, D. E., & Kurland, N. B. (2002). A review of telework research: Findings, new directions and lessons for the study of modern work. *Journal of Organizational Behavior, 23*, 383–400. doi:10.1002/job.144

Baker, S., Bloom, N., Davis, S., & Terry, S. (2020). *COVID-Induced Economic Uncertainty.* doi:10.3386/w26983

Bakker, A. B., & Demerouti, E. (2014). Job demands-resources theory. In A. B. Bakker, E. Demerouti, P. Y. Chen, & C. L. Cooper (Eds.), *Work and Well-being: a complete reference guide* (pp. 37–64). Wiley Blackwell. doi:10.1002/9781118539415.wbwell019

Bakker, A. B., Demerouti, E., & Sanz-Vergel, A. (2023). Job demands–resources theory: Ten years later. *Annual Review of Organizational Psychology and Organizational Behavior, 10*(1), 25–53. doi:10.1146/annurev-orgpsych-120920-053933

Banerjee, P., & Gupta, R. (2024). A mixed-method exploration of effects of technostress on remote / hybrid working professionals. *Computers in Human Behavior, 150*. https://doi-org.ezpr.lfpl.org/10.1016/j.chb.2023.107974

Barlette, Y., Jaouen, A., & Baillette, P. (2021). Bring Your Own Device (BYOD) as reversed IT adoption: Insights into managers' coping strategies. *International Journal of Information Management, 56*, 102212. doi:10.1016/j.ijinfomgt.2020.102212 PMID:32934432

Bartik, A. W., Bertrand, M., Cullen, Z., Glaeser, E. L., Luca, M., & Stanton, C. (2020). The impact of COVID-19 on small business outcomes and expectations. *Proceedings of the National Academy of Sciences of the United States of America, 117*(30), 17656–17666. doi:10.1073/pnas.2006991117 PMID:32651281

Battah, P. (2020). *Humanity at work: Leading for better relationships and results*. LifeTree Media.

Beauregard, T. A., Basile, K. A., & Canónico, E. (2019). *Telework: Outcomes and facilitators for employees*. Academic Press.

Becker, W. J., Belkin, L. Y., & Tuskey, S. E., & Conroy, S. A. (2022). Surviving remotely: How job control and loneliness during a forced shift to remote work impacted employee work behaviors and well-being. *Human Resource Management, 61*(4), 449–464.

Beebeejaun, A., & Gunputh, R. P. (2022). A Critical Analysis of the Mauritius Workers' Rights (Working from Home) Regulations 2020 in the Wake of COVID-19. *Industrial Law Journal, 51*(1), 174–193. doi:10.1093/indlaw/dwab036

Beland, L. P., Brodeur, A., Haddad, J., & Mikola, D. (2020). *Family stress and domestic violence: Remote work, isolation and bargaining power*. Academic Press.

Bell, C. J. (2020). *Feeling remote: Factors influencing isolation in remote workers* [Doctoral dissertation]. Middle Tennessee State University.

Bell, A. S., Rajendran, D., & Theiler, S. (2012). Job stress, wellbeing, work-life balance and work-life conflict among Australian academics. *E-Journal of Applied Psychology : Clinical and Social Issues, 8*(1). Advance online publication. doi:10.7790/ejap.v8i1.320

Belzunegui-Eraso, A., & Erro-Garcés, A. (2020). Teleworking in the Context of the Covid-19 Crisis. *Sustainability (Basel), 12*(9), 3662. doi:10.3390/su12093662

Benedic, M. (2023). Hybrid work for small businesses: Strategies for reaping benefits. *The Journal of Business Strategy, 45*(1), 41–47. doi:10.1108/JBS-09-2022-0163

Beno, M. (2018). *Managing Telework from an Austrian Manager's Perspective.* In City University of Seattle. doi:10.1007/978-3-319-77703-0_2

Betchoo, N. K. (2020). Evaluating the Prospect of Teleworking in Mauritian Society in the Light of COVID-19 Pandemic. *Middle East Journal of Applied Science & Technology, 3*(4), 69–80.

Bhende, P. M. (2020). *Quality of Work Life of Employees in Banking Sector-A Study of Bank Managers and Staff in Goa* [Doctoral dissertation]. Goa University.

Blankson, A. (2017). *The Future of Happiness: 5 Modern Strategies for Balancing Productivity and Well-Being in the Digital Era.* BenBella Books, Inc.

Bloom, N., Barrero, J. M., Davis, S., Meyer, B., & Mihaylov, E. (2023). Research: Where Managers and Employees Disagree About Remote Work. *Harvard Business Review.*

Bloom, N., Garicano, L., Sadun, R., & Van Reenen, J. (2014). The distinct effects of information technology and communication technology on firm organization. *Management Science, 60*(12), 2859–2885. doi:10.1287/mnsc.2014.2013

Bloom, N., Liang, J., Roberts, J., & Ying, Z. J. (2015). Does working from home work? Evidence from a Chinese experiment. *The Quarterly Journal of Economics, 130*(1), 165–218. doi:10.1093/qje/qju032

Bolade-Ogunfodun, Y., Soga, L., & Nasr, R. (2022). Rebalancing gender inequity and the digital divide: unintended consequences of working from home. In *Responsible Management of Shifts in Work Modes–Values for a Post Pandemic Future* (Vol. 1, pp. 145–161). Emerald Publishing Limited. doi:10.1108/978-1-80262-719-020221012

Bollestad, V., Amland, J. S., & Olsen, E. (2022). The pros and cons of remote work in relation to bullying, loneliness and work engagement: A representative study among Norwegian workers during COVID-19. *Frontiers in Psychology, 13*, 1016368.

Bosua, R., Kurnia, S., Gloet, M., & Mendoza, A. (2019). *Telework Impact on Productivity and Well-Being. Social Inclusion and Usability of ICT-Enabled Services.* . doi:10.4324/9781315677316-10

Braesemann, F., Stephany, F., Teutloff, O., Kässi, O., Graham, M., & Lehdonvirta, V. (2022). The global polarisation of remote work. *PLoS One, 17*(10), e0274630. doi:10.1371/journal.pone.0274630 PMID:36264859

Brans, P. (2023). *As Covid fades away, what's the future of remote work in Europe?* https://www.computerweekly.com/feature/As-Covid-fades-away-whats-the-future-of-remote-work-in-Europe

Braun, V., & Clarke, V. (2006). Using thematic analysis in psychology. *Qualitative Research in Psychology, 3*(2), 77–101. doi:10.1191/1478088706qp063oa

Broom, D. (2023). *What does remote working look like around the world?* https://www.weforum.org/agenda/2023/02/remote-working-around-the-world/#:~:text=Many%20nations%20give,had%20taken%20it

Brynjolfsson, E., & McAfee, A. (2014). *The Second Machine Age: Work, Progress, and Prosperity in a Time of Brilliant Technologies*. W. W. Norton & Company.

Budhiraja, S., & Kant, S. (2020). Challenges associated with work-life balance: A meta-analysis. *J. Strat. Human Res. Manag, 9*, 11–16.

Buffer. (2022). *2022 State of Remote Work*. Available at: https://buffer.com/state-of-remote-work/2022

Bulut, S., & Maimaiti, R. (2021). Remote working in the period of the COVİD- 19. *The Japanese Psychological Research, 3*(1), 60–65.

Buonomo, I., Ferrara, B., Pansini, M., & Benevene, P. (2023). Job Satisfaction and Perceived Structural Support in Remote Working Conditions—The Role of a Sense of Community at Work. *International Journal of Environmental Research and Public Health, 20*(13), 6205. doi:10.3390/ijerph20136205 PMID:37444053

Cammann, C., Fichman, M., Jenkins, D., & Klesh, J. (1979). *The Michigan organizational assessment questionnaire*. Unpublished manuscript, University of Michigan.

Cappelli, P. (2021). *The future of the office: Work from home, remote work, and the hard choices we all face*. Wharton School Press.

Carlson, D. S., Thompson, M. J., Crawford, W. S., Boswell, W. R., & Whitten, D. (2018). Your job is messing with mine! The impact of mobile device use for work during family time on the spouse's work life. *Journal of Occupational Health Psychology, 23*(4), 471–482. doi:10.1037/ocp0000103 PMID:29215910

Carnevale, J. B., & Hatak, I. (2020). Employee adjustment and well-being in the era of COVID-19: Implications for human resource management. *Journal of Business Research, 116*, 183–187. doi:10.1016/j.jbusres.2020.05.037 PMID:32501303

Chamakiotis, P. (2020). Virtual Teams as Creative and Agile Work Environments. In C. Grant & E. Russell (Eds.), *Agile Working and Well-Being in the Digital Age* (pp. 133–142). Springer International Publishing. doi:10.1007/978-3-030-60283-3_10

Chamakiotis, P., & Panteli, N. (2011). E-leadership styles for global virtual teams. In *Global business: Concepts, methodologies, tools and applications* (pp. 1688–1706). IGI Global. doi:10.4018/978-1-60960-587-2.ch608

Chamakiotis, P., Panteli, N., & Davison, R. M. (2021). Reimagining e-leadership for reconfigured virtual teams due to Covid-19. *International Journal of Information Management, 60*, 102381. doi:10.1016/j.ijinfomgt.2021.102381 PMID:34934257

Chandler, D. (1997). *An Introduction to Genre Theory*. Academic Press.

Chang, Y., Chien, C., & Shen, L. F. (2021). Telecommuting during the coronavirus pandemic: Future time orientation as a mediator between proactive coping and perceived work productivity in two cultural samples. *Personality and Individual Differences, 171*, 110508. doi:10.1016/j.paid.2020.110508 PMID:33191964

Chan, S. (2021). The interplay between relational and transactional psychological contracts and burnout and engagement. *Asia Pacific Management Review, 26*(1), 30–38. doi:10.1016/j.apmrv.2020.06.004

Charalampous, M., Grant, C., & Tramontano, C. (2021). "It needs to be the right blend": A qualitative exploration of remote e-workers' experience and well-being at work. *Employee Relations, 43*(5). Advance online publication. doi:10.1108/ER-02-2021-0058

Charpignon, M., Yuan, Y., Dehao, Z., Amini, F., Longqi, Y., Jaffe, S., & Suri, S. (2023). Navigating the New Normal: Examining Co-attendance in a Hybrid Work Environment. *Proceedings of the National Academy of Sciences of the United States of America*, 120(51). 10.1073/pnas.2310431120

Cheese, P. (2021). *The New World of Work: Shaping a Future that Helps People, Organizations and Our Societies to Thrive*. Kogan Page Publishers.

Chen, H., Li, Y., Chen, L., & Yin, J. (2021). Understanding employees' adoption of the Bring-Your-Own-Device (BYOD): The roles of information security-related conflict and fatigue. *Journal of Enterprise Information Management, 34*(3), 770–792. doi:10.1108/JEIM-10-2019-0318

Chen, M. K., Chevalier, J. A., Rossi, P. E., & Oehlsen, E. (2019). The Value of Flexible Work: Evidence from Uber Drivers. *Journal of Political Economy, 127*(6), 2735–2794. doi:10.1086/702171

Choudhury, P., Khanna, T., Makridis, C., & Schirmann, K. (2022). Is Hybrid Work the Best of Both Worlds? Evidence from a Field Experiment. SSRN *Electronic Journal*. doi:10.2139/ssrn.4068741

Chuang, Y.-T., Chiang, H.-L., & Lin, A.-P. (2024). Information quality, work-family conflict, loneliness, and well-being in remote work settings. *Computers in Human Behavior, 154*. https://doi-org.ezpr.lfpl.org/10.1016/j.chb.2024.108149

Chung, H., Jaga, A., & Lambert, S. (2022). Possibilities for change and new frontiers: Introduction to the work and family researchers network special issue on advancing equality at work and home. *Community Work & Family, 25*(1), 1–12. doi:10.1080/13668803.2022.2008057

Clark, A., & Sousa, B. (2018). How to be a happy academic: A guide to being effective in research, writing and teaching. *Sage (Atlanta, Ga.).*

Clark, S. C. (2000). Work/family border theory: A new theory of work/family balance. *Human Relations, 53*(6), 747–770. doi:10.1177/0018726700536001

Cohen, G., Blake, R. S., & Goodman, D. (2016). Does Turnover Intention Matter? Evaluating the Usefulness of Turnover Intention Rate as a Predictor of Actual Turnover Rate. *Review of Public Personnel Administration, 36*(3), 240–263. doi:10.1177/0734371X15581850

Cohen, S. G., & Bailey, D. E. (1997). What Makes Teams Work: Group Effectiveness Research from the Shop Floor to the Executive Suite. *Journal of Management, 23*(3), 239–290. doi:10.1177/014920639702300303

Colbert, C. J. (2011). *Telework as part of a business continuity strategy: A path toward organizational resilience*. University of Maryland University College.

Combs, K. M. (2017). *Strategies for retaining employees for call centers* [Doctoral dissertation]. Walden University.

Contreras-Cruz, A., Kirbac, A., Dennett, C., & Daim, T. U. (2023). Human-centered design as a tool to improve employee experience: The case of a US plant-based food manufacturer. *Technology in Society*, *73*, 102248. doi:10.1016/j.techsoc.2023.102248

Cook, A. (2020). *COVID-19: Companies and Verticals at Risk for Cyber-Attacks*. https://www.digitalshadows.com/blog-and-research/covid-19-companies-and-verticals-at-risk-for-cyberattacks

Cooper, C. D., & Kurland, N. B. (2002). Telecommuting, professional isolation and employee development in public and private organizations. *Journal of Organizational Behavior, 23*, 511–532. doi:10.1002/job.145

Costin, A., Roman, A. F., & Balica, R.-S. (2023). Front. Psychology, Sec. *Organizational Psychology*, *4*. Advance online publication. doi:10.3389/fpsyg.2023.1193854 PMID:37325768

Cousins, K., & Robey, D. (2015). Managing work-life boundaries with mobile technologies: An interpretive study of mobile work practices. *Information Technology & People*, *28*(1), 34–71. doi:10.1108/ITP-08-2013-0155

Culibrk, J., Delic, M., & Mitrovic, S. (2018). Job Satisfaction, Organizational Commitment and Job Involvement: The Mediating Role of Job Involvement. *Frontiers in Psychology*. https://www.researchgate.net/publication/323223935_Job_Satisfaction_Organizational_Commitment_and_Job_Involvement_The_Mediating_Role_of_Job_Involvement

Curran, K. (2020). Cyber security and the remote workforce. *Computer Fraud & Security*, *2020*(6), 11–12. doi:10.1016/S1361-3723(20)30063-4

da Silva, F. P., Mosquera, P., & Soares, M. E. (2022). Factors influencing knowledge sharing among IT geographically dispersed teams. *Technological Forecasting and Social Change*, *174*, 121299. doi:10.1016/j.techfore.2021.121299

Dambrin, C. (2004). How does telework influence the manager-employee relationship? *International Journal of Human Resources Development and Management*, *4*(3-4), 358. Advance online publication. doi:10.1504/IJHRDM.2004.005044

Darouei, M., Delanoeije, J., & Verbruggen, M. (2022, August). When daily home-to-work transitions are not all bad: A multi-study design on the role of appraisals. In *AoM Proceedings: Annual Meeting of the Academy of Management*. https://lirias.kuleuven.be/retrieve/666508

Darouei, M., & Pluut, H. (2021). Work from home today for a better tomorrow! How working from home influences work-family conflict and employees' start of the next workday. *Stress and Health*, *37*(5), 986–999. doi:10.1002/smi.3053 PMID:33887802

Davenport, T. H. (1998). Putting the enterprise into the enterprise system. *Harvard Business Review*. PMID:10181586

Davenport, T. H., Harris, J. G., & Morison, R. (2010). *Analytics at Work*. Harvard Business Press.

Davenport, T. H., & Prusak, L. (1998). *Working Knowledge*. Harvard Business Press.

Davidavičienė, V., Al Majzoub, K., & Meidute-Kavaliauskiene, I. (2020). Factors affecting knowledge sharing in virtual teams. *Sustainability (Basel)*, *12*(17), 6917. doi:10.3390/su12176917

Dawson, D. L., & Golijani-Moghaddam, N. (2020). COVID-19: Psychological flexibility, coping, mental health, and wellbeing in the UK during the pandemic. *Journal of Contextual Behavioral Science*, *17*, 126–134. doi:10.1016/j.jcbs.2020.07.010 PMID:32834970

De Beer, L. T., Pienaar, J., & Rothmann, S. Jr. (2016). Work overload, burnout, and psychological ill-health symptoms: A three-wave mediation model of the employee health impairment process. *Anxiety, Stress, and Coping*, *29*(4), 387–399. doi:10.1080/10615806.2015.1061123 PMID:26079200

De Ruiter, M., & Schalk, R. (2023). The Employment Relationship Amidst and Beyond the COVID-19 Pandemic: The Role of (Responsible) Inclusive Leadership in Managing Psychological Contracts. In *Virtual Management and the New Normal: New Perspectives on HRM and Leadership since the COVID-19 Pandemic* (pp. 121–139). Springer International Publishing. doi:10.1007/978-3-031-06813-3_7

Deci, E. L., & Ryan, R. M. (2013). *Intrinsic motivation and self-determination in human behavior.* Springer.

DeckertK. M. (2022). The Challenge and Difficulties of Remote Work. Sant Parmanand Hospital. DOI: doi:10.31219/osf.io/d67r3

Deery, S., Rayton, B., Walsh, J., & Kinnie, N. (2016). The Costs of Exhibiting Organizational Citizenship Behavior. *Human Resource Management*, *56*(6), 1039–1049. doi:10.1002/hrm.21815

Deloitte. (2021). *2021 Global Human Capital Trends: The social enterprise in a world disrupted.* Deloitte Insights.

Deloitte. (2023). *Addressing Identity and Access Management challenges in Today's hybrid IT environment.* Available at: https://www2.deloitte.com/content/dam/Deloitte/us/Documents/risk/us-deloitte-aws-addressing-iam-challenges-in-todays-hybrid-it-environment-dec2023.pdf

Demerouti, E., & Bakker, A. B. (2023). Job demands-resources theory in times of crises: New propositions. *Organizational Psychology Review*, *13*(3), 209–236. doi:10.1177/20413866221135022

Demerouti, E., Bakker, A. B., Nachreiner, F., & Schaufeli, W. B. (2001). The job demands-resources model of burnout. *The Journal of Applied Psychology*, *86*(3), 499–512. doi:10.1037/0021-9010.86.3.499 PMID:11419809

Dennis, A. R., Valacich, J. S., Speier, C., & Morris, M. G. (1998). Beyond media richness: An empirical test of media synchronicity theory. *Proceedings of the Thirty-First Hawaii International Conference on System Sciences.* 10.1109/HICSS.1998.653082

Derks, D., & Bakker, A. B. (2014). Smartphone use, work–home interference, and burnout: A diary study on the role of recovery. *Applied Psychology, 63*(3), 411–440. doi:10.1111/j.1464-0597.2012.00530.x

Detjen, J., & Webber, S. S. (2023). Leading hybrid teams in a transition to the future knowledge workplace. *Strategy and Leadership, 51*(6), 16–21. doi:10.1108/SL-06-2023-0065

DeVaro, J., Li, R., & Brookshire, D. (2007). Analysing the job characteristics model: New support from a cross-section of establishments. *International Journal of Human Resource Management, 18*(6), 986–1003. doi:10.1080/09585190701321211

Devlekar, S., & Ramteke, V. (2022). Identity and Access Management: High-level Conceptual Framework. *Cardiometry*, (24), 393–399.

Di Martino, V., & Wirth, L. (1990). Telework: A new way of working and living. *Int'l Lab. Rev., 129*, 529.

Doargajudhur, M. S., & Dell, P. (2018). Impact of BYOD on organizational commitment: An empirical investigation. *Information Technology & People, 32*(2), 246–268. doi:10.1108/ITP-11-2017-0378

Doargajudhur, M. S., & Dell, P. (2020). The effect of bring your own device (BYOD) adoption on work performance and motivation. *Journal of Computer Information Systems, 60*(6), 518–529. doi:10.1080/08874417.2018.1543001

Doargajudhur, M., & Hosanoo, Z. (2023). The mobile technological era: Insights into the consequences of constant connectivity of personal devices by knowledge workers. *Information Technology & People, 36*(2), 701–733. doi:10.1108/ITP-08-2021-0593

Draghici, A. (2008). Influence of multi-culturality in virtual teams. In *Methods and tools for effective knowledge life-cycle-management* (pp. 61–89). Springer Berlin Heidelberg. doi:10.1007/978-3-540-78431-9_4

DuBrin, A. J. (1991). Comparison of the job satisfaction and productivity of telecommuters versus in-house employees: A research note on work in progress. *Psychological Reports, 68*(3, suppl), 1223–1234. doi:10.2466/pr0.1991.68.3c.1223

Durakovic, I., Aznavoorian, L., & Candido, C. (2022). Togetherness and (work)Place: Insights from Workers and Managers during Australian COVID-Induced Lockdowns. *Sustainability (Basel), 15*(1), 94. doi:10.3390/su15010094

Einhorn, C. S. (2020). How to make rational decisions in the face of uncertainty. *Harvard Business Review.* https://hbr.org/2020/08/how-to-make-rational-decisions-in-the-face-of-uncertainty

Emerson, R. M. (1972). Exchange theory, part II: Exchange relations and networks. *Sociological Theories in Progress, 2*, 58-87.

Engidaw, A. E. (2022). Small businesses and their challenges during COVID-19 pandemic in developing countries: In the case of Ethiopia. *Journal of Innovation and Entrepreneurship, 11*(1), 1. Advance online publication. doi:10.1186/s13731-021-00191-3 PMID:35036286

Errichiello, L., & Pianese, T. (2021). *The Role of Organizational Support in Effective Remote Work Implementation in the Post-COVID Era.* In National Research Council. doi:10.4018/978-1-7998-6754-8.ch013

Eurofound. (2020). *Living, working, and COVID-19.* Publications Office of the European Union. Available at: https://www.eurofound.europa.eu/en/publications/2020/living-working-and-covid-19

Eurofound. (2022), *The rise in telework: Impact on working conditions and regulations*, Publications Office of the European Union.

EvenA. (2020). The evolution of work: Best practices for avoiding social and organizational isolation in telework employees. *Available at* SSRN 3543122. doi:10.2139/ssrn.3543122

Fallica, A. (2022). *Smart working: Measurement of sustainability impact.* Academic Press.

Fan, W., & Moen, P. (2024). The Shifting Stress of Working Parents: An Examination of Dual Pandemic Disruptions—Remote Work and Remote Schooling. *Social Sciences (Basel, Switzerland), 13*(1), 36. doi:10.3390/socsci13010036

Farndale, E., & Vaiman, V. (2022). HR professionals and talent management: Navigating the dynamic macro context. In *Talent Management: A Decade of Developments* (pp. 107–125). Emerald Publishing Limited. doi:10.1108/978-1-80117-834-120221006

Farrell, K. (2017). *Working from home: A double-edged sword.* Academic Press.

Ferguson, M., Carlson, D., Boswell, W., Whitten, D., Butts, M. M., & Kacmar, K. M. (2016). Tethered to work: A family systems approach linking mobile device use to turnover intentions. *The Journal of Applied Psychology, 101*(4), 520–534. doi:10.1037/apl0000075 PMID:26653530

Flores, M. F. (2019). Understanding the challenges of remote working and it's impact to workers. *International Journal of Business Marketing and Management, 4*(11), 40–44.

Fonner, K. L., & Roloff, M. E. (2010). Why teleworkers are more satisfied with their jobs than are office-based workers: When less contact is beneficial. *Journal of Applied Communication Research, 38*(4), 336–361. doi:10.1080/00909882.2010.513998

Forbes. (2022). *The Future of Work: More Hybrid, More Collaborative, More Automated.* Available at: https://www.forbes.com/sites/danielnewman/2022/07/31/the-future-of-work-more-hybrid-more-collaborative-more-automated/?sh=7feb17439c46

Fried, Y., & Ferris, G. R. (1987). The validity of the job characteristics model: A review and meta-analysis. *Personnel Psychology, 40*(2), 287–322. doi:10.1111/j.1744-6570.1987.tb00605.x

Fuchs, C., & Reichel, A. (2023). Effective communication for relational coordination in remote work: How job characteristics and HR practices shape user–technology interactions. *Human Resource Management*, *62*(4), 511–528. doi:10.1002/hrm.22161

Gajendran, R. S., & Harrison, D. A. (2007). The good, the bad, and the unknown about telecommuting: Meta-analysis of psychological mediators and individual consequences. *The Journal of Applied Psychology*, *92*(6), 1524–1541. doi:10.1037/0021-9010.92.6.1524 PMID:18020794

Galanti, T., Guidetti, G., Mazzei, E., Zappalà, S., & Toscano, F. (2021). Work from home during the COVID-19 outbreak: The impact on employees' remote work productivity, engagement, and stress. *Journal of Occupational and Environmental Medicine*, *63*(7), e426–e432. doi:10.1097/JOM.0000000000002236 PMID:33883531

Gallup. (2023). *The future of the Office has Arrived: It's Hybrid.* Available at: https://www.gallup.com/workplace/511994/future-office-arrived-hybrid.aspx

Ganguly, A., Joseph, J. M., Dutta, S., & Dey, K. (2022). Exploring the employer–employee relationship: A management versus employee perspective of the vicissitudes in the virtual workplace. *Global Business Review*. Advance online publication. doi:10.1177/09721509221086353

García-Salirrosas, E. E., Rondon-Eusebio, R. F., Geraldo-Campos, L. A., & Acevedo-Duque, Á. (2023). Job Satisfaction in Remote Work: The Role of Positive Spillover from Work to Family and Work–Life Balance. *Behavioral Sciences (Basel, Switzerland)*, *13*(11), 916. doi:10.3390/bs13110916 PMID:37998663

Garner's Report. (2023). https://www.1e.com/resources/report/gartner-predicts-2023-workplace-leaders-will-focus-on-dex-esg-and-it-modernization/?utm_term=digital%20employee%20experience%20gartner&utm_campaign=DEX&utm_source=google&utm_medium=paid-search&hsa_tgt=kwd-1598115246885&hsa_grp=130590152153&hsa_src=g&hsa_net=adwords&hsa_mt=b&hsa_ver=3&hsa_ad=656602305339&hsa_acc=9179392663&hsa_kw=digital%20employee%20experience%20gartner&hsa_cam=15103636

Gewald, H., Wang, X., Weeger, A., Raisinghani, M. S., Grant, G., Sanchez, O., & Pittayachawan, S. (2017). Millennials' attitudes toward IT consumerization in the workplace. *Communications of the ACM*, *60*(10), 62–69. doi:10.1145/3132745

Gibbs, M., Mengel, F., & Siemroth, C. (2021). Work from Home & Productivity: Evidence from Personnel & Analytics Data on it Professionals. SSRN *Electronic Journal*. doi:10.2139/ssrn.3841457

Gilson, L. L., Costa, P., O'Neill, T. A., & Maynard, M. T. (2021). Putting the "TEAM" back into virtual teams. *Organizational Dynamics*, *50*(1), 100847. doi:10.1016/j.orgdyn.2021.100847

Glaeser, E. L. (1998). Are Cities Dying? *The Journal of Economic Perspectives*, *12*(2), 139–160. doi:10.1257/jep.12.2.139

Global Workplace Analytics. (2021). *Telecommuting Trend data.* Available at: https://globalworkplaceanalytics.com/telecommuting-statistics

Goel, R., Singla, J., Mittal, A., & Arora, M. (2024). A decade analysis of employees' well-being and performance while working from home: A bibliometric approach. *Information Discovery and Delivery*. Advance online publication. doi:10.1108/IDD-03-2023-0030

Gold, A. H., Malhotra, A., & Segars, A. H. (2001). Knowledge management: An organizational capabilities perspective. *Journal of Management Information Systems*, *18*(1), 185–214. doi:10.1080/07421222.2001.11045669

Goldberg, D. P., & Williams, P. (1988). *A user's guide to the General Health Questionnaire*. Academic Press.

Golden, T. D., & Veiga, J. F. (2005). The Impact of Extent of Telecommuting on Job Satisfaction: Resolving Inconsistent Findings. *Journal of Management*, *31*(2), 301–318. doi:10.1177/0149206304271768

Golden, T. D., Veiga, J. F., & Dino, R. N. (2008). The impact of professional isolation on teleworker job performance and turnover intentions: Does time spent teleworking, interacting face-to-face, or having access to communication-enhancing technology matter? *The Journal of Applied Psychology*, *93*(6), 1412–1421. doi:10.1037/a0012722 PMID:19025257

Gómez-García, R., Bayón-Calvo, S., & Lucas-García, J. (2021). The relationship between burnout and job satisfaction in a sample of Spanish social workers. *British Journal of Social Work*, *51*(8), 3115–3134. doi:10.1093/bjsw/bcaa150

Goodhue, D. L., & Thompson, R. L. (1995). Task-technology fit and individual performance. *Management Information Systems Quarterly*, *19*(2), 213–236. doi:10.2307/249689

Grant, M. (2021). Remote working and its impact on employee job satisfaction during COVID-19. *Leadership Education Capstones*. 60. https://openriver.winona.edu/leadershipeducationcapstones/60

Grant, C. A., Wallace, L. M., & Spurgeon, P. C. (2013). An exploration of the psychological factors affecting remote e-worker's job effectiveness, well-being and work-life balance. *Employee Relations*, *35*(5), 527–546. doi:10.1108/ER-08-2012-0059

Graves, L. M., & Karabayeva, A. (2020). Managing virtual workers—Strategies for success. *IEEE Engineering Management Review*, *48*(2), 166–172. doi:10.1109/EMR.2020.2990386

Greenhaus, J. H., Collins, K. M., & Shaw, J. D. (2003). The relation between work-family balance and quality of life. *Journal of Vocational Behavior*, *63*(3), 510–531. doi:10.1016/S0001-8791(02)00042-8

Greenhaus, J. H., Ziegert, J. C., & Allen, T. D. (2012). When family-supportive supervision matters: Relations between multiple sources of support and work–family balance. *Journal of Vocational Behavior*, *80*(2), 266–275. doi:10.1016/j.jvb.2011.10.008

Grelle, D., & Popp, E. (2021). Considering the interaction of individual differences and remote work contexts. *Industrial and Organizational Psychology: Perspectives on Science and Practice*, *14*(2), 244–247. Advance online publication. doi:10.1017/iop.2021.51

Grenville-Cleave, B., Guðmundsdóttir, D., Huppert, F., King, V., Roffey, D., Roffey, S., & de Vries, M. (2021). *Creating the world we want to live in: How positive psychology can build a brighter future*. Routledge. doi:10.4324/9781003031789

Gropel, P. (2006). Balancing work and personal life: Relation to needs and life satisfaction. *Československá Psychologie*, *1*, 71–83.

Grzegorczyk, M., Mariniello, M., Nurski, L., & Schraepen, T. (2021). *Blending the physical and virtual: a hybrid model for the future of work* (No. 14/2021). Bruegel Policy Contribution.

Ha, N. (2021). Workplace isolation in the growth trend of remote working: a literature review. *Rev economic Bus Stud, 14*(1), 97-113.

Haan, K. (2023). *Remote work statistics and trends in 2023*. https://www.forbes.com/advisor/business/remote-work-statistics/

Haan, K. (2023). Remote Work Statistics and Trends. *Forbes*. https://www.forbes.com/advisor/business/remote-work-statistics/

Hackman, J. R., & Oldham, G. R. (1975). Development of the job diagnostic survey. *The Journal of Applied Psychology*, *60*(2), 159–170. doi:10.1037/h0076546

Hackman, J. R., & Oldham, G. R. (1976). Motivation through the design of work: Test of a theory. *Organizational Behavior and Human Performance*, *16*(2), 250–279. doi:10.1016/0030-5073(76)90016-7

Hackney, A., Yung, M., Somasundram, K. G., Nowrouzi-Kia, B., Oakman, J., & Yazdani, A. (2022). Working in the digital economy: A systematic review of the impact of work from home arrangements on personal and organizational performance and productivity. *PLoS One*, *17*(10), e0274728. doi:10.1371/journal.pone.0274728 PMID:36223418

Haider, S., Jabeen, S., & Ahmad, J. (2018). Moderated mediation between work life balance and employee job performance: The role of psychological wellbeing and satisfaction with coworkers. *Journal of Work and Organizational Psychology*, *34*(1), 29–37. doi:10.5093/jwop2018a4

Hair, J. F. Jr, Matthews, L. M., Matthews, R. L., & Sarstedt, M. (2017). PLS-SEM or CB-SEM: Updated guidelines on which method to use. *International Journal of Multivariate Data Analysis*, *1*(2), 107–123. doi:10.1504/IJMDA.2017.087624

Hair, J. F., Ringle, C. M., & Sarstedt, M. (2011). PLS-SEM: Indeed a silver bullet. *Journal of Marketing Theory and Practice*, *19*(2), 139–152. doi:10.2753/MTP1069-6679190202

Hajal, G. E. (2022). Teleworking and the jobs of tomorrow. *Research in Hospitality Management*, *12*(1), 21–27. doi:10.1080/22243534.2022.2080953

Hamlin, C. R. (2018). *ICT and Diminished Work-life Balance among Civilian Law Enforcement Professionals: A Multiple Case Study* [Doctoral dissertation]. Northcentral University.

Hardie, E., Kashima, E., & Pridmore, P. (2005). The influence of relational, individual and collective self-aspects on stress, uplifts and health. *Self and Identity*, *4*(1), 1–24. doi:10.1080/13576500444000146

Hardin, A., Davison, R. M., Schneider, C., Looney, C. A., & Sarker, S. (in press). Contextualising collective efficacy in virtual team research: The essential role of collaborative technologies in the virtual team efficacy conceptual framework. *Information Systems Journal*, 1–30. doi:10.1111/isj.12484

Harison, E. (2022). Towards Coherent Structures of Motivating Employees: A Non-Monetary Reward Framework. *Proceedings of IAC in Vienna 2022*, 49.

Harvard Business Review. (2021a). *Remote Work Should Be (Mostly) Asynchronous*. Available at: https://hbr.org/2021/12/remote-work-should-be-mostly-asynchronous

Harvard Business Review. (2021b). The New Reality of Remote Work. *Harvard Business Review*.

Haughton, A., & Ivey, W. (2023). A Growth Comparison of Mauritius and Jamaica. In *Overcoming Productivity Challenges in Small Countries: Lessons from Jamaica* (pp. 229–258). Springer Nature Switzerland. doi:10.1007/978-3-031-23301-2_9

Heissler, C., Kern, M., & Ohly, S. (2022). When thinking about work makes employees reach for their devices: A longitudinal autoregressive diary study. *Journal of Business and Psychology*, *37*(5), 999–1016. doi:10.1007/s10869-021-09781-0

Hill, E. J., Hawkins, A. J., Ferris, M., & Weitzman, M. (2001). Finding an extra day a week: The positive influence of perceived job flexibility on work and family life balance. *Family Relations*, *50*(1), 49–58. doi:10.1111/j.1741-3729.2001.00049.x

Hill, E. J., McClanahan, K. J., & Driggs, T. (2023). Telework. In *Encyclopedia of quality of life and well-being research* (pp. 1–3). Springer International Publishing. doi:10.1007/978-3-031-17299-1_2990

Hill, N. S., Axtell, C., Raghuram, S., & Nurmi, N. (2022). Unpacking virtual work's dual effects on employee well-being: An integrative review and future research agenda. *Journal of Management*.

Hills, J. W., & Jensen, J. F. (1998). Telepresence technology in medicine: Principles and applications. *Proceedings of the IEEE*, *86*(3), 569–580. doi:10.1109/5.662880

Hirsch, P. B. (2021). Sustaining corporate culture in a world of hybrid work. *The Journal of Business Strategy*, *42*(5), 358–361. doi:10.1108/JBS-06-2021-0100

Hjálmsdóttir, A., & Bjarnadóttir, V. S. (2021). "I have turned into a foreman here at home": Families and work–life balance in times of COVID-19 in a gender equality paradise. *Gender, Work and Organization*, *28*(1), 268–283. doi:10.1111/gwao.12552 PMID:33041540

Hoang, A. T., Nižetić, S., Olcer, A. I., Ong, H. C., Chen, W. H., Chong, C. T., Thomas, S., Bandh, S. A., & Nguyen, X. P. (2021). Impacts of COVID-19 pandemic on the global energy system and the shift progress to renewable energy: Opportunities, challenges, and policy implications. *Energy Policy*, *154*, 112322. doi:10.1016/j.enpol.2021.112322 PMID:34566236

Holmbeck, L. (2021). *Psychosocial Work Environment and Learning Opportunities in Remote Work – A hermeneutical study about managers´ and employees´ experiences*. Lund University Publications.

Hopkins, J., & Bardoel, A. (2023). The future is hybrid: How organisations are designing and supporting sustainable hybrid work models in post-pandemic Australia. *Sustainability (Basel), 15*(4), 3086. doi:10.3390/su15043086

Hosley, C. F. (2010). *The perceived effects of technology on product management team collaboration* (Publication No. 3425728) [Doctoral dissertation, University of Phoenix]. University of Phoenix ProQuest Dissertations Publishing.

Huang, K. Y., & Chuang, Y. R. (2016). A task–technology fit view of job search website impact on performance effects: An empirical analysis from Taiwan. *Cogent Business & Management, 3*(1), 1253943. doi:10.1080/23311975.2016.1253943

Hubble, R. (2017). *Re-Engineer Your Workday: Create Time for Life*. Balboa Press.

Huo, M. L., & Jiang, Z. (2023). Work–life conflict and job performance: The mediating role of employee wellbeing and the moderating role of trait extraversion. *Personality and Individual Differences, 205*, 112109. doi:10.1016/j.paid.2023.112109

Hwang, H., Malhotra, N. K., Kim, Y., Tomiuk, M. A., & Hong, S. (2010). A comparative study on parameter recovery of three approaches to structural equation modeling. *JMR, Journal of Marketing Research, 47*(4), 699–712. doi:10.1509/jmkr.47.4.699

Ilag, B. N. (2021). Tools and Technology for Effective Remote Work. *International Journal of Computer Applications, 174*(21), 13–16. doi:10.5120/ijca2021921109

Ilmola-Sheppard, L., Strelkovskii, N., Rovenskaya, E., Abramzon, S., & Bar, R. (2020). *A systems description of the national well-being system*. Version 1.0.

Ingram, K. E. (2012). *Antecedents and consequences of relational ambivalence: A longitudinal and daily diary study investigation*. Academic Press.

International Labour Organization. (2020). *An Employers' Guide on Working from Home in Response to the Outbreak of COVID-19*. Retrieved from: https://www.ilo.org/wcmsp5/groups/public/---ed_dialogue/---act_emp/documents/publication/wcms_745024.pdf

Iqbal, K. M., Khalid, F., & Barykin, S. Y. (2021). Hybrid Workplace: The Future of Work. In B. Khan, M. Kuofie, & S. Suman (Eds.), *Handbook of Research on Future Opportunities for Technology Management Education* (pp. 28–48). IGI Global. doi:10.4018/978-1-7998-8327-2.ch003

Ireland, S. G. (2022). A Study on the Need for Investment in Employee Wellbeing in Small and Medium Enterprises (SMEs) In Bangalore, India. Academic Press.

Izvercian, M., Seran, S., & Ivascu, L. (2016). Job Satisfaction Variables: A Grounded Theory Approach. *Procedia - Social and Behavioral Sciences*. https://www.researchgate.net/publication/304106042_Job_Satisfaction_Variables_A_Grounded_Theory_Approach

Jalagat, R., & Jalagat, A. (2019). Rationalizing remote working concept and its implications on employee productivity. *Global Journal of Advanced Research, 6*(3), 95-100.

Jämsen, R., Sivunen, A., & Blomqvist, K. (2022). Employees' perceptions of relational communication in full-time remote work in the public sector. *Computers in Human Behavior, 132*, 107240.

Jayaweera, T., Bal, M., Chudzikowski, K., & De Jong, S. (2021). The impact of economic factors on the relationships between psychological contract breach and work outcomes: A meta-analysis. *Employee Relations, 43*(3), 667–686. doi:10.1108/ER-03-2020-0095

Jiang, L., Pan, Z., Luo, Y., Guo, Z., & Kou, D. (2023). More flexible and more innovative: The impact of flexible work arrangements on the innovation behavior of knowledge employees. *Frontiers in Psychology, 14*, 1053242. Advance online publication. doi:10.3389/fpsyg.2023.1053242 PMID:37179861

Jindal, A., Agarwal, S., Garg, P., & Rastogi, R. (2013). Role of job design for achieving work life balance. In *Conference on Excellence in Research and Education* (pp. 203-217).

Johari, J., Tan, F. Y., & Zulkarnain, Z. I. T. (2018). Autonomy, workload, work-life balance and job performance among teachers. *International Journal of Educational Management, 32*(1), 107–120. doi:10.1108/IJEM-10-2016-0226

Johnson, N., Pinsonneault, A., & Boisvert, M. (2001). *The Impacts of Telecommuting on Organizations and Individuals*. https://www.researchgate.net/publication/315386065_The_Impacts_of_Telecommuting_on_Organizations_and_Individuals

Jones, Y. (2022). *Factors Contributing to Workplace Isolation in Remote Workers* [Doctoral dissertation]. Trevecca Nazarene University.

Jones, F., Burke, R. J., & Westman, M. (Eds.). (2013). *Work-life balance: A psychological perspective*. Psychology Press. doi:10.4324/9780203536810

Julsrud, T. E. (1996). Teleworking–the vision. *Feature Status, 16*.

Jung, Y.-W., Lim, Y., & Kim, M. (2017). Possibilities and limitations of online document tools for design collaboration: The case of Google Docs. *Proceedings of the 2017 ACM Conference on Computer Supported Cooperative Work and Social Computing*.

Kaliannan, M., Perumal, K., & Dorasamy, M. (2016). Developing a work-life balance model towards improving job satisfaction among medical doctors across different generations. *Journal of Developing Areas, 50*(5), 343–351. doi:10.1353/jda.2016.0035

Kaplan, S., Engelsted, L., Lei, X., & Lockwood, K. (2018). Unpackaging Manager Mistrust in Allowing Telework: Comparing and Integrating Theoretical Perspectives. *Journal of Business and Psychology, 33*(3), 365–382. doi:10.1007/s10869-017-9498-5

Kasbuntoro, D. I., Maemunah, S., Mahfud, I., Fahlevi, M., & Parashakti, R. D. (2020). Work-life balance and job satisfaction: A case study of employees on banking companies in Jakarta. *International Journal of Control and Automation, 13*(4), 439–451.

Katz & Krueger. (2019). Understanding Trends in Alternative Work Arrangements in the United States. *RSF: The Russell Sage Foundation Journal of the Social Sciences, 5*(5), 132. doi:10.7758/rsf.2019.5.5.07

Kelly, M. (2011). *Off Balance: Getting Beyond the Work-Life Balance Myth to Personal and Professional Satisfaction*. Penguin.

Kimbu, A. N., Adam, I., Dayour, F., & de Jong, A. (2023). COVID-19-induced redundancy and socio-psychological well-being of tourism employees: Implications for organizational recovery in a resource-scarce context. *Journal of Travel Research*, *62*(1), 55–74. doi:10.1177/00472875211054571 PMID:36605829

Kirchmeyer, C. (2000). Work-life initiatives: Greed or benevolence regarding workers' time? In C. L. Cooper & D. M. Rousseau (Eds.), *Trends in organizational behavior, Vol. 7. Time in organizational behavior* (pp. 79–93). John Wiley & Sons Ltd. https://psycnet.apa.org/record/2001-00452-005

Klein, H. J., Molloy, J. C., & Brinsfield, C. T. (2012). Reconceptualizing workplace commitment to redress a stretched construct: Revisiting assumptions and removing confounds. *Academy of Management Review*, *37*(1), 130–151.

Kline, R. B. (2023). *Principles and practice of structural equation modeling*. Guilford publications.

Kłopotek, M. (2017). The advantages and disadvantages of remote working from the perspective of young employees. *Organizacja i Zarządzanie: kwartalnik naukowy,* (4), 39-49.

Kniffin, K. M., Narayanan, J., Anseel, F., Antonakis, J., Ashford, S. P., Bakker, A. B., Bamberger, P., Bapuji, H., Bhave, D. P., Choi, V. K., Creary, S. J., Demerouti, E., Flynn, F. J., Gelfand, M. J., Greer, L. L., Johns, G., Kesebir, S., Klein, P. G., Lee, S. Y., ... Vugt, M. (2021). COVID-19 and the workplace: Implications, issues, and insights for future research and action. *The American Psychologist*, *76*(1), 63–77. doi:10.1037/amp0000716 PMID:32772537

Köffer, S., Junglas, I., Chiperi, C., & Niehaves, B. (2014). Dual use of mobile IT and work-to-life conflict in the context of IT consumerization. *Thirty fifth international conference on Information Systems Completed Research Paper.*

Koh, Y., Soepriyanto, G., Aljuaid, M., & Hasan, F. (2022). The effect of transformational leadership and remote working on employee performance during COVID-19 pandemic. *Frontiers in Psychology*, *13*, 919631. doi:10.3389/fpsyg.2022.919631 PMID:36033029

Kollmann, T., Stöckmann, C., & Kensbock, J. M. (2019). I can't get no sleep—The differential impact of entrepreneurial stressors on work-home interference and insomnia among experienced versus novice entrepreneurs. *Journal of Business Venturing*, *34*(4), 692–708. doi:10.1016/j.jbusvent.2018.08.001

Kondo, M., & Sonenshein, S. (2020). *Joy at work: Organizing your professional life*. Pan Macmillan.

Kossek, E. E., Perrigino, M. B., & Lautsch, B. A. (2023). Work-life flexibility policies from a boundary control and implementation perspective: A review and research framework. *Journal of Management*, *49*(6), 2062–2108. doi:10.1177/01492063221140354

Kowske, B. J., Rasch, R., & Wiley, J. (2010). Millennials' (Lack of) Attitude Problem: An Empirical Examination of Generational Effects on Work Attitudes. *Journal of Business and Psychology*, *25*(2), 265–279. doi:10.1007/s10869-010-9171-8

Krehl, E. H., & Büttgen, M. (2022). Uncovering the complexities of remote leadership and the usage of digital tools during the COVID-19 pandemic: A qualitative diary study. *German Journal of Human Resource Management*, *36*(3), 325–352. doi:10.1177/23970022221083697

Kritika. (2023). *Demystifying cyber crimes*. IGI Global. https://www.igi-global.com/chapter/demystifying-cyber-crimes/330260

Kroemer, A. D., & Kroemer, K. H. (2016). *Office ergonomics: Ease and efficiency at work*. CRC Press. doi:10.1201/9781315368603

Krombholz, K., Hobel, H., Huber, M., & Weippl, E. (2015). Advanced social engineering attacks. *Journal of Information Security and Applications, 22*, 113-122.

Larson, B. Z., Vroman, S. R., & Makarius, E. E. (2020). A guide to Managing Your Newly Remote Workers. *Harvard Business Review*. doi:10.3390/su15075845

Laumer, S., Maier, C., & Weitzel, T. (2017). Information quality, user satisfaction, and the manifestation of workarounds: A qualitative and quantitative study of enterprise content management system users. *European Journal of Information Systems*, *26*(4), 333–360. doi:10.1057/s41303-016-0029-7

Launer, J. (2020). Burnout in the age of COVID-19. *Postgraduate Medical Journal*, *96*(1136), 367–368. doi:10.1136/postgradmedj-2020-137980 PMID:32457131

Lazarus, R. S., & Folkman, S. (1984). *Stress, appraisal, and coping*. Springer Publishing Company Inc.

Leclercq-Vandelannoitte, A. (2017). An ethical perspective on emerging forms of ubiquitous IT-based control. *Journal of Business Ethics*, *142*(1), 139–154. doi:10.1007/s10551-015-2708-z

Lee, R. T., & Ashforth, B. E. (1996). A meta-analytic examination of the correlates of the three dimensions of job burnout. *The Journal of Applied Psychology*, *81*(2), 123–133. doi:10.1037/0021-9010.81.2.123 PMID:8603909

Lerman, D. C., O'Brien, M. J., Neely, L., Call, N. A., Tsami, L., Schieltz, K. M., Berg, W. K., Graber, J., Huang, P., Kopelman, T., & Cooper-Brown, L. J. (2020). Remote Coaching of Caregivers via Telehealth: Challenges and Potential Solutions. *Journal of Behavioral Education*, *29*(2), 195–221. doi:10.1007/s10864-020-09378-2 PMID:36093285

Li, K. C., Chen, X., & Susilo, W. (Eds.). (2019). *Advances in cyber security: Principles, techniques, and applications*. Springer. doi:10.1007/978-981-13-1483-4

LinkedIn. (2022). *Has remote work Peaked?* Available at: https://www.linkedin.com/pulse/has-remote-work-peaked-get-hired-by-linkedin-news/

Locke, E. A. (1976). The nature and causes of job satisfaction. Handbook of Industrial and Organizational Psychology.

Locke, E. A. (1969). What is job satisfaction? *Organizational Behavior and Human Performance*, *4*(4), 309–336. doi:10.1016/0030-5073(69)90013-0

Lorenzen, E. M. (2008). *Work/life Issues in America: A Public Policy Analysis* [Doctoral dissertation]. California State University, Northridge.

Lowry, P. B., Roberts, T. L., Romano, N. C. Jr, Cheney, P. D., & Hightower, R. T. (2006). The Impact of Group Size and Social Presence on Small-Group Communication: Does Computer-Mediated Communication Make a Difference? *Small Group Research*, *37*(6), 631–661. doi:10.1177/1046496406294322

Luft, J., & Shields, M. D. (2014). Subjectivity in developing and validating causal explanations in positivist accounting research. *Accounting, Organizations and Society*, *39*(7), 550–558. doi:10.1016/j.aos.2013.09.001

Lu, G., Du, R. Y., & Peng, X. (2022). The impact of schedule consistency on shift worker productivity: An empirical investigation. *Manufacturing & Service Operations Management*, *24*(5), 2780–2796. doi:10.1287/msom.2022.1132

Lund, S., Madgavkar, A., Manyika, J., & Smit, S. (2020). *What's next for remote work: An analysis of 2,000 tasks, 800 jobs, and nine countries*. McKinsey & Company. https://www.mckinsey.com/featured-insights/future-of-work/whats-next-for-remote-work-an-analysis-of-2000-tasks-800-jobs-and-nine-countries

MacArthur, D. G., Balasubramanian, S., Frankish, A., Huang, N., Morris, J., Walter, K., Jostins, L., Habegger, L., Pickrell, J. K., Montgomery, S. B., Albers, C. A., Zhang, Z. D., Conrad, D. F., Lunter, G., Zheng, H., Ayub, Q., DePristo, M. A., Banks, E., Hu, M., ... Tyler-Smith, C. (2012). A systematic survey of loss-of-function variants in human protein-coding genes. *Science*, *335*(6070), 823–828. doi:10.1126/science.1215040 PMID:22344438

Madan, P., Tripathi, S., Khalique, F., & Puri, G. (Eds.). (2022). *Re-envisioning Organizations Through Transformational Change: A Practitioners Guide to Work, Workforce, and Workplace*. CRC Press. doi:10.4324/9781003267751

Mahendran, A. V., Panatik, S. A., Rajab, A., & Nordin, N. (2019). The Influence of Work-life Balance on Burnout among Nurses. *Indian Journal of Public Health Research & Development*, *10*(11), 3338. doi:10.5958/0976-5506.2019.04097.X

Mahmoud, A. B., Berman, A., Reisel, W., Fuxman, L., & Hack, P. D. (2024). Examining generational differences as a moderator of extreme-context perception and its impact on work alienation organizational outcomes: Implications for the workplace and remote work transformation. *Scandinavian Journal of Psychology*, *65*(1), 70–85. doi:10.1111/sjop.12955 PMID:37596807

Mahoney, C. B., Lea, J., Schumann, P. L., & Jillson, I. A. (2020). Turnover, burnout, and job satisfaction of certified registered nurse anesthetists in the United States: Role of job characteristics and personality. *AANA Journal*, *88*(1), 39–48. PMID:32008617

Mallett, O., Marks, A., & Skountridaki, L. (2020). Where does work belong anymore? The implications of intensive homebased working. *Gender in Management*, *35*(7/8), 657–665. doi:10.1108/GM-06-2020-0173

Marcus, J. S. (2023). COVID-19 and the Shift to Remote Work. In Beyond the Pandemic? Exploring the Impact of COVID-19 on Telecommunications and the Internet. Emerald Publishing Limited. doi:10.1108/978-1-80262-049-820231003

Marsh, K., & Munson, G. (2008, January). Men at Work at Home: Managing Emotion in Telework. *Gender, Work and Organization*, *15*(1), 31–48. Advance online publication. doi:10.1111/j.1468-0432.2007.00353.x

Martin, B., & Mesler, R. (2012). Is Telework Effective for Organizations? A Meta-Analysis of Empirical Research on Perceptions of Telework and Organizational Outcomes. *Management Research Review*. https://www.researchgate.net/publication/242340515_Is_Telework_Effective_for_Organizations_A_Meta-Analysis_of_Empirical_Research_on_Perceptions_of_Telework_and_Organizational_Outcomes

Marx, K. (1867). Capital: A critique of political economy. Volume 1, Part 1: The process of capitalist production. New York, NY: Cosimo.

Maslach, C., Jackson, S. E., & Leiter, M. P. (1997). *Maslach burnout inventory*. Scarecrow Education.

Maslach, C., Schaufeli, W. B., & Leiter, M. P. (2001). Job burnout. *Annual Review of Psychology*, *52*(1), 397–422. doi:10.1146/annurev.psych.52.1.397 PMID:11148311

Maslow, A. H. (1943). A theory of human motivation. *Psychological Review*, *50*(4), 370–396. doi:10.1037/h0054346

Mathur, A., Purohit, A., & Mathur, R. (2013). Camparative Analysis Of Overall Work Life Balance Of Medical Professionals. *International Journal of Management Research and Development*, *3*(2).

Matsuo, T., Kobayashi, D., Taki, F., Sakamoto, F., Uehara, Y., Mori, N., & Fukui, T. (2020). Prevalence of health care worker burnout during the coronavirus disease 2019 (COVID-19) pandemic in Japan. *JAMA Network Open*, *3*(8), e2017271–e2017271. doi:10.1001/jamanetworkopen.2020.17271 PMID:32749466

Mazmanian, M., Orlikowski, W. J., & Yates, J. (2013). The autonomy paradox: The implications of mobile email devices for knowledge professionals. *Organization Science*, *24*(5), 1337–1357. doi:10.1287/orsc.1120.0806

McDaniel, B. T., O'Connor, K., & Drouin, M. (2021). Work-related technoference at home and feelings of work spillover, overload, life satisfaction and job satisfaction. *International Journal of Workplace Health Management*, *14*(5), 526–541. doi:10.1108/IJWHM-11-2020-0197

McKinsey & Company. (2021). *What executives are saying about the future of hybrid work*. McKinsey & Company. Available at: https://www.mckinsey.com/capabilities/people-and-organizational-performance/our-insights/what-executives-are-saying-about-the-future-of-hybrid-work

Merrill, J. S. (2021). *Remote Work, Employee Engagement, and the Implications of COVID-19* [Doctoral dissertation]. The College of St. Scholastica.

Meyer, J. P., & Allen, N. J. (1997). *Commitment in the workplace: Theory, research, and application*. Sage publications. doi:10.4135/9781452231556

Mirkovic, J., Dietrich, S., Dittrich, D., & Reiher, P. (2004). *Internet denial of service: attack and defense mechanisms (Radia Perlman Computer Networking and Security)*. Prentice Hall PTR.

Mitchell, J. W. (2024). *The State of Hybrid Workplaces in 2024*. https://www.forbes.com/sites/forbesbooksauthors/2024/01/24/the-state-of-hybrid-workplaces-in-2024/?sh=14f14adba408Mitchell

Mitchell, T. (2022). *COVID-19 Pandemic Continues To Reshape Work in America*. Pew Research Center. Retrieved from: https://www.pewresearch.org/social-trends/2022/02/16/covid-19-pandemic-continues-to-reshape-work-in-america/

Miyake, F., Odgerel, C. O., Hino, A., Ikegami, K., Nagata, T., Tateishi, S., ... Ishimaru, T. (2022). Job stress and loneliness among desk workers during the COVID-19 pandemic in Japan: Focus on remote working. *Environmental Health and Preventive Medicine*, *27*, 33–33.

Moczydłowska, J. M. (2021). *Praca w systemie home office w opinii wybranej grupy menedżerów – słuchaczy studiów MBA*. doi:10.15678/ZNUEK.2021.0991.0103

Mokhtarian, P. L., & Salomon, I. (2001). How derived is the demand for travel? Some conceptual and measurement considerations. *Transportation Research Part A, Policy and Practice*, *35*(8), 695–719. doi:10.1016/S0965-8564(00)00013-6

Molek-Winiarska, D. (Ed.). (2022). Shaping Employee Experience in the Changing Social and Organisation Conditions. Wydawnictwo Uniwersytetu Ekonomicznego we Wrocławiu. doi:10.15611/2022.989.4

Moore, K. A. (2017). 21. Mindfulness at work. *Research handbook on work and well-being*, 453.

Morganson, V., Major, D., Oborn, K., Verive, J., & Heelan, M. (2010). Comparing Telework Locations and Traditional Work Arrangements: Differences in Work-Life Balance Support, Job Satisfaction, and Inclusion. *Journal of Managerial Psychology*, *25*(6), 578–595. Advance online publication. doi:10.1108/02683941011056941

Morgeson, F. P., & Humphrey, S. E. (2006). The Work Design Questionnaire (WDQ): Developing and validating a comprehensive measure for assessing job design and the nature of work. *The Journal of Applied Psychology*, *91*(6), 1321–1339. doi:10.1037/0021-9010.91.6.1321 PMID:17100487

Moschella, D., Neal, D., Opperman, P., & Taylor, J. (2004). *The Consumerization of Information Technology. El Segundo*. CSC Research White Paper. Available at: https://www.smaele.nl/documents/Taylor-Consumerization-2004.pdf

Mowday, R. T., Steers, R. M., & Porter, L. W. (1979). The measurement of organizational commitment. *Journal of Vocational Behavior*, *14*(2), 224–247. doi:10.1016/0001-8791(79)90072-1

Mukherjee, S., & Narang, D. (2023). Digital economy and work-from-home: The rise of home offices amidst the COVID-19 outbreak in India. *Journal of the Knowledge Economy*, *14*(2), 924–945. doi:10.1007/s13132-022-00896-0

Mulcahy, D., & Andreeva, T. (2023). *Employer Perspectives on Employee Work Location: Collaboration, Culture and Control*. Working Paper 05/23. https://www.bruegel.org

Mullins, L. B., Scutelnicu, G., & Charbonneau, É. (2022). A Qualitative Study of Pandemic-Induced Telework: Federal Workers Thrive, Working Parents Struggle. *Public Administration Quarterly*, *46*(3), 258–281. doi:10.37808/paq.46.3.4

Murgia, A., Bozzon, R., Digennaro, P., Mezihorak, P., Mondon-Navazo, M., & Borghi, P. (2020). Hybrid Areas of Work Between Employment and Self-Employment: Emerging Challenges and Future Research Directions. *Frontiers in Sociology*, *4*, 86. Advance online publication. doi:10.3389/fsoc.2019.00086 PMID:33869406

Mutebi, N., & Hobbs, A. (2022). *The impact of remote and hybrid working on workers and organisations*. https://researchbriefings.files.parliament.uk/documents/POST-PB-0049/POST-PB-0049.pdf

Nadinloyi, K. B., Sadeghi, H., & Hajloo, N. (2013). Relationship between job satisfaction and employees mental health. *Procedia: Social and Behavioral Sciences*, *84*, 293–297. doi:10.1016/j.sbspro.2013.06.554

Nair, A. V., McGregor, A., & Caputi, P. (2020). The impact of challenge and hindrance demands on burnout, work engagement, and presenteeism. A cross-sectional study using the job demands–resources model. *Journal of Occupational and Environmental Medicine*, *62*(8), e392–e397. doi:10.1097/JOM.0000000000001908 PMID:32404829

Nemţeanu, M. S., & Dabija, D. C. (2023). Negative impact of telework, job insecurity, and work–life conflict on employee behaviour. *International Journal of Environmental Research and Public Health*, *20*(5), 4182. doi:10.3390/ijerph20054182 PMID:36901192

Next, I. T. C. (2021). *ITC NEXT: Reimagining Businesses for a Redefined Future*. Retrieved from- https://www.itcportal.com/about-itc/ChairmanSpeakContent.aspx?id=2394&type=B&news=2021

Nguyen, M. H. (2021). Factors influencing home-based telework in Hanoi (Vietnam) during and after the COVID-19 era. *Transportation*, *48*(6), 3207–3238. doi:10.1007/s11116-021-10169-5 PMID:33518829

Nickson, A., Gair, S., & Miles, D. (2016). Supporting isolated workers in their work with families in rural and remote Australia: Exploring peer group supervision. *Children Australia*, *41*(4), 265–274.

Niehaves, B., Köffer, S., & Ortbach, K. (2013). The effect of private IT use on work performance-towards an IT consumerization theory. *11th International Conference on Wirtschaftsinformatik*, 39-53

Niehaves, B., Köffer, S., Ortbach, K., & Katschewitz, S. (2012). Towards an IT consumerization theory: a theory and practice review. *18th Americas Conference on Information Systems,* 1-17.

Nilles, J. M., Carlson, F. R., Gray, P., & Hanneman, G. (1976). Telecommuting-an alternative to urban transportation congestion. *IEEE Transactions on Systems, Man, and Cybernetics, SMC-6*(2), 77–84. doi:10.1109/TSMC.1976.5409177

Ninaus, K., Diehl, S., & Terlutter, R. (2021). Employee perceptions of information and communication technologies in work life, perceived burnout, job satisfaction and the role of work-family balance. *Journal of Business Research, 136*, 652–666. doi:10.1016/j.jbusres.2021.08.007

Nitzl, C. (2016). The use of partial least squares structural equation modelling (PLS-SEM) in management accounting research: Directions for future theory development. *Journal of Accounting Literature*, *37*(1), 19–35. doi:10.1016/j.acclit.2016.09.003

Njaramba, L. (2023). *New Ways of Working and Burnout in the Post-Pandemic Era: A Case of National Health Insurance Fund (NHIF), Kenya* [Master's thesis].

Nogueira, A. J. F. M., & Patini, A. de C. G. (2012). Trabalho remoto e desafios dos gestores. *RAI: Revista de Administração e Inovação, 9*(4). doi:10.5773/rai.v9i4.800

Noh, E., & Lee, K. H. (2022). The Empirical Research on Professional Isolation During Remote Work in Context of COVID-19 and Organizational Citizenship Behavior: Sequential Mediation Effect of Psychological Empowerment and Affective Organizational Commitment. *Korean Journal of Industrial and Organizational Psychology*, *35*(3), 513–558.

Ochoa, J. J. (2023). *Occupational Burnout and Work-Life Balance Strategies Among Chaplains* [Doctoral dissertation]. University of Arizona Global Campus.

Oldham, G. R., & Fried, Y. (2016). Job design research and theory: Past, present and future. *Organizational Behavior and Human Decision Processes, 136*, 20–35. doi:10.1016/j.obhdp.2016.05.002

Olson, M. H. (1983). Remote office work: Changing work patterns in space and time. *Communications of the ACM*, *26*(3), 182–187.

Olson, M. H., & Primps, S. B. (1984). Working at home with computers: Work and non-work issues. *The Journal of Social Issues*, *40*(3), 97–112.

Onder, C. C. (2016). *Unfolding of telecommuting's effects in organizations: Performance, commitment, and mechanisms of action* [Doctoral dissertation]. Colorado State University.

Onnis, L. (2017). *What can we learn about improving workforce retention from five words.* Academic Press.

Onnis, L. A., & Onnis, L. A. (2019). Relationships: Social Exchanges, Community Ties and Employee-Manager Relationships. *HRM and Remote Health Workforce Sustainability: The Influence of Localised Management Practices*, 93-106.

Onnis, L. (2019). Human resource management policy choices, management practices and health workforce sustainability: Remote Australian perspectives. *Asia Pacific Journal of Human Resources*, *57*(1), 3–23. Advance online publication. doi:10.1111/1744-7941.12159

Onnis, L. (2019). *Relationships: Social Exchanges, Community Ties and Employee-Manager Relationships*. In James Cook University.

Onyeka, N. F. (2022). *Exploring Female Perspectives of Work-Life Balance and the Role of Organizational Culture* [Doctoral dissertation]. Capella University.

Organ, D. W. (1988). *Organizational citizenship behavior: The good soldier syndrome*. Lexington Books/DC Heath and Com.

Orzeł, B., & Wolniak, R. (2022). Digitization in the Design and Construction Industry—Remote Work in the Context of Sustainability: A Study from Poland. *Sustainability (Basel)*, *14*(3), 1332. doi:10.3390/su14031332

Palanisamy, R., Norman, A. A., & Mat Kiah, M. L. (2022). BYOD policy compliance: Risks and strategies in organizations. *Journal of Computer Information Systems*, *62*(1), 61–72. doi:10.1080/08874417.2019.1703225

Parham, S., & Rauf, M. A. (2020). COVID-19 and obligatory remote working in HEIs: An exploratory study of faculties' work-life balance, well-being and productivity during the pandemic. *International Journal of Economics. Commerce and Management*, *8*(12), 384–400.

Parker, S. K., Andrei, D. M., & Van den Broeck, A. (2019). Poor work design begets poor work design: Capacity and willingness antecedents of individual work design behavior. *The Journal of Applied Psychology*, *104*(7), 907–928. doi:10.1037/apl0000383 PMID:30640488

Pearman, C. D. (2023). *Tai Chi Enhanced Workplace Health and Wellness: A Systematic Literature Review for a Program Design, Implementation, and Evaluation* [Doctoral dissertation]. University of Arizona Global Campus.

Pee, L. G., & Lee, J. (2015). Intrinsically motivating employees' online knowledge sharing: Understanding the effects of job design. *International Journal of Information Management*, *35*(6), 679–690. doi:10.1016/j.ijinfomgt.2015.08.002

Peeters, M. C., Montgomery, A. J., Bakker, A. B., & Schaufeli, W. B. (2005). Balancing work and home: How job and home demands are related to burnout. *International Journal of Stress Management*, *12*(1), 43–61. doi:10.1037/1072-5245.12.1.43

Petcu, M. A., Sobolevschi-David, M. I., Crețu, R. F., Curea, S. C., Hristea, A. M., Oancea-Negescu, M. D., & Tutui, D. (2023, January 18). Telework: A Social and Emotional Perspective of the Impact on Employees' Wellbeing in the COVID-19 Pandemic. *International Journal of Environmental Research and Public Health*, *20*(3), 1811. doi:10.3390/ijerph20031811 PMID:36767179

Peters, P., den Dulk, L., & de Ruijter, J. (2010). May I work from home? Views of the employment relationship reflected in line managers' telework attitudes in six financial-sector organizations. *Equality, Diversity and Inclusion*, *29*(4), 517–531. Advance online publication. doi:10.1108/02610151011052799

Petrovski, P., & Gleeson, G. (1997). The relationship between job satisfaction and psychological health in people with an intellectual disability in competitive employment. *Journal of Intellectual & Developmental Disability*, *22*(3), 199–211. doi:10.1080/13668259700033411

Poleg, D. (2021). What is the Future of Offices When Workers Have a Choice? *The New York Times*. https://www.nytimes.com/2021/01/04/upshot/work-office-from-home.html

Popovici, V., & Popovici, A. (2020). *Remote Work Revolution: Current Opportunities and Challenges for Organizations.* Ovidius University Annals: Economic Sciences Series.

Powell, R. E., Henstenburg, J. M., Cooper, G., Hollander, J. E., & Rising, K. L. (2017). Patient Perceptions of Telehealth Primary Care Video Visits. *Annals of Family Medicine*, *15*(3), 225–229. doi:10.1370/afm.2095 PMID:28483887

Pradhan, R. K., & Kumar, U. (Eds.). (2021). *Emotion, well-being, and resilience: Theoretical perspectives and practical applications*. CRC Press.

Prasad, K. D. V., Vaidya, R., & Rani, R. (2023). Remote working and occupational stress: Effects on IT-enabled industry employees in Hyderabad Metro, India. *Frontiers in Psychology*, *14*, 998. doi:10.3389/fpsyg.2023.1069402 PMID:37063549

Prasad, R., & Rohokale, V. (2020). *Cyber security: the lifeline of information and communication technology*. Springer International Publishing. doi:10.1007/978-3-030-31703-4

PwC. (2021). *Global Crisis Survey 2021: India Insights*. Available at: https://www.pwc.in/assets/pdfs/services/crisis-management/pwcs-global-crisis-survey-2021.pdf

Qi, C. (2021). Will BYOD induce work-to-life conflict during COVID-19 pandemic period? Exploring the importance of Work Connectivity Behavior After-hours (WCBA) enabled by BYOD. In PACIS (p. 222). Academic Press.

Raišienė, A. G., Rapuano, V., Varkulevičiūtė, K., & Stachová, K. (2020). Working from home—Who is happy? A survey of Lithuania's employees during the COVID-19 quarantine period. *Sustainability (Basel)*, *12*(13), 5332. doi:10.3390/su12135332

Rakha, N. A. (2023). Ensuring Cyber-security in Remote Workforce: Legal Implications and International Best Practices. *International Journal of Law and Policy*, *1*(3).

Ray, R. L., Singh, V. P., Singh, S. K., Acharya, B. S., & He, Y. (2022). What is the impact of COVID-19 pandemic on global carbon emissions? *The Science of the Total Environment*, *816*, 151503. doi:10.1016/j.scitotenv.2021.151503 PMID:34752864

Regulation, P. (2016). Regulation (EU) 2016/679 of the European Parliament and of the Council. *Regulation*, *679*, 2016.

Reina, R., Ventura, M., Cristofaro, C. L., & Vesperi, W. (2021). Digitalize work in pandemic time: Practices for remote working and job redesign in public organization. *HR Analytics and Digital HR Practices: Digitalization post COVID-19*, 73–99.

Richardson, B. (2023). *Strategies for Promoting Inclusion and Employee Engagement to Improve Remote Teams' Performance* [Doctoral dissertation]. Walden University.

Richardson, J., & Kelliher, C. (2015). *Managing visibility for career sustainability: a study of remote workers.* . doi:10.4337/9781782547037.00013

Ringle, C. M., Wende, S., & Will, A. (2009). The finite mixture partial least squares approach: Methodology and application. Handbook of partial least squares: Concepts, methods, and applications. Berlin: Springer.

Risikko, J. (2021). *Telecommuting during a crisis: the development of manager-employee trust from the manager's point of view* [Master's thesis].

Rook, K. S. (1984). The negative side of social interaction: Impact on psychological well-being. *Journal of Personality and Social Psychology*, *46*(5), 1097–1108. doi:10.1037/0022-3514.46.5.1097 PMID:6737206

Roth, S. C., Hinton, E. G., & Jivanelli, B. (2023). Impact of COVID-19 on Parents/Guardians in the Library Profession: A Narrative Review and Shared Experiences. *Library Leadership & Management (Online)*, *37*(2), 1–72.

Rotundo, M., & Sackett, P. R. (2002). The relative importance of task, citizenship, and counterproductive performance to global ratings of job performance: A policy-capturing approach. *The Journal of Applied Psychology*, *87*(1), 66–80. doi:10.1037/0021-9010.87.1.66 PMID:11916217

Rousseau, D. M. (1989). Psychological and implied contracts in organizations. *Employee Responsibilities and Rights Journal*, *2*(2), 121–139. doi:10.1007/BF01384942

Rubin, K. I. (2022). *The Great Balancing Act: Women Seeking Work-Life Balance during COVID-19* [Doctoral dissertation]. University of Southern California.

Ruderman, M. N., Clerkin, C., & Deal, J. J. (2017). The long-hours culture. *The Routledge Companion to Wellbeing at Work: Taylor & Francis*, 207-221.

Rudolph, C. W., Allan, B., Clark, M., Hertel, G., Hirschi, A., Kunze, F., Shockley, K., Shoss, M., Sonnentag, S., & Zacher, H. (2021). Pandemics: Implications for research and practice in industrial and organizational psychology. *Industrial and Organizational Psychology: Perspectives on Science and Practice*, *14*(1-2), 1–35. doi:10.1017/iop.2020.48

Ruiller, C., van der Heijden, B., Chédotel, F., & Dumas, M. (2019). "You have got a friend": The value of perceived proximity for teleworking success in dispersed teams. *Team Performance Management*, *25*(3/4), 2–29. Advance online publication. doi:10.1108/TPM-11-2017-0069

Ruotsalainen, S., Jantunen, S., & Sinervo, T. (2020). Which factors are related to Finnish home care workers' job satisfaction, stress, psychological distress and perceived quality of care?-a mixed method study. *BMC Health Services Research*, *20*(1), 1–13. doi:10.1186/s12913-020-05733-1 PMID:32988396

Ryan, R. M., & Deci, E. L. (2000). Self-determination theory and the facilitation of intrinsic motivation, social development, and well-being. *The American Psychologist*, *55*(1), 68–78. doi:10.1037/0003-066X.55.1.68 PMID:11392867

Saez, K. (2023). Problematizing workplace learning for the work-from-home and hybrid workforce following. *COVID*, 19.

Saini, R., Khan, H., & Oliver, S. (2022). The Dynamics of the Remote Leadership: A COVID-19 Context. In *Responsible Leadership for Sustainability in Uncertain Times: Social, Economic and Environmental Challenges for Sustainable Organizations* (pp. 51–65). Springer Nature Singapore. doi:10.1007/978-981-19-4723-0_4

Sannassee, R. V., Seetanah, B., & Lamport, M. J. (2014). Export diversification and economic growth: the case of Mauritius. In Connecting to Global Markets (pp. 11-23). WTO iLibrary.

Schall, M. A. (2019). The Relationship Between Remote Work and Job Satisfaction: The Mediating Roles of Perceived Autonomy, Work-Family Conflict, and Telecommuting Intensity. *Master's Theses., 5017*. Advance online publication. doi:10.31979/etd.2x82-58pg

Schauder, M. J. (2015). *Employee–Organization Relationship*. doi:10.1002/9781118521373.wbeaa020

Schaufeli, W. B., Dierendonck, D. V., & Gorp, K. V. (1996). Burnout and reciprocity: Towards a dual-level social exchange model. *Work and Stress, 10*(3), 225–237. doi:10.1080/02678379608256802

Schaufeli, W., & Enzmann, D. (2020). *The burnout companion to study and practice: A critical analysis*. CRC Press. doi:10.1201/9781003062745

Schieman, S., & Young, M. C. (2013). Are communications about work outside regular working hours associated with work-to-family conflict, psychological distress and sleep problems? *Work and Stress, 27*(3), 244–261. doi:10.1080/02678373.2013.817090

Schmidt, J. B., Montoya-Weiss, M. M., & Massey, A. P. (2001). New product development decision-making effectiveness: Comparing individuals, face-to-face teams, and virtual teams. *Decision Sciences, 32*(4), 575–600. doi:10.1111/j.1540-5915.2001.tb00973.x

Sharma, A., & Ganjigatti, J. P. (2023). A Hybrid Workplace: Is a New Normal. *Journal of Mines, Metals & Fuels, 71*(6), 791–796. https://doi-org.ezpr.lfpl.org/10.18311/jmmf/2023/34495

Sharma, A., & Singh, B. (2022). Measuring Impact of E-commerce on Small Scale Business: A Systematic Review. *Journal of Corporate Governance and International Business Law, 5*(1).

Shepherd, M. M., & Martz, W. B. (2006). Media Richness Theory and the Distance Education Environment. *Journal of Computer Information Systems, 47*, 114–122. doi:10.1080/08874417.2006.11645945

Sheveleva, A., & Rogov, E. (2021). *Organization of remote work in the context of digitalization*. doi:10.1051/e3sconf/202127312042

Shifrin, N. V., & Michel, J. S. (2022). Flexible work arrangements and employee health: A meta-analytic review. *Work and Stress, 36*(1), 60–85. doi:10.1080/02678373.2021.1936287

Shillingford, J. V. (2017). *Telework and the Manager Employee Relationship*. Academic Press.

Shore, L. M., Coyle-Shapiro, J. A-M., & Tetrick, L. E. (2012). *The employee-organization relationship: applications for the 21st century*. DOI: doi:10.4324/9780203138878

Siengthai, S., & Pila-Ngarm, P. (2016). The interaction effect of job redesign and job satisfaction on employee performance. *Evidence-based HRM: a Global Forum for Empirical Scholarship*. 10.1108/EBHRM-01-2015-0001

Siganidou, S. (2020). *Work-related employee stress in 4* & 5* hotels in Greece. A study about the city of Thessaloniki*. Academic Press.

Sigeman, H., & Adolfsson, M. (2021). *Maintaining well-being when working remotely: Work habits of employees at a geographical IT office in Sweden during the COVID-19 pandemic, their perceived well-being and productivity*. Academic Press.

Silvester, J., & Konstantinou, E. (2010). *Lighting, well-being and performance at work*. City University.

Simon, H. A. (1957). *Models of man: Social and rational; mathematical essays on rational human behavior in society setting*. Wiley.

Sims, H. P. Jr, Szilagyi, A. D., & Keller, R. T. (1976). The measurement of job characteristics. *Academy of Management Journal, 19*(2), 195–212. doi:10.2307/255772 PMID:1029346

Singh, B. (2023a). Blockchain Technology in Renovating Healthcare: Legal and Future Perspectives. In Revolutionizing Healthcare Through Artificial Intelligence and Internet of Things Applications (pp. 177-186). IGI Global.

Singh, B. (2019). Profiling Public Healthcare: A Comparative Analysis Based on the Multidimensional Healthcare Management and Legal Approach. *Indian Journal of Health and Medical Law*, *2*(2), 1–5.

Singh, B. (2020). Global science and jurisprudential approach concerning healthcare and illness. *Indian Journal of Health and Medical Law*, *3*(1), 7–13.

Singh, B. (2022a). Relevance of Agriculture-Nutrition Linkage for Human Healthcare: A Conceptual Legal Framework of Implication and Pathways. *Justice and Law Bulletin*, *1*(1), 44–49.

Singh, B. (2022b). COVID-19 Pandemic and Public Healthcare: Endless Downward Spiral or Solution via Rapid Legal and Health Services Implementation with Patient Monitoring Program. *Justice and Law Bulletin*, *1*(1), 1–7.

Singh, B. (2023b). Federated Learning for Envision Future Trajectory Smart Transport System for Climate Preservation and Smart Green Planet: Insights into Global Governance and SDG-9 (Industry, Innovation and Infrastructure). *National Journal of Environmental Law*, *6*(2), 6–17.

Singh, B. (2023c). Tele-Health Monitoring Lensing Deep Neural Learning Structure: Ambient Patient Wellness via Wearable Devices for Real-Time Alerts and Interventions. *Indian Journal of Health and Medical Law*, *6*(2), 12–16.

Singh, B. (2024). Legal Dynamics Lensing Metaverse Crafted for Videogame Industry and E-Sports: Phenomenological Exploration Catalyst Complexity and Future. *Journal of Intellectual Property Rights Law*, *7*(1), 8–14.

Singh, P., Bala, H., Dey, B. L., & Filieri, R. (2022). Enforced remote working: The impact of digital platform-induced stress and remote working experience on technology exhaustion and subjective wellbeing. *Journal of Business Research, 151*, 269–286. doi:10.1016/j.jbusres.2022.07.002 PMID:35847196

Siponen, M., & Vance, A. (2010). Neutralization: New insights into the problem of employee information systems security policy violations. *Management Information Systems Quarterly, 34*(3), 487–502. doi:10.2307/25750688

Slack. (2021). *The great executive-employee disconnect.* Retrieved from: https://slack.com/intl/en-in/blog/news/the-great-executive-employee-disconnect

Slavković, M., Sretenović, S., & Bugarčić, M. (2022). Remote working for sustainability of organization during the covid-19 pandemic: The mediator-moderator role of social support. *Sustainability (Basel), 14*(1), 70. doi:10.3390/su14010070

Smith, A. (1776). *An inquiry into the nature and causes of the wealth of nations: Volume One*. Printed for W. Strahan and T. Cadell.

Smith, A. (1976). *An inquiry into the nature and causes of the wealth of nations* (R. H. Campbell, A. S. Skinner, & W. B. Todd, Eds.).

Smith, S. A., Patmos, A., & Pitts, M. J. (2018). Communication and teleworking: A study of communication channel satisfaction, personality, and job satisfaction for teleworking employees. *International Journal of Business Communication, 55*(1), 44–68. doi:10.1177/2329488415589101

Soelton, M., Hardianti, D., Kuncoro, S., & Jumadi, J. (2020, February). Factors affecting burnout in manufacturing industries. In *4th International Conference on Management, Economics and Business (ICMEB 2019)* (pp. 46-52). Atlantis Press.

Soga, L., Bolade-Ogunfodun, Y., & Laker, B. (2021). Design your work environment to manage unintended tech consequences. *MIT Sloan Management Review.*

Song, S., & Song, J. (2022). A Study on Workplace Isolation in Remote Working: The Analysis of Workplace Isolation Perception and The Effect on Work Engagement. *Korean Journal of Industrial and Organizational Psychology, 35*(3), 559–581.

Song, Y., & Gao, J. (2020). Does telework stress employees out? A study on working at home and subjective well-being for wage/salary workers. *Journal of Happiness Studies, 21*(7), 2649–2668. doi:10.1007/s10902-019-00196-6

Staniulienė, S., & Leonavičiūtė, A. (2022). Self-organization of remote employees in Lithuanian companies. *Problems and Perspectives in Management, 20*(2), 486–496. doi:10.21511/ppm.20(2).2022.40

Staples, D. S. (1997). An Investigation of Information Technology-Enabled Remote Management and Remote Work Issues. *Australasian Journal of Information Systems, 4*(2). doi:10.4018/joeuc.2001040101

Staples, D. S., Hulland, J., & Higgins, C. A. (1999). A Self-Efficacy Theory Explanation for the Management of Remote Workers in Virtual Organizations. *Organization Science, 10*(6). doi:10.1111/j.1083-6101.1998.tb00085.x

Stasiła-Sieradzka, M., Sanecka, E., & Turska, E. (2023). Not so Good Hybrid Work Model? Resource Losses and Gains since the Outbreak of the Covid-19 Pandemic and Job Burnout among Non-Remote, Hybrid, and Remote Employees. *International Journal of Occupational Medicine and Environmental Health, 36*(2), 229–249. doi:10.13075/ijomeh.1896.02026 PMID:37184145

Straus, E., Uhlig, L., Kühnel, J., & Korunka, C. (2023). Remote workers' well-being, perceived productivity, and engagement: Which resources should HRM improve during COVID-19? A longitudinal diary study. *International Journal of Human Resource Management, 34*(15), 2960–2990. doi:10.1080/09585192.2022.2075235

Straus, S. G. (1996). Getting a clue: The effects of communication media and information distribution on participation and performance in computer-mediated and face-to-face groups. *Small Group Research, 27*(1), 115–142. doi:10.1177/1046496496271006

Sui, M., Yu, Z., & Zhou, M. (2023). The Impact of Psychological Contract, Physical and Mental Health on Burnout in Grassroots Civil Servants: Evidence from China. *Psychology Research and Behavior Management, 16*, 3461–3476. doi:10.2147/PRBM.S423011 PMID:37664138

Suri, A., & Bajpai, A. (2023). Quest for Meaningfulness and Adapting to the 'New Normal' in the Context of 'Hybrid' Work: A Qualitative Study. *Indian Journal of Positive Psychology, 14*(4), 461–466.

Tao, Y., You, S., Zhu, J., & You, F. (2024). Energy, climate, and environmental sustainability of trend toward occupational-dependent hybrid work: Overview, research challenges, and outlook. *Journal of Cleaner Production, 442*, 141083. doi:10.1016/j.jclepro.2024.141083

Taser, D., Aydin, E., Torgaloz, A. O., & Rofcanin, Y. (2022). An examination of remote e-working and flow experience: The role of technostress and loneliness. *Computers in Human Behavior, 127*, 107020.

Taskin, L., & Bridoux, F. (2010). Telework: A challenge to knowledge transfer in organizations. *International Journal of Human Resource Management, 21*(10), 2503–2520. Advance online publication. doi:10.1080/09585192.2010.516600

ten Brummelhuis, L. L., ter Hoeven, C. L., & Toniolo-Barrios, M. (2021). Staying in the loop: Is constant connectivity to work good or bad for work performance? *Journal of Vocational Behavior, 128*, 103589. doi:10.1016/j.jvb.2021.103589

Tenorio, J. L., & Vu, H. T. (2021). *A Cross-sectional Study during the COVID-19 Pandemic: Work Events and Affective Reactions as Predictors of Loss of Productivity and Intention to Leave* [Master's thesis]. Handelshøyskolen BI.

Thames, L., & Schaefer, D. (2017). *Cybersecurity for industry 4.0.* Springer. doi:10.1007/978-3-319-50660-9

Thompson, R. J., Payne, S. C., & Taylor, A. B. (2015). Applicant attraction to flexible work arrangements: Separating the influence of flextime and flexplace. *Journal of Occupational and Organizational Psychology*, *88*(4), 726–749. doi:10.1111/joop.12095

Timsal, A., & Awais, M. (2016). Flexibility or ethical dilemma: An overview of the work from home policies in modern organizations around the world. *Human Resource Management International Digest*, *24*(7), 12–15. doi:10.1108/HRMID-03-2016-0027

Tiwari, S., & Shah, M. (2022). Importance Of Employee Relations for The Management of Organization. *Central European Management Journal*, *30*(4), 2508–2512. doi:10.57030/23364890.cemj.30.4.265

Tomasi, D. (2020). Coronavirus disease (COVID-19). A socioepidemiological analysis. Bennington, VT: Vermont Academy of Arts and Sciences LV.

Torten, R., Reaiche, C., & Caraballo, E. L. (2016). Teleworking in the new milleneum. *Journal of Developing Areas*, *50*(5), 317–326. doi:10.1353/jda.2016.0060

Toscano, F., & Zappalà, S. (2020). Social isolation and stress as predictors of productivity perception and remote work satisfaction during the COVID-19 pandemic: The role of concern about the virus in a moderated double mediation. *Sustainability*, *12*(23), 9804.

Turel, O., & Connelly, C. E. (2012). Team spirit: The influence of psychological collectivism on the usage of E-Collaboration Tools. *Group Decision and Negotiation*, *21*(5), 703–725. doi:10.1007/s10726-011-9245-7

Tuzovic, S., & Kabadayi, S. (2021). The influence of social distancing on employee well-being: A conceptual framework and research agenda. *Journal of Service Management*, *32*(2), 145–160. doi:10.1108/JOSM-05-2020-0140

U.S. Census Bureau. (2021). *Census Bureau Estimates Show Average One-Way Travel Time to Work Rises to All-Time High*. Available at: Census.gov. https://www.census.gov/newsroom/press-releases/2021/one-way-travel-time-to-work-rises.html

Upwork. (2021). *Future Workforce Report 2021: How Remote Work is Changing Businesses Forever.* Retrieved from: https://www.upwork.com/research/future-workforce-report

Van Zoonen, W., & Sivunen, A. E. (2022). The impact of remote work and Mediated communication frequency on isolation and psychological distress. *European Journal of Work and Organizational Psychology*, *31*(4), 610–621.

Vance, A., Siponen, M., & Pahnila, S. (2012). Motivating IS security compliance: Insights from habit and protection motivation theory. *Information & Management*, *49*(3-4), 190–198. doi:10.1016/j.im.2012.04.002

Veld, M., Paauwe, J., & Boselie, P. (2010). HRM and strategic climates in hospitals: Does the message come across at the ward level? *Human Resource Management Journal*, *20*(4), 339–356. doi:10.1111/j.1748-8583.2010.00139.x

Viollaz, M. (2022). Does working from home work in developing countries? *IZA World of Labor: Evidence-Based Policy Making*. Advance online publication. doi:10.15185/izawol.504

Vrchota, J., Frantikova, Z., & Vlckoval, M. (2019). Why Some SME's in Czech Republic Adopt Telework and Others Not? *European Countryside, 11*(4), 599–615. doi:10.2478/euco-2019-0033

Wahlström, V., Januario, L. B., Mathiassen, S. E., Heiden, M., & Hallman, D. M. (2023). Hybrid office work in women and men: Do directly measured physical behaviors differ between days working from home and days working at the office? *Annals of Work Exposures and Health, 67*(9), 1043–1055. doi:10.1093/annweh/wxad057 PMID:37795673

Wall, K. D. (1993). A model of decision making under bounded rationality. *Journal of Economic Behavior & Organization, 20*(3), 331–352. doi:10.1016/0167-2681(93)90030-S

Wang, J., Hong, Y., Wang, J., Xu, J., Tang, Y., Han, Q. L., & Kurths, J. (2022). Cooperative and competitive multi-agent systems: From optimization to games. *IEEE/CAA Journal of Automatica Sinica, 9*(5), 763-783.

Wang, B., Liu, Y., Qian, J., & Parker, S. K. (2021). Achieving effective remote working during the COVID-19 pandemic: A work design perspective. *Applied Psychology, 70*(1), 16–59.

Wan, W. (2022). *Making Kindness Our Business*. Marshall Cavendish International Asia Pte Ltd.

Weber, M. B., & De Fino, M. (2022). *Virtual technical services: a handbook*. Rowman & Littlefield. doi:10.5771/9781538152645

Weeger, A., Wang, X., & Gewald, H. (2016). IT consumerization: BYOD-program acceptance and its impact on employer attractiveness. *Journal of Computer Information Systems, 56*(1), 1–10. doi:10.1080/08874417.2015.11645795

Wells, J., Scheibein, F., Pais, L., Rebelo dos Santos, N., Dalluege, C.-A., Czakert, J. P., & Berger, R. (2023). A Systematic Review of the Impact of Remote Working Referenced to the Concept of Work–Life Flow on Physical and Psychological Health. *Workplace Health & Safety, 71*(11), 507–521. doi:10.1177/21650799231176397 PMID:37387511

Wheatley, D., Hardill, I., & Buglass, S. (Eds.). (2021). *Handbook of research on remote work and worker well-being in the post-COVID-19 era*. IGI Global. doi:10.4018/978-1-7998-6754-8

Wilson, C. G., Bond, C. E., & Shipley, T. F. (2019). How can geologic decision-making under uncertainty be improved? *Solid Earth, 10*(5), 1469–1488. doi:10.5194/se-10-1469-2019

Wiradendi Wolor, C., Solikhah, S., Fidhyallah, N. F., & Lestari, D. P. (2020). Effectiveness of e-training, e-leadership, and work life balance on employee performance during COVID-19. *Journal of Asian Finance, Economics and Business, 7*(10).

Wong, S. I., Bunjak, A., Černe, M., & Fieseler, C. (2021). Fostering creative performance of platform Crowdworkers: The digital feedback dilemma. *International Journal of Electronic Commerce, 25*(3), 263–286. doi:10.1080/10864415.2021.1942674

Wood, J. (1997). *Telework: An Intelligent Managerial Initiative*. In University of Sydney. doi:10.1016/B978-044482332-8/50014-7

World Bank. (2021). *From Remote Work to Smart Work in the Post-COVID Era: An Illustrative Case for Leapfrogging into Smart Government.* Retrieved From- https://www.worldbank.org/en/events/2021/02/02/from-remote-work-to-smart-work-in-the-post-covid-era-an-illustrative-case-for-leapfrogging-into-smart-government

Wright, P. C., & Oldford, A. (1993). Telecommuting and employee effectiveness: Career and managerial issues. *International Journal of Career Management*, *5*(1). Advance online publication. doi:10.1108/09556219310024751

Yang, L., Holtz, D., Jaffe, S., Suri, S., Sinha, S., Weston, J., Joyce, C., Shah, N., Sherman, K., Hetch, B., & Teevan, J. (2022). The effects of remote work on collaboration among information workers. *Nature Human Behaviour*, *6*(1), 43–54. doi:10.1038/s41562-021-01196-4 PMID:34504299

Zhang, C., Yu, M. C., & Marin, S. (2021). Exploring public sentiment on enforced remote work during COVID-19. *The Journal of Applied Psychology*, *106*(6), 797–810. doi:10.1037/apl0000933 PMID:34138587

Related References

To continue our tradition of advancing information science and technology research, we have compiled a list of recommended IGI Global readings. These references will provide additional information and guidance to further enrich your knowledge and assist you with your own research and future publications.

Abdul Razak, R., & Mansor, N. A. (2021). Instagram Influencers in Social Media-Induced Tourism: Rethinking Tourist Trust Towards Tourism Destination. In M. Dinis, L. Bonixe, S. Lamy, & Z. Breda (Eds.), *Impact of New Media in Tourism* (pp. 135-144). IGI Global. https://doi.org/10.4018/978-1-7998-7095-1.ch009

Abir, T., & Khan, M. Y. (2022). Importance of ICT Advancement and Culture of Adaptation in the Tourism and Hospitality Industry for Developing Countries. In C. Ramos, S. Quinteiro, & A. Gonçalves (Eds.), *ICT as Innovator Between Tourism and Culture* (pp. 30–41). IGI Global. https://doi.org/10.4018/978-1-7998-8165-0.ch003

Abtahi, M. S., Behboudi, L., & Hasanabad, H. M. (2017). Factors Affecting Internet Advertising Adoption in Ad Agencies. *International Journal of Innovation in the Digital Economy*, *8*(4), 18–29. doi:10.4018/IJIDE.2017100102

Afenyo-Agbe, E., & Mensah, I. (2022). Principles, Benefits, and Barriers to Community-Based Tourism: Implications for Management. In I. Mensah & E. Afenyo-Agbe (Eds.), *Prospects and Challenges of Community-Based Tourism and Changing Demographics* (pp. 1–29). IGI Global. doi:10.4018/978-1-7998-7335-8.ch001

Agbo, V. M. (2022). Distributive Justice Issues in Community-Based Tourism. In I. Mensah & E. Afenyo-Agbe (Eds.), *Prospects and Challenges of Community-Based Tourism and Changing Demographics* (pp. 107–129). IGI Global. https://doi.org/10.4018/978-1-7998-7335-8.ch005

Agrawal, S. (2017). The Impact of Emerging Technologies and Social Media on Different Business(es): Marketing and Management. In O. Rishi & A. Sharma (Eds.), *Maximizing Business Performance and Efficiency Through Intelligent Systems* (pp. 37–49). Hershey, PA: IGI Global. doi:10.4018/978-1-5225-2234-8.ch002

Ahmad, A., & Johari, S. (2022). Georgetown as a Gastronomy Tourism Destination: Visitor Awareness Towards Revisit Intention of Nasi Kandar Restaurant. In M. Valeri (Ed.), *New Governance and Management in Touristic Destinations* (pp. 71–83). IGI Global. https://doi.org/10.4018/978-1-6684-3889-3.ch005

Alkhatib, G., & Bayouq, S. T. (2021). A TAM-Based Model of Technological Factors Affecting Use of E-Tourism. *International Journal of Tourism and Hospitality Management in the Digital Age*, *5*(2), 50–67. https://doi.org/10.4018/IJTHMDA.20210701.oa1

Altinay Ozdemir, M. (2021). Virtual Reality (VR) and Augmented Reality (AR) Technologies for Accessibility and Marketing in the Tourism Industry. In C. Eusébio, L. Teixeira, & M. Carneiro (Eds.), *ICT Tools and Applications for Accessible Tourism* (pp. 277-301). IGI Global. https://doi.org/10.4018/978-1-7998-6428-8.ch013

Anantharaman, R. N., Rajeswari, K. S., Angusamy, A., & Kuppusamy, J. (2017). Role of Self-Efficacy and Collective Efficacy as Moderators of Occupational Stress Among Software Development Professionals. *International Journal of Human Capital and Information Technology Professionals*, *8*(2), 45–58. doi:10.4018/IJHCITP.2017040103

Aninze, F., El-Gohary, H., & Hussain, J. (2018). The Role of Microfinance to Empower Women: The Case of Developing Countries. *International Journal of Customer Relationship Marketing and Management*, *9*(1), 54–78. doi:10.4018/IJCRMM.2018010104

Antosova, G., Sabogal-Salamanca, M., & Krizova, E. (2021). Human Capital in Tourism: A Practical Model of Endogenous and Exogenous Territorial Tourism Planning in Bahía Solano, Colombia. In V. Costa, A. Moura, & M. Mira (Eds.), *Handbook of Research on Human Capital and People Management in the Tourism Industry* (pp. 282–302). IGI Global. https://doi.org/10.4018/978-1-7998-4318-4.ch014

Arsenijević, O. M., Orčić, D., & Kastratović, E. (2017). Development of an Optimization Tool for Intangibles in SMEs: A Case Study from Serbia with a Pilot Research in the Prestige by Milka Company. In M. Vemić (Ed.), *Optimal Management Strategies in Small and Medium Enterprises* (pp. 320–347). Hershey, PA: IGI Global. doi:10.4018/978-1-5225-1949-2.ch015

Aryanto, V. D., Wismantoro, Y., & Widyatmoko, K. (2018). Implementing Eco-Innovation by Utilizing the Internet to Enhance Firm's Marketing Performance: Study of Green Batik Small and Medium Enterprises in Indonesia. *International Journal of E-Business Research*, *14*(1), 21–36. doi:10.4018/IJEBR.2018010102

Asero, V., & Billi, S. (2022). New Perspective of Networking in the DMO Model. In M. Valeri (Ed.), *New Governance and Management in Touristic Destinations* (pp. 105–118). IGI Global. https://doi.org/10.4018/978-1-6684-3889-3.ch007

Atiku, S. O., & Fields, Z. (2017). Multicultural Orientations for 21st Century Global Leadership. In N. Baporikar (Ed.), *Management Education for Global Leadership* (pp. 28–51). Hershey, PA: IGI Global. doi:10.4018/978-1-5225-1013-0.ch002

Atiku, S. O., & Fields, Z. (2018). Organisational Learning Dimensions and Talent Retention Strategies for the Service Industries. In N. Baporikar (Ed.), *Global Practices in Knowledge Management for Societal and Organizational Development* (pp. 358–381). Hershey, PA: IGI Global. doi:10.4018/978-1-5225-3009-1.ch017

Atsa'am, D. D., & Kuset Bodur, E. (2021). Pattern Mining on How Organizational Tenure Affects the Psychological Capital of Employees Within the Hospitality and Tourism Industry: Linking Employees' Organizational Tenure With PsyCap. *International Journal of Tourism and Hospitality Management in the Digital Age, 5*(2), 17–28. https://doi.org/10.4018/IJTHMDA.2021070102

Ávila, L., & Teixeira, L. (2018). The Main Concepts Behind the Dematerialization of Business Processes. In M. Khosrow-Pour, D.B.A. (Ed.), Encyclopedia of Information Science and Technology, Fourth Edition (pp. 888-898). Hershey, PA: IGI Global. https://doi.org/doi:10.4018/978-1-5225-2255-3.ch076

Ayorekire, J., Mugizi, F., Obua, J., & Ampaire, G. (2022). Community-Based Tourism and Local People's Perceptions Towards Conservation: The Case of Queen Elizabeth Conservation Area, Uganda. In I. Mensah & E. Afenyo-Agbe (Eds.), *Prospects and Challenges of Community-Based Tourism and Changing Demographics* (pp. 56–82). IGI Global. https://doi.org/10.4018/978-1-7998-7335-8.ch003

Baleiro, R. (2022). Tourist Literature and the Architecture of Travel in Olga Tokarczuk and Patti Smith. In R. Baleiro & R. Pereira (Eds.), *Global Perspectives on Literary Tourism and Film-Induced Tourism* (pp. 202-216). IGI Global. https://doi.org/10.4018/978-1-7998-8262-6.ch011

Barat, S. (2021). Looking at the Future of Medical Tourism in Asia. *International Journal of Tourism and Hospitality Management in the Digital Age, 5*(1), 19–33. https://doi.org/10.4018/IJTHMDA.2021010102

Barbosa, C. A., Magalhães, M., & Nunes, M. R. (2021). Travel Instagramability: A Way of Choosing a Destination? In M. Dinis, L. Bonixe, S. Lamy, & Z. Breda (Eds.), *Impact of New Media in Tourism* (pp. 173-190). IGI Global. https://doi.org/10.4018/978-1-7998-7095-1.ch011

Bari, M. W., & Khan, Q. (2021). Pakistan as a Destination of Religious Tourism. In E. Alaverdov & M. Bari (Eds.), *Global Development of Religious Tourism* (pp. 1-10). IGI Global. https://doi.org/10.4018/978-1-7998-5792-1.ch001

Bartens, Y., Chunpir, H. I., Schulte, F., & Voß, S. (2017). Business/IT Alignment in Two-Sided Markets: A COBIT 5 Analysis for Media Streaming Business Models. In S. De Haes & W. Van Grembergen (Eds.), *Strategic IT Governance and Alignment in Business Settings* (pp. 82–111). Hershey, PA: IGI Global. doi:10.4018/978-1-5225-0861-8.ch004

Bashayreh, A. M. (2018). Organizational Culture and Organizational Performance. In W. Lee & F. Sabetzadeh (Eds.), *Contemporary Knowledge and Systems Science* (pp. 50–69). Hershey, PA: IGI Global. doi:10.4018/978-1-5225-5655-8.ch003

Bechthold, L., Lude, M., & Prügl, R. (2021). Crisis Favors the Prepared Firm: How Organizational Ambidexterity Relates to Perceptions of Organizational Resilience. In A. Zehrer, G. Glowka, K. Schwaiger, & V. Ranacher-Lackner (Eds.), *Resiliency Models and Addressing Future Risks for Family Firms in the Tourism Industry* (pp. 178–205). IGI Global. https://doi.org/10.4018/978-1-7998-7352-5.ch008

Bedford, D. A. (2018). Sustainable Knowledge Management Strategies: Aligning Business Capabilities and Knowledge Management Goals. In N. Baporikar (Ed.), *Global Practices in Knowledge Management for Societal and Organizational Development* (pp. 46–73). Hershey, PA: IGI Global. doi:10.4018/978-1-5225-3009-1.ch003

Bekjanov, D., & Matyusupov, B. (2021). Influence of Innovative Processes in the Competitiveness of Tourist Destination. In J. Soares (Ed.), *Innovation and Entrepreneurial Opportunities in Community Tourism* (pp. 243–263). IGI Global. https://doi.org/10.4018/978-1-7998-4855-4.ch014

Bharwani, S., & Musunuri, D. (2018). Reflection as a Process From Theory to Practice. In M. Khosrow-Pour, D.B.A. (Ed.), Encyclopedia of Information Science and Technology, Fourth Edition (pp. 1529-1539). Hershey, PA: IGI Global. doi:10.4018/978-1-5225-2255-3.ch132

Bhatt, G. D., Wang, Z., & Rodger, J. A. (2017). Information Systems Capabilities and Their Effects on Competitive Advantages: A Study of Chinese Companies. *Information Resources Management Journal*, *30*(3), 41–57. doi:10.4018/IRMJ.2017070103

Bhushan, M., & Yadav, A. (2017). Concept of Cloud Computing in ESB. In R. Bhadoria, N. Chaudhari, G. Tomar, & S. Singh (Eds.), *Exploring Enterprise Service Bus in the Service-Oriented Architecture Paradigm* (pp. 116–127). Hershey, PA: IGI Global. doi:10.4018/978-1-5225-2157-0.ch008

Bhushan, S. (2017). System Dynamics Base-Model of Humanitarian Supply Chain (HSCM) in Disaster Prone Eco-Communities of India: A Discussion on Simulation and Scenario Results. *International Journal of System Dynamics Applications*, *6*(3), 20–37. doi:10.4018/IJSDA.2017070102

Binder, D., & Miller, J. W. (2021). A Generations' Perspective on Employer Branding in Tourism. In V. Costa, A. Moura, & M. Mira (Eds.), *Handbook of Research on Human Capital and People Management in the Tourism Industry* (pp. 152–174). IGI Global. https://doi.org/10.4018/978-1-7998-4318-4.ch008

Birch Freeman, A. A., Mensah, I., & Antwi, K. B. (2022). Smiling vs. Frowning Faces: Community Participation for Sustainable Tourism in Ghanaian Communities. In I. Mensah & E. Afenyo-Agbe (Eds.), *Prospects and Challenges of Community-Based Tourism and Changing Demographics* (pp. 83–106). IGI Global. https://doi.org/10.4018/978-1-7998-7335-8.ch004

Biswas, A., & De, A. K. (2017). On Development of a Fuzzy Stochastic Programming Model with Its Application to Business Management. In S. Trivedi, S. Dey, A. Kumar, & T. Panda (Eds.), *Handbook of Research on Advanced Data Mining Techniques and Applications for Business Intelligence* (pp. 353–378). Hershey, PA: IGI Global. doi:10.4018/978-1-5225-2031-3.ch021

Boragnio, A., & Faracce Macia, C. (2021). "Taking Care of Yourself at Home": Use of E-Commerce About Food and Care During the COVID-19 Pandemic in the City of Buenos Aires. In M. Korstanje (Ed.), *Socio-Economic Effects and Recovery Efforts for the Rental Industry: Post-COVID-19 Strategies* (pp. 45–71). IGI Global. https://doi.org/10.4018/978-1-7998-7287-0.ch003

Borges, V. D. (2021). Happiness: The Basis for Public Policy in Tourism. In A. Perinotto, V. Mayer, & J. Soares (Eds.), *Rebuilding and Restructuring the Tourism Industry: Infusion of Happiness and Quality of Life* (pp. 1–25). IGI Global. https://doi.org/10.4018/978-1-7998-7239-9.ch001

Bücker, J., & Ernste, K. (2018). Use of Brand Heroes in Strategic Reputation Management: The Case of Bacardi, Adidas, and Daimler. In A. Erdemir (Ed.), *Reputation Management Techniques in Public Relations* (pp. 126–150). Hershey, PA: IGI Global. doi:10.4018/978-1-5225-3619-2.ch007

Buluk Eşitti, B. (2021). COVID-19 and Alternative Tourism: New Destinations and New Tourism Products. In M. Demir, A. Dalgıç, & F. Ergen (Eds.), *Handbook of Research on the Impacts and Implications of COVID-19 on the Tourism Industry* (pp. 786–805). IGI Global. https://doi.org/10.4018/978-1-7998-8231-2.ch038

Bureš, V. (2018). Industry 4.0 From the Systems Engineering Perspective: Alternative Holistic Framework Development. In R. Brunet-Thornton & F. Martinez (Eds.), *Analyzing the Impacts of Industry 4.0 in Modern Business Environments* (pp. 199–223). Hershey, PA: IGI Global. doi:10.4018/978-1-5225-3468-6.ch011

Buzady, Z. (2017). Resolving the Magic Cube of Effective Case Teaching: Benchmarking Case Teaching Practices in Emerging Markets – Insights from the Central European University Business School, Hungary. In D. Latusek (Ed.), *Case Studies as a Teaching Tool in Management Education* (pp. 79–103). Hershey, PA: IGI Global. doi:10.4018/978-1-5225-0770-3.ch005

Camillo, A. (2021). *Legal Matters, Risk Management, and Risk Prevention: From Forming a Business to Legal Representation.* IGI Global. doi:10.4018/978-1-7998-4342-9.ch004

Căpusneanu, S., & Topor, D. I. (2018). Business Ethics and Cost Management in SMEs: Theories of Business Ethics and Cost Management Ethos. In I. Oncioiu (Ed.), *Ethics and Decision-Making for Sustainable Business Practices* (pp. 109–127). Hershey, PA: IGI Global. doi:10.4018/978-1-5225-3773-1.ch007

Chan, R. L., Mo, P. L., & Moon, K. K. (2018). Strategic and Tactical Measures in Managing Enterprise Risks: A Study of the Textile and Apparel Industry. In K. Strang, M. Korstanje, & N. Vajjhala (Eds.), *Research, Practices, and Innovations in Global Risk and Contingency Management* (pp. 1–19). Hershey, PA: IGI Global. doi:10.4018/978-1-5225-4754-9.ch001

Charlier, S. D., Burke-Smalley, L. A., & Fisher, S. L. (2018). Undergraduate Programs in the U.S: A Contextual and Content-Based Analysis. In J. Mendy (Ed.), *Teaching Human Resources and Organizational Behavior at the College Level* (pp. 26–57). Hershey, PA: IGI Global. doi:10.4018/978-1-5225-2820-3.ch002

Chumillas, J., Güell, M., & Quer, P. (2022). The Use of ICT in Tourist and Educational Literary Routes: The Role of the Guide. In C. Ramos, S. Quinteiro, & A. Gonçalves (Eds.), *ICT as Innovator Between Tourism and Culture* (pp. 15–29). IGI Global. https://doi.org/10.4018/978-1-7998-8165-0.ch002

Dahlberg, T., Kivijärvi, H., & Saarinen, T. (2017). IT Investment Consistency and Other Factors Influencing the Success of IT Performance. In S. De Haes & W. Van Grembergen (Eds.), *Strategic IT Governance and Alignment in Business Settings* (pp. 176–208). Hershey, PA: IGI Global. doi:10.4018/978-1-5225-0861-8.ch007

Damnjanović, A. M. (2017). Knowledge Management Optimization through IT and E-Business Utilization: A Qualitative Study on Serbian SMEs. In M. Vemić (Ed.), *Optimal Management Strategies in Small and Medium Enterprises* (pp. 249–267). Hershey, PA: IGI Global. doi:10.4018/978-1-5225-1949-2.ch012

Daneshpour, H. (2017). Integrating Sustainable Development into Project Portfolio Management through Application of Open Innovation. In M. Vemić (Ed.), *Optimal Management Strategies in Small and Medium Enterprises* (pp. 370–387). Hershey, PA: IGI Global. doi:10.4018/978-1-5225-1949-2.ch017

Daniel, A. D., & Reis de Castro, V. (2018). Entrepreneurship Education: How to Measure the Impact on Nascent Entrepreneurs. In A. Carrizo Moreira, J. Guilherme Leitão Dantas, & F. Manuel Valente (Eds.), *Nascent Entrepreneurship and Successful New Venture Creation* (pp. 85–110). Hershey, PA: IGI Global. doi:10.4018/978-1-5225-2936-1.ch004

David, R., Swami, B. N., & Tangirala, S. (2018). Ethics Impact on Knowledge Management in Organizational Development: A Case Study. In N. Baporikar (Ed.), *Global Practices in Knowledge Management for Societal and Organizational Development* (pp. 19–45). Hershey, PA: IGI Global. doi:10.4018/978-1-5225-3009-1.ch002

De Uña-Álvarez, E., & Villarino-Pérez, M. (2022). Fostering Ecocultural Resources, Identity, and Tourism in Inland Territories (Galicia, NW Spain). In G. Fernandes (Ed.), *Challenges and New Opportunities for Tourism in Inland Territories: Ecocultural Resources and Sustainable Initiatives* (pp. 1-16). IGI Global. https://doi.org/10.4018/978-1-7998-7339-6.ch001

Delias, P., & Lakiotaki, K. (2018). Discovering Process Horizontal Boundaries to Facilitate Process Comprehension. *International Journal of Operations Research and Information Systems, 9*(2), 1–31. doi:10.4018/IJORIS.2018040101

Denholm, J., & Lee-Davies, L. (2018). Success Factors for Games in Business and Project Management. In *Enhancing Education and Training Initiatives Through Serious Games* (pp. 34–68). Hershey, PA: IGI Global. doi:10.4018/978-1-5225-3689-5.ch002

Deshpande, M. (2017). Best Practices in Management Institutions for Global Leadership: Policy Aspects. In N. Baporikar (Ed.), *Management Education for Global Leadership* (pp. 1–27). Hershey, PA: IGI Global. doi:10.4018/978-1-5225-1013-0.ch001

Deshpande, M. (2018). Policy Perspectives for SMEs Knowledge Management. In N. Baporikar (Ed.), *Knowledge Integration Strategies for Entrepreneurship and Sustainability* (pp. 23–46). Hershey, PA: IGI Global. doi:10.4018/978-1-5225-5115-7.ch002

Dezdar, S. (2017). ERP Implementation Projects in Asian Countries: A Comparative Study on Iran and China. *International Journal of Information Technology Project Management, 8*(3), 52–68. doi:10.4018/IJITPM.2017070104

Domingos, D., Respício, A., & Martinho, R. (2017). Reliability of IoT-Aware BPMN Healthcare Processes. In C. Reis & M. Maximiano (Eds.), *Internet of Things and Advanced Application in Healthcare* (pp. 214–248). Hershey, PA: IGI Global. doi:10.4018/978-1-5225-1820-4.ch008

Dosumu, O., Hussain, J., & El-Gohary, H. (2017). An Exploratory Study of the Impact of Government Policies on the Development of Small and Medium Enterprises in Developing Countries: The Case of Nigeria. *International Journal of Customer Relationship Marketing and Management*, *8*(4), 51–62. doi:10.4018/IJCRMM.2017100104

Durst, S., Bruns, G., & Edvardsson, I. R. (2017). Retaining Knowledge in Smaller Building and Construction Firms. *International Journal of Knowledge and Systems Science*, *8*(3), 1–12. doi:10.4018/IJKSS.2017070101

Edvardsson, I. R., & Durst, S. (2017). Outsourcing, Knowledge, and Learning: A Critical Review. *International Journal of Knowledge-Based Organizations*, *7*(2), 13–26. doi:10.4018/IJKBO.2017040102

Edwards, J. S. (2018). Integrating Knowledge Management and Business Processes. In M. Khosrow-Pour, D.B.A. (Ed.), Encyclopedia of Information Science and Technology, Fourth Edition (pp. 5046-5055). Hershey, PA: IGI Global. doi:10.4018/978-1-5225-2255-3.ch437

Eichelberger, S., & Peters, M. (2021). Family Firm Management in Turbulent Times: Opportunities for Responsible Tourism. In A. Zehrer, G. Glowka, K. Schwaiger, & V. Ranacher-Lackner (Eds.), *Resiliency Models and Addressing Future Risks for Family Firms in the Tourism Industry* (pp. 103–124). IGI Global. https://doi.org/10.4018/978-1-7998-7352-5.ch005

Eide, D., Hjalager, A., & Hansen, M. (2022). Innovative Certifications in Adventure Tourism: Attributes and Diffusion. In R. Augusto Costa, F. Brandão, Z. Breda, & C. Costa (Eds.), *Planning and Managing the Experience Economy in Tourism* (pp. 161-175). IGI Global. https://doi.org/10.4018/978-1-7998-8775-1.ch009

Ejiogu, A. O. (2018). Economics of Farm Management. In *Agricultural Finance and Opportunities for Investment and Expansion* (pp. 56–72). Hershey, PA: IGI Global. doi:10.4018/978-1-5225-3059-6.ch003

Ekanem, I., & Abiade, G. E. (2018). Factors Influencing the Use of E-Commerce by Small Enterprises in Nigeria. *International Journal of ICT Research in Africa and the Middle East*, *7*(1), 37–53. doi:10.4018/IJICTRAME.2018010103

Ekanem, I., & Alrossais, L. A. (2017). Succession Challenges Facing Family Businesses in Saudi Arabia. In P. Zgheib (Ed.), *Entrepreneurship and Business Innovation in the Middle East* (pp. 122–146). Hershey, PA: IGI Global. doi:10.4018/978-1-5225-2066-5.ch007

El Faquih, L., & Fredj, M. (2017). Ontology-Based Framework for Quality in Configurable Process Models. *Journal of Electronic Commerce in Organizations*, *15*(2), 48–60. doi:10.4018/JECO.2017040104

Faisal, M. N., & Talib, F. (2017). Building Ambidextrous Supply Chains in SMEs: How to Tackle the Barriers? *International Journal of Information Systems and Supply Chain Management, 10*(4), 80–100. doi:10.4018/IJISSCM.2017100105

Fernandes, T. M., Gomes, J., & Romão, M. (2017). Investments in E-Government: A Benefit Management Case Study. *International Journal of Electronic Government Research, 13*(3), 1–17. doi:10.4018/IJEGR.2017070101

Figueira, L. M., Honrado, G. R., & Dionísio, M. S. (2021). Human Capital Management in the Tourism Industry in Portugal. In V. Costa, A. Moura, & M. Mira (Eds.), *Handbook of Research on Human Capital and People Management in the Tourism Industry* (pp. 1–19). IGI Global. doi:10.4018/978-1-7998-4318-4.ch001

Gao, S. S., Oreal, S., & Zhang, J. (2018). Contemporary Financial Risk Management Perceptions and Practices of Small-Sized Chinese Businesses. In I. Management Association (Ed.), Global Business Expansion: Concepts, Methodologies, Tools, and Applications (pp. 917-931). Hershey, PA: IGI Global. doi:10.4018/978-1-5225-5481-3.ch041

Garg, R., & Berning, S. C. (2017). Indigenous Chinese Management Philosophies: Key Concepts and Relevance for Modern Chinese Firms. In B. Christiansen & G. Koc (Eds.), *Transcontinental Strategies for Industrial Development and Economic Growth* (pp. 43–57). Hershey, PA: IGI Global. doi:10.4018/978-1-5225-2160-0.ch003

Gencer, Y. G. (2017). Supply Chain Management in Retailing Business. In U. Akkucuk (Ed.), *Ethics and Sustainability in Global Supply Chain Management* (pp. 197–210). Hershey, PA: IGI Global. doi:10.4018/978-1-5225-2036-8.ch011

Gera, R., Arora, S., & Malik, S. (2021). Emotional Labor in the Tourism Industry: Strategies, Antecedents, and Outcomes. In V. Costa, A. Moura, & M. Mira (Eds.), *Handbook of Research on Human Capital and People Management in the Tourism Industry* (pp. 73–91). IGI Global. https://doi.org/10.4018/978-1-7998-4318-4.ch004

Giacosa, E. (2018). The Increasing of the Regional Development Thanks to the Luxury Business Innovation. In L. Carvalho (Ed.), *Handbook of Research on Entrepreneurial Ecosystems and Social Dynamics in a Globalized World* (pp. 260–273). Hershey, PA: IGI Global. doi:10.4018/978-1-5225-3525-6.ch011

Glowka, G., Tusch, M., & Zehrer, A. (2021). The Risk Perception of Family Business Owner-Manager in the Tourism Industry: A Qualitative Comparison of the Intra-Firm Senior and Junior Generation. In A. Zehrer, G. Glowka, K. Schwaiger, & V. Ranacher-Lackner (Eds.), *Resiliency Models and Addressing Future Risks for Family Firms in the Tourism Industry* (pp. 126–153). IGI Global. https://doi.org/10.4018/978-1-7998-7352-5.ch006

Glykas, M., & George, J. (2017). Quality and Process Management Systems in the UAE Maritime Industry. *International Journal of Productivity Management and Assessment Technologies, 5*(1), 20–39. doi:10.4018/IJPMAT.2017010102

Glykas, M., Valiris, G., Kokkinaki, A., & Koutsoukou, Z. (2018). Banking Business Process Management Implementation. *International Journal of Productivity Management and Assessment Technologies, 6*(1), 50–69. doi:10.4018/IJPMAT.2018010104

Gomes, J., & Romão, M. (2017). The Balanced Scorecard: Keeping Updated and Aligned with Today's Business Trends. *International Journal of Productivity Management and Assessment Technologies*, *5*(2), 1–15. doi:10.4018/IJPMAT.2017070101

Gomes, J., & Romão, M. (2017). Aligning Information Systems and Technology with Benefit Management and Balanced Scorecard. In S. De Haes & W. Van Grembergen (Eds.), *Strategic IT Governance and Alignment in Business Settings* (pp. 112–131). Hershey, PA: IGI Global. doi:10.4018/978-1-5225-0861-8.ch005

Goyal, A. (2021). Communicating and Building Destination Brands With New Media. In M. Dinis, L. Bonixe, S. Lamy, & Z. Breda (Eds.), *Impact of New Media in Tourism* (pp. 1-20). IGI Global. https://doi.org/10.4018/978-1-7998-7095-1.ch001

Grefen, P., & Turetken, O. (2017). Advanced Business Process Management in Networked E-Business Scenarios. *International Journal of E-Business Research, 13*(4), 70–104. doi:10.4018/IJEBR.2017100105

Guasca, M., Van Broeck, A. M., & Vanneste, D. (2021). Tourism and the Social Reintegration of Colombian Ex-Combatants. In J. da Silva, Z. Breda, & F. Carbone (Eds.), *Role and Impact of Tourism in Peacebuilding and Conflict Transformation* (pp. 66-86). IGI Global. https://doi.org/10.4018/978-1-7998-5053-3.ch005

Haider, A., & Saetang, S. (2017). Strategic IT Alignment in Service Sector. In S. Rozenes & Y. Cohen (Eds.), *Handbook of Research on Strategic Alliances and Value Co-Creation in the Service Industry* (pp. 231–258). Hershey, PA: IGI Global. doi:10.4018/978-1-5225-2084-9.ch012

Hajilari, A. B., Ghadaksaz, M., & Fasghandis, G. S. (2017). Assessing Organizational Readiness for Implementing ERP System Using Fuzzy Expert System Approach. *International Journal of Enterprise Information Systems, 13*(1), 67–85. doi:10.4018/IJEIS.2017010105

Haldorai, A., Ramu, A., & Murugan, S. (2018). Social Aware Cognitive Radio Networks: Effectiveness of Social Networks as a Strategic Tool for Organizational Business Management. In H. Bansal, G. Shrivastava, G. Nguyen, & L. Stanciu (Eds.), *Social Network Analytics for Contemporary Business Organizations* (pp. 188–202). Hershey, PA: IGI Global. doi:10.4018/978-1-5225-5097-6.ch010

Hall, O. P. Jr. (2017). Social Media Driven Management Education. *International Journal of Knowledge-Based Organizations*, *7*(2), 43–59. doi:10.4018/IJKBO.2017040104

Hanifah, H., Halim, H. A., Ahmad, N. H., & Vafaei-Zadeh, A. (2017). Innovation Culture as a Mediator Between Specific Human Capital and Innovation Performance Among Bumiputera SMEs in Malaysia. In N. Ahmad, T. Ramayah, H. Halim, & S. Rahman (Eds.), *Handbook of Research on Small and Medium Enterprises in Developing Countries* (pp. 261–279). Hershey, PA: IGI Global. doi:10.4018/978-1-5225-2165-5.ch012

Hartlieb, S., & Silvius, G. (2017). Handling Uncertainty in Project Management and Business Development: Similarities and Differences. In Y. Raydugin (Ed.), *Handbook of Research on Leveraging Risk and Uncertainties for Effective Project Management* (pp. 337–362). Hershey, PA: IGI Global. doi:10.4018/978-1-5225-1790-0.ch016

Hass, K. B. (2017). Living on the Edge: Managing Project Complexity. In Y. Raydugin (Ed.), *Handbook of Research on Leveraging Risk and Uncertainties for Effective Project Management* (pp. 177–201). Hershey, PA: IGI Global. doi:10.4018/978-1-5225-1790-0.ch009

Hawking, P., & Carmine Sellitto, C. (2017). Developing an Effective Strategy for Organizational Business Intelligence. In M. Tavana (Ed.), *Enterprise Information Systems and the Digitalization of Business Functions* (pp. 222–237). Hershey, PA: IGI Global. doi:10.4018/978-1-5225-2382-6.ch010

Hawking, P., & Sellitto, C. (2017). A Fast-Moving Consumer Goods Company and Business Intelligence Strategy Development. *International Journal of Enterprise Information Systems, 13*(2), 22–33. doi:10.4018/IJEIS.2017040102

Hawking, P., & Sellitto, C. (2017). Business Intelligence Strategy: Two Case Studies. *International Journal of Business Intelligence Research, 8*(2), 17–30. doi:10.4018/IJBIR.2017070102

Hee, W. J., Jalleh, G., Lai, H., & Lin, C. (2017). E-Commerce and IT Projects: Evaluation and Management Issues in Australian and Taiwanese Hospitals. *International Journal of Public Health Management and Ethics, 2*(1), 69–90. doi:10.4018/IJPHME.2017010104

Hernandez, A. A. (2018). Exploring the Factors to Green IT Adoption of SMEs in the Philippines. *Journal of Cases on Information Technology, 20*(2), 49–66. doi:10.4018/JCIT.2018040104

Hollman, A., Bickford, S., & Hollman, T. (2017). Cyber InSecurity: A Post-Mortem Attempt to Assess Cyber Problems from IT and Business Management Perspectives. *Journal of Cases on Information Technology, 19*(3), 42–70. doi:10.4018/JCIT.2017070104

Ibrahim, F., & Zainin, N. M. (2021). Exploring the Technological Impacts: The Case of Museums in Brunei Darussalam. *International Journal of Tourism and Hospitality Management in the Digital Age, 5*(1), 1–18. https://doi.org/10.4018/IJTHMDA.2021010101

Igbinakhase, I. (2017). Responsible and Sustainable Management Practices in Developing and Developed Business Environments. In Z. Fields (Ed.), *Collective Creativity for Responsible and Sustainable Business Practice* (pp. 180–207). Hershey, PA: IGI Global. doi:10.4018/978-1-5225-1823-5.ch010

Iwata, J. J., & Hoskins, R. G. (2017). Managing Indigenous Knowledge in Tanzania: A Business Perspective. In P. Jain & N. Mnjama (Eds.), *Managing Knowledge Resources and Records in Modern Organizations* (pp. 198–214). Hershey, PA: IGI Global. doi:10.4018/978-1-5225-1965-2.ch012

Jain, P. (2017). Ethical and Legal Issues in Knowledge Management Life-Cycle in Business. In P. Jain & N. Mnjama (Eds.), *Managing Knowledge Resources and Records in Modern Organizations* (pp. 82–101). Hershey, PA: IGI Global. doi:10.4018/978-1-5225-1965-2.ch006

James, S., & Hauli, E. (2017). Holistic Management Education at Tanzanian Rural Development Planning Institute. In N. Baporikar (Ed.), *Management Education for Global Leadership* (pp. 112–136). Hershey, PA: IGI Global. doi:10.4018/978-1-5225-1013-0.ch006

Janošková, M., Csikósová, A., & Čulková, K. (2018). Measurement of Company Performance as Part of Its Strategic Management. In R. Leon (Ed.), *Managerial Strategies for Business Sustainability During Turbulent Times* (pp. 309–335). Hershey, PA: IGI Global. doi:10.4018/978-1-5225-2716-9.ch017

Jean-Vasile, A., & Alecu, A. (2017). Theoretical and Practical Approaches in Understanding the Influences of Cost-Productivity-Profit Trinomial in Contemporary Enterprises. In A. Jean Vasile & D. Nicolò (Eds.), *Sustainable Entrepreneurship and Investments in the Green Economy* (pp. 28–62). Hershey, PA: IGI Global. doi:10.4018/978-1-5225-2075-7.ch002

Joia, L. A., & Correia, J. C. (2018). CIO Competencies From the IT Professional Perspective: Insights From Brazil. *Journal of Global Information Management, 26*(2), 74–103. doi:10.4018/JGIM.2018040104

Juma, A., & Mzera, N. (2017). Knowledge Management and Records Management and Competitive Advantage in Business. In P. Jain & N. Mnjama (Eds.), *Managing Knowledge Resources and Records in Modern Organizations* (pp. 15–28). Hershey, PA: IGI Global. doi:10.4018/978-1-5225-1965-2.ch002

K., I., & A, V. (2018). Monitoring and Auditing in the Cloud. In K. Munir (Ed.), *Cloud Computing Technologies for Green Enterprises* (pp. 318-350). Hershey, PA: IGI Global. https://doi.org/ doi:10.4018/978-1-5225-3038-1.ch013

Kabra, G., Ghosh, V., & Ramesh, A. (2018). Enterprise Integrated Business Process Management and Business Intelligence Framework for Business Process Sustainability. In A. Paul, D. Bhattacharyya, & S. Anand (Eds.), *Green Initiatives for Business Sustainability and Value Creation* (pp. 228–238). Hershey, PA: IGI Global. doi:10.4018/978-1-5225-2662-9.ch010

Kaoud, M. (2017). Investigation of Customer Knowledge Management: A Case Study Research. *International Journal of Service Science, Management, Engineering, and Technology, 8*(2), 12–22. doi:10.4018/IJSSMET.2017040102

Katuu, S. (2018). A Comparative Assessment of Enterprise Content Management Maturity Models. In N. Gwangwava & M. Mutingi (Eds.), *E-Manufacturing and E-Service Strategies in Contemporary Organizations* (pp. 93–118). Hershey, PA: IGI Global. doi:10.4018/978-1-5225-3628-4.ch005

Khan, M. Y., & Abir, T. (2022). The Role of Social Media Marketing in the Tourism and Hospitality Industry: A Conceptual Study on Bangladesh. In C. Ramos, S. Quinteiro, & A. Gonçalves (Eds.), *ICT as Innovator Between Tourism and Culture* (pp. 213–229). IGI Global. https://doi.org/10.4018/978-1-7998-8165-0.ch013

Kinnunen, S., Ylä-Kujala, A., Marttonen-Arola, S., Kärri, T., & Baglee, D. (2018). Internet of Things in Asset Management: Insights from Industrial Professionals and Academia. *International Journal of Service Science, Management, Engineering, and Technology, 9*(2), 104–119. doi:10.4018/IJSSMET.2018040105

Klein, A. Z., Sabino de Freitas, A., Machado, L., Freitas, J. C. Jr, Graziola, P. G. Jr, & Schlemmer, E. (2017). Virtual Worlds Applications for Management Education. In L. Tomei (Ed.), *Exploring the New Era of Technology-Infused Education* (pp. 279–299). Hershey, PA: IGI Global. doi:10.4018/978-1-5225-1709-2.ch017

Kővári, E., Saleh, M., & Steinbachné Hajmásy, G. (2022). The Impact of Corporate Digital Responsibility (CDR) on Internal Stakeholders' Satisfaction in Hungarian Upscale Hotels. In M. Valeri (Ed.), *New Governance and Management in Touristic Destinations* (pp. 35–51). IGI Global. https://doi.org/10.4018/978-1-6684-3889-3.ch003

Kożuch, B., & Jabłoński, A. (2017). Adopting the Concept of Business Models in Public Management. In M. Lewandowski & B. Kożuch (Eds.), *Public Sector Entrepreneurship and the Integration of Innovative Business Models* (pp. 10–46). Hershey, PA: IGI Global. doi:10.4018/978-1-5225-2215-7.ch002

Kumar, J., Adhikary, A., & Jha, A. (2017). Small Active Investors' Perceptions and Preferences Towards Tax Saving Mutual Fund Schemes in Eastern India: An Empirical Note. *International Journal of Asian Business and Information Management*, *8*(2), 35–45. doi:10.4018/IJABIM.2017040103

Latusi, S., & Fissore, M. (2021). Pilgrimage Routes to Happiness: Comparing the Camino de Santiago and Via Francigena. In A. Perinotto, V. Mayer, & J. Soares (Eds.), *Rebuilding and Restructuring the Tourism Industry: Infusion of Happiness and Quality of Life* (pp. 157–182). IGI Global. https://doi.org/10.4018/978-1-7998-7239-9.ch008

Lavassani, K. M., & Movahedi, B. (2017). Applications Driven Information Systems: Beyond Networks toward Business Ecosystems. *International Journal of Innovation in the Digital Economy*, *8*(1), 61–75. doi:10.4018/IJIDE.2017010104

Lazzareschi, V. H., & Brito, M. S. (2017). Strategic Information Management: Proposal of Business Project Model. In G. Jamil, A. Soares, & C. Pessoa (Eds.), *Handbook of Research on Information Management for Effective Logistics and Supply Chains* (pp. 59–88). Hershey, PA: IGI Global. doi:10.4018/978-1-5225-0973-8.ch004

Lechuga Sancho, M. P., & Martín Navarro, A. (2022). Evolution of the Literature on Social Responsibility in the Tourism Sector: A Systematic Literature Review. In G. Fernandes (Ed.), *Challenges and New Opportunities for Tourism in Inland Territories: Ecocultural Resources and Sustainable Initiatives* (pp. 169–186). IGI Global. https://doi.org/10.4018/978-1-7998-7339-6.ch010

Lederer, M., Kurz, M., & Lazarov, P. (2017). Usage and Suitability of Methods for Strategic Business Process Initiatives: A Multi Case Study Research. *International Journal of Productivity Management and Assessment Technologies*, *5*(1), 40–51. doi:10.4018/IJPMAT.2017010103

Lee, I. (2017). A Social Enterprise Business Model and a Case Study of Pacific Community Ventures (PCV). In V. Potocan, M. Ünğan, & Z. Nedelko (Eds.), *Handbook of Research on Managerial Solutions in Non-Profit Organizations* (pp. 182–204). Hershey, PA: IGI Global. doi:10.4018/978-1-5225-0731-4.ch009

Leon, L. A., Seal, K. C., Przasnyski, Z. H., & Wiedenman, I. (2017). Skills and Competencies Required for Jobs in Business Analytics: A Content Analysis of Job Advertisements Using Text Mining. *International Journal of Business Intelligence Research, 8*(1), 1–25. doi:10.4018/IJBIR.2017010101

Levy, C. L., & Elias, N. I. (2017). SOHO Users' Perceptions of Reliability and Continuity of Cloud-Based Services. In M. Moore (Ed.), *Cybersecurity Breaches and Issues Surrounding Online Threat Protection* (pp. 248–287). Hershey, PA: IGI Global. doi:10.4018/978-1-5225-1941-6.ch011

Levy, M. (2018). Change Management Serving Knowledge Management and Organizational Development: Reflections and Review. In N. Baporikar (Ed.), *Global Practices in Knowledge Management for Societal and Organizational Development* (pp. 256–270). Hershey, PA: IGI Global. doi:10.4018/978-1-5225-3009-1.ch012

Lewandowski, M. (2017). Public Organizations and Business Model Innovation: The Role of Public Service Design. In M. Lewandowski & B. Kożuch (Eds.), *Public Sector Entrepreneurship and the Integration of Innovative Business Models* (pp. 47–72). Hershey, PA: IGI Global. doi:10.4018/978-1-5225-2215-7.ch003

Lhannaoui, H., Kabbaj, M. I., & Bakkoury, Z. (2017). A Survey of Risk-Aware Business Process Modelling. *International Journal of Risk and Contingency Management, 6*(3), 14–26. doi:10.4018/IJRCM.2017070102

Li, J., Sun, W., Jiang, W., Yang, H., & Zhang, L. (2017). How the Nature of Exogenous Shocks and Crises Impact Company Performance?: The Effects of Industry Characteristics. *International Journal of Risk and Contingency Management, 6*(4), 40–55. doi:10.4018/IJRCM.2017100103

Lopez-Fernandez, M., Perez-Perez, M., Serrano-Bedia, A., & Cobo-Gonzalez, A. (2021). Small and Medium Tourism Enterprise Survival in Times of Crisis: "El Capricho de Gaudí. In D. Toubes & N. Araújo-Vila (Eds.), *Risk, Crisis, and Disaster Management in Small and Medium-Sized Tourism Enterprises* (pp. 103–129). IGI Global. doi:10.4018/978-1-7998-6996-2.ch005

Mahajan, A., Maidullah, S., & Hossain, M. R. (2022). Experience Toward Smart Tour Guide Apps in Travelling: An Analysis of Users' Reviews on Audio Odigos and Trip My Way. In R. Augusto Costa, F. Brandão, Z. Breda, & C. Costa (Eds.), *Planning and Managing the Experience Economy in Tourism* (pp. 255-273). IGI Global. https://doi.org/10.4018/978-1-7998-8775-1.ch014

Malega, P. (2017). Small and Medium Enterprises in the Slovak Republic: Status and Competitiveness of SMEs in the Global Markets and Possibilities of Optimization. In M. Vemić (Ed.), *Optimal Management Strategies in Small and Medium Enterprises* (pp. 102–124). Hershey, PA: IGI Global. doi:10.4018/978-1-5225-1949-2.ch006

Malewska, K. M. (2017). Intuition in Decision-Making on the Example of a Non-Profit Organization. In V. Potocan, M. Üngan, & Z. Nedelko (Eds.), *Handbook of Research on Managerial Solutions in Non-Profit Organizations* (pp. 378–399). Hershey, PA: IGI Global. doi:10.4018/978-1-5225-0731-4.ch018

Maroofi, F. (2017). Entrepreneurial Orientation and Organizational Learning Ability Analysis for Innovation and Firm Performance. In N. Baporikar (Ed.), *Innovation and Shifting Perspectives in Management Education* (pp. 144–165). Hershey, PA: IGI Global. doi:10.4018/978-1-5225-1019-2.ch007

Marques, M., Moleiro, D., Brito, T. M., & Marques, T. (2021). Customer Relationship Management as an Important Relationship Marketing Tool: The Case of the Hospitality Industry in Estoril Coast. In M. Dinis, L. Bonixe, S. Lamy, & Z. Breda (Eds.), Impact of New Media in Tourism (pp. 39-56). IGI Global. https://doi.org/ doi:10.4018/978-1-7998-7095-1.ch003

Martins, P. V., & Zacarias, M. (2017). A Web-based Tool for Business Process Improvement. *International Journal of Web Portals*, *9*(2), 68–84. doi:10.4018/IJWP.2017070104

Matthies, B., & Coners, A. (2017). Exploring the Conceptual Nature of e-Business Projects. *Journal of Electronic Commerce in Organizations*, *15*(3), 33–63. doi:10.4018/JECO.2017070103

Mayer, V. F., Fraga, C. C., & Silva, L. C. (2021). Contributions of Neurosciences to Studies of Well-Being in Tourism. In A. Perinotto, V. Mayer, & J. Soares (Eds.), *Rebuilding and Restructuring the Tourism Industry: Infusion of Happiness and Quality of Life* (pp. 108–128). IGI Global. https://doi.org/10.4018/978-1-7998-7239-9.ch006

McKee, J. (2018). Architecture as a Tool to Solve Business Planning Problems. In M. Khosrow-Pour, D.B.A. (Ed.), Encyclopedia of Information Science and Technology, Fourth Edition (pp. 573-586). Hershey, PA: IGI Global. doi:10.4018/978-1-5225-2255-3.ch050

McMurray, A. J., Cross, J., & Caponecchia, C. (2018). The Risk Management Profession in Australia: Business Continuity Plan Practices. In N. Bajgoric (Ed.), *Always-On Enterprise Information Systems for Modern Organizations* (pp. 112–129). Hershey, PA: IGI Global. doi:10.4018/978-1-5225-3704-5.ch006

Meddah, I. H., & Belkadi, K. (2018). Mining Patterns Using Business Process Management. In R. Hamou (Ed.), *Handbook of Research on Biomimicry in Information Retrieval and Knowledge Management* (pp. 78–89). Hershey, PA: IGI Global. doi:10.4018/978-1-5225-3004-6.ch005

Melian, A. G., & Camprubí, R. (2021). The Accessibility of Museum Websites: The Case of Barcelona. In C. Eusébio, L. Teixeira, & M. Carneiro (Eds.), *ICT Tools and Applications for Accessible Tourism* (pp. 234–255). IGI Global. https://doi.org/10.4018/978-1-7998-6428-8.ch011

Mendes, L. (2017). TQM and Knowledge Management: An Integrated Approach Towards Tacit Knowledge Management. In D. Jaziri-Bouagina & G. Jamil (Eds.), *Handbook of Research on Tacit Knowledge Management for Organizational Success* (pp. 236–263). Hershey, PA: IGI Global. doi:10.4018/978-1-5225-2394-9.ch009

Menezes, V. D., & Cavagnaro, E. (2021). Communicating Sustainable Initiatives in the Hotel Industry: The Case of the Hotel Jakarta Amsterdam. In F. Brandão, Z. Breda, R. Costa, & C. Costa (Eds.), *Handbook of Research on the Role of Tourism in Achieving Sustainable Development Goals* (pp. 224-234). IGI Global. https://doi.org/10.4018/978-1-7998-5691-7.ch013

Mitas, O., Bastiaansen, M., & Boode, W. (2022). If You're Happy, I'm Happy: Emotion Contagion at a Tourist Information Center. In R. Augusto Costa, F. Brandão, Z. Breda, & C. Costa (Eds.), *Planning and Managing the Experience Economy in Tourism* (pp. 122-140). IGI Global. https://doi.org/10.4018/978-1-7998-8775-1.ch007

Mnjama, N. M. (2017). Preservation of Recorded Information in Public and Private Sector Organizations. In P. Jain & N. Mnjama (Eds.), *Managing Knowledge Resources and Records in Modern Organizations* (pp. 149–167). Hershey, PA: IGI Global. doi:10.4018/978-1-5225-1965-2.ch009

Mokoqama, M., & Fields, Z. (2017). Principles of Responsible Management Education (PRME): Call for Responsible Management Education. In Z. Fields (Ed.), *Collective Creativity for Responsible and Sustainable Business Practice* (pp. 229–241). Hershey, PA: IGI Global. doi:10.4018/978-1-5225-1823-5.ch012

Monteiro, A., Lopes, S., & Carbone, F. (2021). Academic Mobility: Bridging Tourism and Peace Education. In J. da Silva, Z. Breda, & F. Carbone (Eds.), *Role and Impact of Tourism in Peacebuilding and Conflict Transformation* (pp. 275-301). IGI Global. https://doi.org/10.4018/978-1-7998-5053-3.ch016

Muniapan, B. (2017). Philosophy and Management: The Relevance of Vedanta in Management. In P. Ordóñez de Pablos (Ed.), *Managerial Strategies and Solutions for Business Success in Asia* (pp. 124–139). Hershey, PA: IGI Global. doi:10.4018/978-1-5225-1886-0.ch007

Murad, S. E., & Dowaji, S. (2017). Using Value-Based Approach for Managing Cloud-Based Services. In A. Turuk, B. Sahoo, & S. Addya (Eds.), *Resource Management and Efficiency in Cloud Computing Environments* (pp. 33–60). Hershey, PA: IGI Global. doi:10.4018/978-1-5225-1721-4.ch002

Mutahar, A. M., Daud, N. M., Thurasamy, R., Isaac, O., & Abdulsalam, R. (2018). The Mediating of Perceived Usefulness and Perceived Ease of Use: The Case of Mobile Banking in Yemen. *International Journal of Technology Diffusion*, *9*(2), 21–40. doi:10.4018/IJTD.2018040102

Naidoo, V. (2017). E-Learning and Management Education at African Universities. In N. Baporikar (Ed.), *Management Education for Global Leadership* (pp. 181–201). Hershey, PA: IGI Global. doi:10.4018/978-1-5225-1013-0.ch009

Naidoo, V., & Igbinakhase, I. (2018). Opportunities and Challenges of Knowledge Retention in SMEs. In N. Baporikar (Ed.), *Knowledge Integration Strategies for Entrepreneurship and Sustainability* (pp. 70–94). Hershey, PA: IGI Global. doi:10.4018/978-1-5225-5115-7.ch004

Naumov, N., & Costandachi, G. (2021). Creativity and Entrepreneurship: Gastronomic Tourism in Mexico. In J. Soares (Ed.), *Innovation and Entrepreneurial Opportunities in Community Tourism* (pp. 90–108). IGI Global. https://doi.org/10.4018/978-1-7998-4855-4.ch006

Nayak, S., & Prabhu, N. (2017). Paradigm Shift in Management Education: Need for a Cross Functional Perspective. In N. Baporikar (Ed.), *Management Education for Global Leadership* (pp. 241–255). Hershey, PA: IGI Global. doi:10.4018/978-1-5225-1013-0.ch012

Nedelko, Z., & Potocan, V. (2017). Management Solutions in Non-Profit Organizations: Case of Slovenia. In V. Potocan, M. Üngan, & Z. Nedelko (Eds.), *Handbook of Research on Managerial Solutions in Non-Profit Organizations* (pp. 1–22). Hershey, PA: IGI Global. doi:10.4018/978-1-5225-0731-4.ch001

Nedelko, Z., & Potocan, V. (2017). Priority of Management Tools Utilization among Managers: International Comparison. In V. Wang (Ed.), *Encyclopedia of Strategic Leadership and Management* (pp. 1083–1094). Hershey, PA: IGI Global. doi:10.4018/978-1-5225-1049-9.ch075

Nedelko, Z., Raudeliūnienė, J., & Črešnar, R. (2018). Knowledge Dynamics in Supply Chain Management. In N. Baporikar (Ed.), *Knowledge Integration Strategies for Entrepreneurship and Sustainability* (pp. 150–166). Hershey, PA: IGI Global. doi:10.4018/978-1-5225-5115-7.ch008

Nguyen, H. T., & Hipsher, S. A. (2018). Innovation and Creativity Used by Private Sector Firms in a Resources-Constrained Environment. In S. Hipsher (Ed.), *Examining the Private Sector's Role in Wealth Creation and Poverty Reduction* (pp. 219–238). Hershey, PA: IGI Global. doi:10.4018/978-1-5225-3117-3.ch010

Obicci, P. A. (2017). Risk Sharing in a Partnership. In *Risk Management Strategies in Public-Private Partnerships* (pp. 115–152). Hershey, PA: IGI Global. doi:10.4018/978-1-5225-2503-5.ch004

Obidallah, W. J., & Raahemi, B. (2017). Managing Changes in Service Oriented Virtual Organizations: A Structural and Procedural Framework to Facilitate the Process of Change. *Journal of Electronic Commerce in Organizations, 15*(1), 59–83. doi:10.4018/JECO.2017010104

Ojo, O. (2017). Impact of Innovation on the Entrepreneurial Success in Selected Business Enterprises in South-West Nigeria. *International Journal of Innovation in the Digital Economy, 8*(2), 29–38. doi:10.4018/IJIDE.2017040103

Okdinawati, L., Simatupang, T. M., & Sunitiyoso, Y. (2017). Multi-Agent Reinforcement Learning for Value Co-Creation of Collaborative Transportation Management (CTM). *International Journal of Information Systems and Supply Chain Management, 10*(3), 84–95. doi:10.4018/IJISSCM.2017070105

Olivera, V. A., & Carrillo, I. M. (2021). Organizational Culture: A Key Element for the Development of Mexican Micro and Small Tourist Companies. In J. Soares (Ed.), *Innovation and Entrepreneurial Opportunities in Community Tourism* (pp. 227–242). IGI Global. doi:10.4018/978-1-7998-4855-4.ch013

Ossorio, M. (2022). Corporate Museum Experiences in Enogastronomic Tourism. In R. Augusto Costa, F. Brandão, Z. Breda, & C. Costa (Eds.), Planning and Managing the Experience Economy in Tourism (pp. 107-121). IGI Global. https://doi.org/ doi:10.4018/978-1-7998-8775-1.ch006

Ossorio, M. (2022). Enogastronomic Tourism in Times of Pandemic. In G. Fernandes (Ed.), *Challenges and New Opportunities for Tourism in Inland Territories: Ecocultural Resources and Sustainable Initiatives* (pp. 241–255). IGI Global. https://doi.org/10.4018/978-1-7998-7339-6.ch014

Özekici, Y. K. (2022). ICT as an Acculturative Agent and Its Role in the Tourism Context: Introduction, Acculturation Theory, Progress of the Acculturation Theory in Extant Literature. In C. Ramos, S. Quinteiro, & A. Gonçalves (Eds.), *ICT as Innovator Between Tourism and Culture* (pp. 42–66). IGI Global. https://doi.org/10.4018/978-1-7998-8165-0.ch004

Pal, K. (2018). Building High Quality Big Data-Based Applications in Supply Chains. In A. Kumar & S. Saurav (Eds.), *Supply Chain Management Strategies and Risk Assessment in Retail Environments* (pp. 1–24). Hershey, PA: IGI Global. doi:10.4018/978-1-5225-3056-5.ch001

Palos-Sanchez, P. R., & Correia, M. B. (2018). Perspectives of the Adoption of Cloud Computing in the Tourism Sector. In J. Rodrigues, C. Ramos, P. Cardoso, & C. Henriques (Eds.), *Handbook of Research on Technological Developments for Cultural Heritage and eTourism Applications* (pp. 377–400). Hershey, PA: IGI Global. doi:10.4018/978-1-5225-2927-9.ch018

Papadopoulou, G. (2021). Promoting Gender Equality and Women Empowerment in the Tourism Sector. In F. Brandão, Z. Breda, R. Costa, & C. Costa (Eds.), Handbook of Research on the Role of Tourism in Achieving Sustainable Development Goals (pp. 152-174). IGI Global. https://doi.org/ doi:10.4018/978-1-7998-5691-7.ch009

Papp-Váry, Á. F., & Tóth, T. Z. (2022). Analysis of Budapest as a Film Tourism Destination. In R. Baleiro & R. Pereira (Eds.), *Global Perspectives on Literary Tourism and Film-Induced Tourism* (pp. 257-279). IGI Global. https://doi.org/10.4018/978-1-7998-8262-6.ch014

Patiño, B. E. (2017). New Generation Management by Convergence and Individual Identity: A Systemic and Human-Oriented Approach. In N. Baporikar (Ed.), *Innovation and Shifting Perspectives in Management Education* (pp. 119–143). Hershey, PA: IGI Global. doi:10.4018/978-1-5225-1019-2.ch006

Patro, C. S. (2021). Digital Tourism: Influence of E-Marketing Technology. In M. Dinis, L. Bonixe, S. Lamy, & Z. Breda (Eds.), *Impact of New Media in Tourism* (pp. 234-254). IGI Global. https://doi.org/10.4018/978-1-7998-7095-1.ch014

Pawliczek, A., & Rössler, M. (2017). Knowledge of Management Tools and Systems in SMEs: Knowledge Transfer in Management. In A. Bencsik (Ed.), *Knowledge Management Initiatives and Strategies in Small and Medium Enterprises* (pp. 180–203). Hershey, PA: IGI Global. doi:10.4018/978-1-5225-1642-2.ch009

Pejic-Bach, M., Omazic, M. A., Aleksic, A., & Zoroja, J. (2018). Knowledge-Based Decision Making: A Multi-Case Analysis. In R. Leon (Ed.), *Managerial Strategies for Business Sustainability During Turbulent Times* (pp. 160–184). Hershey, PA: IGI Global. doi:10.4018/978-1-5225-2716-9.ch009

Perano, M., Hysa, X., & Calabrese, M. (2018). Strategic Planning, Cultural Context, and Business Continuity Management: Business Cases in the City of Shkoder. In A. Presenza & L. Sheehan (Eds.), *Geopolitics and Strategic Management in the Global Economy* (pp. 57–77). Hershey, PA: IGI Global. doi:10.4018/978-1-5225-2673-5.ch004

Pereira, R., Mira da Silva, M., & Lapão, L. V. (2017). IT Governance Maturity Patterns in Portuguese Healthcare. In S. De Haes & W. Van Grembergen (Eds.), *Strategic IT Governance and Alignment in Business Settings* (pp. 24–52). Hershey, PA: IGI Global. doi:10.4018/978-1-5225-0861-8.ch002

Pérez-Uribe, R. I., Torres, D. A., Jurado, S. P., & Prada, D. M. (2018). Cloud Tools for the Development of Project Management in SMEs. In R. Perez-Uribe, C. Salcedo-Perez, & D. Ocampo-Guzman (Eds.), *Handbook of Research on Intrapreneurship and Organizational Sustainability in SMEs* (pp. 95–120). Hershey, PA: IGI Global. doi:10.4018/978-1-5225-3543-0.ch005

Petrisor, I., & Cozmiuc, D. (2017). Global Supply Chain Management Organization at Siemens in the Advent of Industry 4.0. In L. Saglietto & C. Cezanne (Eds.), *Global Intermediation and Logistics Service Providers* (pp. 123–142). Hershey, PA: IGI Global. doi:10.4018/978-1-5225-2133-4.ch007

Pierce, J. M., Velliaris, D. M., & Edwards, J. (2017). A Living Case Study: A Journey Not a Destination. In N. Silton (Ed.), *Exploring the Benefits of Creativity in Education, Media, and the Arts* (pp. 158–178). Hershey, PA: IGI Global. doi:10.4018/978-1-5225-0504-4.ch008

Pipia, S., & Pipia, S. (2021). Challenges of Religious Tourism in the Conflict Region: An Example of Jerusalem. In E. Alaverdov & M. Bari (Eds.), *Global Development of Religious Tourism* (pp. 135-148). IGI Global. https://doi.org/10.4018/978-1-7998-5792-1.ch009

Poulaki, P., Kritikos, A., Vasilakis, N., & Valeri, M. (2022). The Contribution of Female Creativity to the Development of Gastronomic Tourism in Greece: The Case of the Island of Naxos in the South Aegean Region. In M. Valeri (Ed.), *New Governance and Management in Touristic Destinations* (pp. 246–258). IGI Global. https://doi.org/10.4018/978-1-6684-3889-3.ch015

Radosavljevic, M., & Andjelkovic, A. (2017). Multi-Criteria Decision Making Approach for Choosing Business Process for the Improvement: Upgrading of the Six Sigma Methodology. In J. Stanković, P. Delias, S. Marinković, & S. Rochhia (Eds.), *Tools and Techniques for Economic Decision Analysis* (pp. 225–247). Hershey, PA: IGI Global. doi:10.4018/978-1-5225-0959-2.ch011

Radovic, V. M. (2017). Corporate Sustainability and Responsibility and Disaster Risk Reduction: A Serbian Overview. In M. Camilleri (Ed.), *CSR 2.0 and the New Era of Corporate Citizenship* (pp. 147–164). Hershey, PA: IGI Global. doi:10.4018/978-1-5225-1842-6.ch008

Raghunath, K. M., Devi, S. L., & Patro, C. S. (2018). Impact of Risk Assessment Models on Risk Factors: A Holistic Outlook. In K. Strang, M. Korstanje, & N. Vajjhala (Eds.), *Research, Practices, and Innovations in Global Risk and Contingency Management* (pp. 134–153). Hershey, PA: IGI Global. doi:10.4018/978-1-5225-4754-9.ch008

Raman, A., & Goyal, D. P. (2017). Extending IMPLEMENT Framework for Enterprise Information Systems Implementation to Information System Innovation. In M. Tavana (Ed.), *Enterprise Information Systems and the Digitalization of Business Functions* (pp. 137–177). Hershey, PA: IGI Global. doi:10.4018/978-1-5225-2382-6.ch007

Rao, Y., & Zhang, Y. (2017). The Construction and Development of Academic Library Digital Special Subject Databases. In L. Ruan, Q. Zhu, & Y. Ye (Eds.), *Academic Library Development and Administration in China* (pp. 163–183). Hershey, PA: IGI Global. doi:10.4018/978-1-5225-0550-1.ch010

Ravasan, A. Z., Mohammadi, M. M., & Hamidi, H. (2018). An Investigation Into the Critical Success Factors of Implementing Information Technology Service Management Frameworks. In K. Jakobs (Ed.), *Corporate and Global Standardization Initiatives in Contemporary Society* (pp. 200–218). Hershey, PA: IGI Global. doi:10.4018/978-1-5225-5320-5.ch009

Rezaie, S., Mirabedini, S. J., & Abtahi, A. (2018). Designing a Model for Implementation of Business Intelligence in the Banking Industry. *International Journal of Enterprise Information Systems, 14*(1), 77–103. doi:10.4018/IJEIS.2018010105

Richards, V., Matthews, N., Williams, O. J., & Khan, Z. (2021). The Challenges of Accessible Tourism Information Systems for Tourists With Vision Impairment: Sensory Communications Beyond the Screen. In C. Eusébio, L. Teixeira, & M. Carneiro (Eds.), *ICT Tools and Applications for Accessible Tourism* (pp. 26–54). IGI Global. https://doi.org/10.4018/978-1-7998-6428-8.ch002

Rodrigues de Souza Neto, V., & Marques, O. (2021). Rural Tourism Fostering Welfare Through Sustainable Development: A Conceptual Approach. In A. Perinotto, V. Mayer, & J. Soares (Eds.), *Rebuilding and Restructuring the Tourism Industry: Infusion of Happiness and Quality of Life* (pp. 38–57). IGI Global. https://doi.org/10.4018/978-1-7998-7239-9.ch003

Romano, L., Grimaldi, R., & Colasuonno, F. S. (2017). Demand Management as a Success Factor in Project Portfolio Management. In L. Romano (Ed.), *Project Portfolio Management Strategies for Effective Organizational Operations* (pp. 202–219). Hershey, PA: IGI Global. doi:10.4018/978-1-5225-2151-8.ch008

Rubio-Escuderos, L., & García-Andreu, H. (2021). Competitiveness Factors of Accessible Tourism E-Travel Agencies. In C. Eusébio, L. Teixeira, & M. Carneiro (Eds.), *ICT Tools and Applications for Accessible Tourism* (pp. 196–217). IGI Global. https://doi.org/10.4018/978-1-7998-6428-8.ch009

Rucci, A. C., Porto, N., Darcy, S., & Becka, L. (2021). Smart and Accessible Cities?: Not Always – The Case for Accessible Tourism Initiatives in Buenos Aries and Sydney. In C. Eusébio, L. Teixeira, & M. Carneiro (Eds.), *ICT Tools and Applications for Accessible Tourism* (pp. 115–145). IGI Global. https://doi.org/10.4018/978-1-7998-6428-8.ch006

Ruhi, U. (2018). Towards an Interdisciplinary Socio-Technical Definition of Virtual Communities. In M. Khosrow-Pour, D.B.A. (Ed.), Encyclopedia of Information Science and Technology, Fourth Edition (pp. 4278-4295). Hershey, PA: IGI Global. doi:10.4018/978-1-5225-2255-3.ch371

Ryan, L., Catena, M., Ros, P., & Stephens, S. (2021). Designing Entrepreneurial Ecosystems to Support Resource Management in the Tourism Industry. In V. Costa, A. Moura, & M. Mira (Eds.), *Handbook of Research on Human Capital and People Management in the Tourism Industry* (pp. 265–281). IGI Global. https://doi.org/10.4018/978-1-7998-4318-4.ch013

Sabuncu, I. (2021). Understanding Tourist Perceptions and Expectations During Pandemic Through Social Media Big Data. In M. Demir, A. Dalgıç, & F. Ergen (Eds.), *Handbook of Research on the Impacts and Implications of COVID-19 on the Tourism Industry* (pp. 330–350). IGI Global. https://doi.org/10.4018/978-1-7998-8231-2.ch016

Safari, M. R., & Jiang, Q. (2018). The Theory and Practice of IT Governance Maturity and Strategies Alignment: Evidence From Banking Industry. *Journal of Global Information Management, 26*(2), 127–146. doi:10.4018/JGIM.2018040106

Sahoo, J., Pati, B., & Mohanty, B. (2017). Knowledge Management as an Academic Discipline: An Assessment. In B. Gunjal (Ed.), *Managing Knowledge and Scholarly Assets in Academic Libraries* (pp. 99–126). Hershey, PA: IGI Global. doi:10.4018/978-1-5225-1741-2.ch005

Saini, D. (2017). Relevance of Teaching Values and Ethics in Management Education. In N. Baporikar (Ed.), *Management Education for Global Leadership* (pp. 90–111). Hershey, PA: IGI Global. doi:10.4018/978-1-5225-1013-0.ch005

Sambhanthan, A. (2017). Assessing and Benchmarking Sustainability in Organisations: An Integrated Conceptual Model. *International Journal of Systems and Service-Oriented Engineering, 7*(4), 22–43. doi:10.4018/IJSSOE.2017100102

Sambhanthan, A., & Potdar, V. (2017). A Study of the Parameters Impacting Sustainability in Information Technology Organizations. *International Journal of Knowledge-Based Organizations, 7*(3), 27–39. doi:10.4018/IJKBO.2017070103

Sánchez-Fernández, M. D., & Manríquez, M. R. (2018). The Entrepreneurial Spirit Based on Social Values: The Digital Generation. In P. Isaias & L. Carvalho (Eds.), *User Innovation and the Entrepreneurship Phenomenon in the Digital Economy* (pp. 173–193). Hershey, PA: IGI Global. doi:10.4018/978-1-5225-2826-5.ch009

Sanchez-Ruiz, L., & Blanco, B. (2017). Process Management for SMEs: Barriers, Enablers, and Benefits. In M. Vemić (Ed.), *Optimal Management Strategies in Small and Medium Enterprises* (pp. 293–319). Hershey, PA: IGI Global. doi:10.4018/978-1-5225-1949-2.ch014

Sanz, L. F., Gómez-Pérez, J., & Castillo-Martinez, A. (2018). Analysis of the European ICT Competence Frameworks. In V. Ahuja & S. Rathore (Eds.), *Multidisciplinary Perspectives on Human Capital and Information Technology Professionals* (pp. 225–245). Hershey, PA: IGI Global. doi:10.4018/978-1-5225-5297-0.ch012

Sarvepalli, A., & Godin, J. (2017). Business Process Management in the Classroom. *Journal of Cases on Information Technology, 19*(2), 17–28. doi:10.4018/JCIT.2017040102

Saxena, G. G., & Saxena, A. (2021). Host Community Role in Medical Tourism Development. In M. Singh & S. Kumaran (Eds.), *Growth of the Medical Tourism Industry and Its Impact on Society: Emerging Research and Opportunities* (pp. 105–127). IGI Global. https://doi.org/10.4018/978-1-7998-3427-4.ch006

Saygili, E. E., Ozturkoglu, Y., & Kocakulah, M. C. (2017). End Users' Perceptions of Critical Success Factors in ERP Applications. *International Journal of Enterprise Information Systems, 13*(4), 58–75. doi:10.4018/IJEIS.2017100104

Saygili, E. E., & Saygili, A. T. (2017). Contemporary Issues in Enterprise Information Systems: A Critical Review of CSFs in ERP Implementations. In M. Tavana (Ed.), *Enterprise Information Systems and the Digitalization of Business Functions* (pp. 120–136). Hershey, PA: IGI Global. doi:10.4018/978-1-5225-2382-6.ch006

Schwaiger, K. M., & Zehrer, A. (2021). The COVID-19 Pandemic and Organizational Resilience in Hospitality Family Firms: A Qualitative Approach. In A. Zehrer, G. Glowka, K. Schwaiger, & V. Ranacher-Lackner (Eds.), *Resiliency Models and Addressing Future Risks for Family Firms in the Tourism Industry* (pp. 32–49). IGI Global. https://doi.org/10.4018/978-1-7998-7352-5.ch002

Scott, N., & Campos, A. C. (2022). Cognitive Science of Tourism Experiences. In R. Augusto Costa, F. Brandão, Z. Breda, & C. Costa (Eds.), Planning and Managing the Experience Economy in Tourism (pp. 1-21). IGI Global. https://doi.org/ doi:10.4018/978-1-7998-8775-1.ch001

Seidenstricker, S., & Antonino, A. (2018). Business Model Innovation-Oriented Technology Management for Emergent Technologies. In M. Khosrow-Pour, D.B.A. (Ed.), Encyclopedia of Information Science and Technology, Fourth Edition (pp. 4560-4569). Hershey, PA: IGI Global. doi:10.4018/978-1-5225-2255-3.ch396

Selvi, M. S. (2021). Changes in Tourism Sales and Marketing Post COVID-19. In M. Demir, A. Dalgıç, & F. Ergen (Eds.), *Handbook of Research on the Impacts and Implications of COVID-19 on the Tourism Industry* (pp. 437–460). IGI Global. doi:10.4018/978-1-7998-8231-2.ch021

Senaratne, S., & Gunarathne, A. D. (2017). Excellence Perspective for Management Education from a Global Accountants' Hub in Asia. In N. Baporikar (Ed.), *Management Education for Global Leadership* (pp. 158–180). Hershey, PA: IGI Global. doi:10.4018/978-1-5225-1013-0.ch008

Sensuse, D. I., & Cahyaningsih, E. (2018). Knowledge Management Models: A Summative Review. *International Journal of Information Systems in the Service Sector, 10*(1), 71–100. doi:10.4018/IJISSS.2018010105

Seth, M., Goyal, D., & Kiran, R. (2017). Diminution of Impediments in Implementation of Supply Chain Management Information System for Enhancing its Effectiveness in Indian Automobile Industry. *Journal of Global Information Management, 25*(3), 1–20. doi:10.4018/JGIM.2017070101

Seyal, A. H., & Rahman, M. N. (2017). Investigating Impact of Inter-Organizational Factors in Measuring ERP Systems Success: Bruneian Perspectives. In M. Tavana (Ed.), *Enterprise Information Systems and the Digitalization of Business Functions* (pp. 178–204). Hershey, PA: IGI Global. doi:10.4018/978-1-5225-2382-6.ch008

Shaqrah, A. A. (2018). Analyzing Business Intelligence Systems Based on 7s Model of McKinsey. *International Journal of Business Intelligence Research, 9*(1), 53–63. doi:10.4018/IJBIR.2018010104

Sharma, A. J. (2017). Enhancing Sustainability through Experiential Learning in Management Education. In N. Baporikar (Ed.), *Management Education for Global Leadership* (pp. 256–274). Hershey, PA: IGI Global. doi:10.4018/978-1-5225-1013-0.ch013

Shetty, K. P. (2017). Responsible Global Leadership: Ethical Challenges in Management Education. In N. Baporikar (Ed.), *Innovation and Shifting Perspectives in Management Education* (pp. 194–223). Hershey, PA: IGI Global. doi:10.4018/978-1-5225-1019-2.ch009

Sinthupundaja, J., & Kohda, Y. (2017). Effects of Corporate Social Responsibility and Creating Shared Value on Sustainability. *International Journal of Sustainable Entrepreneurship and Corporate Social Responsibility, 2*(1), 27–38. doi:10.4018/IJSECSR.2017010103

Škarica, I., & Hrgović, A. V. (2018). Implementation of Total Quality Management Principles in Public Health Institutes in the Republic of Croatia. *International Journal of Productivity Management and Assessment Technologies, 6*(1), 1–16. doi:10.4018/IJPMAT.2018010101

Skokic, V. (2021). How Small Hotel Owners Practice Resilience: Longitudinal Study Among Small Family Hotels in Croatia. In A. Zehrer, G. Glowka, K. Schwaiger, & V. Ranacher-Lackner (Eds.), *Resiliency Models and Addressing Future Risks for Family Firms in the Tourism Industry* (pp. 50–73). IGI Global. doi:10.4018/978-1-7998-7352-5.ch003

Smuts, H., Kotzé, P., Van der Merwe, A., & Loock, M. (2017). Framework for Managing Shared Knowledge in an Information Systems Outsourcing Context. *International Journal of Knowledge Management, 13*(4), 1–30. doi:10.4018/IJKM.2017100101

Sousa, M. J., Cruz, R., Dias, I., & Caracol, C. (2017). Information Management Systems in the Supply Chain. In G. Jamil, A. Soares, & C. Pessoa (Eds.), *Handbook of Research on Information Management for Effective Logistics and Supply Chains* (pp. 469–485). Hershey, PA: IGI Global. doi:10.4018/978-1-5225-0973-8.ch025

Spremic, M., Turulja, L., & Bajgoric, N. (2018). Two Approaches in Assessing Business Continuity Management Attitudes in the Organizational Context. In N. Bajgoric (Ed.), *Always-On Enterprise Information Systems for Modern Organizations* (pp. 159–183). Hershey, PA: IGI Global. doi:10.4018/978-1-5225-3704-5.ch008

Steenkamp, A. L. (2018). Some Insights in Computer Science and Information Technology. In *Examining the Changing Role of Supervision in Doctoral Research Projects: Emerging Research and Opportunities* (pp. 113–133). Hershey, PA: IGI Global. doi:10.4018/978-1-5225-2610-0.ch005

Stipanović, C., Rudan, E., & Zubović, V. (2022). Reaching the New Tourist Through Creativity: Sustainable Development Challenges in Croatian Coastal Towns. In M. Valeri (Ed.), *New Governance and Management in Touristic Destinations* (pp. 231–245). IGI Global. https://doi.org/10.4018/978-1-6684-3889-3.ch014

Tabach, A., & Croteau, A. (2017). Configurations of Information Technology Governance Practices and Business Unit Performance. *International Journal of IT/Business Alignment and Governance*, *8*(2), 1–27. doi:10.4018/IJITBAG.2017070101

Talaue, G. M., & Iqbal, T. (2017). Assessment of e-Business Mode of Selected Private Universities in the Philippines and Pakistan. *International Journal of Online Marketing*, *7*(4), 63–77. doi:10.4018/IJOM.2017100105

Tam, G. C. (2017). Project Manager Sustainability Competence. In *Managerial Strategies and Green Solutions for Project Sustainability* (pp. 178–207). Hershey, PA: IGI Global. doi:10.4018/978-1-5225-2371-0.ch008

Tambo, T. (2018). Fashion Retail Innovation: About Context, Antecedents, and Outcome in Technological Change Projects. In I. Management Association (Ed.), Fashion and Textiles: Breakthroughs in Research and Practice (pp. 233-260). Hershey, PA: IGI Global. https://doi.org/ doi:10.4018/978-1-5225-3432-7.ch010

Tantau, A. D., & Frăţilă, L. C. (2018). Information and Management System for Renewable Energy Business. In *Entrepreneurship and Business Development in the Renewable Energy Sector* (pp. 200–244). Hershey, PA: IGI Global. doi:10.4018/978-1-5225-3625-3.ch006

Teixeira, N., Pardal, P. N., & Rafael, B. G. (2018). Internationalization, Financial Performance, and Organizational Challenges: A Success Case in Portugal. In L. Carvalho (Ed.), *Handbook of Research on Entrepreneurial Ecosystems and Social Dynamics in a Globalized World* (pp. 379–423). Hershey, PA: IGI Global. doi:10.4018/978-1-5225-3525-6.ch017

Teixeira, P., Teixeira, L., Eusébio, C., Silva, S., & Teixeira, A. (2021). The Impact of ICTs on Accessible Tourism: Evidence Based on a Systematic Literature Review. In C. Eusébio, L. Teixeira, & M. Carneiro (Eds.), *ICT Tools and Applications for Accessible Tourism* (pp. 1–25). IGI Global. doi:10.4018/978-1-7998-6428-8.ch001

Trad, A., & Kalpić, D. (2018). The Business Transformation Framework, Agile Project and Change Management. In M. Khosrow-Pour, D.B.A. (Ed.), Encyclopedia of Information Science and Technology, Fourth Edition (pp. 620-635). Hershey, PA: IGI Global. https://doi.org/ doi:10.4018/978-1-5225-2255-3.ch054

Trad, A., & Kalpić, D. (2018). The Business Transformation and Enterprise Architecture Framework: The Financial Engineering E-Risk Management and E-Law Integration. In B. Sergi, F. Fidanoski, M. Ziolo, & V. Naumovski (Eds.), *Regaining Global Stability After the Financial Crisis* (pp. 46–65). Hershey, PA: IGI Global. doi:10.4018/978-1-5225-4026-7.ch003

Trengereid, V. (2022). Conditions of Network Engagement: The Quest for a Common Good. In R. Augusto Costa, F. Brandão, Z. Breda, & C. Costa (Eds.), *Planning and Managing the Experience Economy in Tourism* (pp. 69-84). IGI Global. https://doi.org/10.4018/978-1-7998-8775-1.ch004

Turulja, L., & Bajgoric, N. (2018). Business Continuity and Information Systems: A Systematic Literature Review. In N. Bajgoric (Ed.), *Always-On Enterprise Information Systems for Modern Organizations* (pp. 60–87). Hershey, PA: IGI Global. doi:10.4018/978-1-5225-3704-5.ch004

Vargas-Hernández, J. G. (2017). Professional Integrity in Business Management Education. In N. Baporikar (Ed.), *Management Education for Global Leadership* (pp. 70–89). Hershey, PA: IGI Global. doi:10.4018/978-1-5225-1013-0.ch004

Varnacı Uzun, F. (2021). The Destination Preferences of Foreign Tourists During the COVID-19 Pandemic and Attitudes Towards: Marmaris, Turkey. In M. Demir, A. Dalgıç, & F. Ergen (Eds.), *Handbook of Research on the Impacts and Implications of COVID-19 on the Tourism Industry* (pp. 285–306). IGI Global. https://doi.org/10.4018/978-1-7998-8231-2.ch014

Vasista, T. G., & AlAbdullatif, A. M. (2017). Role of Electronic Customer Relationship Management in Demand Chain Management: A Predictive Analytic Approach. *International Journal of Information Systems and Supply Chain Management, 10*(1), 53–67. doi:10.4018/IJISSCM.2017010104

Vieru, D., & Bourdeau, S. (2017). Survival in the Digital Era: A Digital Competence-Based Multi-Case Study in the Canadian SME Clothing Industry. *International Journal of Social and Organizational Dynamics in IT, 6*(1), 17–34. doi:10.4018/IJSODIT.2017010102

Vijayan, G., & Kamarulzaman, N. H. (2017). An Introduction to Sustainable Supply Chain Management and Business Implications. In M. Khan, M. Hussain, & M. Ajmal (Eds.), *Green Supply Chain Management for Sustainable Business Practice* (pp. 27–50). Hershey, PA: IGI Global. doi:10.4018/978-1-5225-0635-5.ch002

Vlachvei, A., & Notta, O. (2017). Firm Competitiveness: Theories, Evidence, and Measurement. In A. Vlachvei, O. Notta, K. Karantininis, & N. Tsounis (Eds.), *Factors Affecting Firm Competitiveness and Performance in the Modern Business World* (pp. 1–42). Hershey, PA: IGI Global. doi:10.4018/978-1-5225-0843-4.ch001

Wang, C., Schofield, M., Li, X., & Ou, X. (2017). Do Chinese Students in Public and Private Higher Education Institutes Perform at Different Level in One of the Leadership Skills: Critical Thinking?: An Exploratory Comparison. In V. Wang (Ed.), *Encyclopedia of Strategic Leadership and Management* (pp. 160–181). Hershey, PA: IGI Global. doi:10.4018/978-1-5225-1049-9.ch013

Wang, J. (2017). Multi-Agent based Production Management Decision System Modelling for the Textile Enterprise. *Journal of Global Information Management, 25*(4), 1–15. doi:10.4018/JGIM.2017100101

Wiedemann, A., & Gewald, H. (2017). Examining Cross-Domain Alignment: The Correlation of Business Strategy, IT Management, and IT Business Value. *International Journal of IT/Business Alignment and Governance, 8*(1), 17–31. doi:10.4018/IJITBAG.2017010102

Wolf, R., & Thiel, M. (2018). Advancing Global Business Ethics in China: Reducing Poverty Through Human and Social Welfare. In S. Hipsher (Ed.), *Examining the Private Sector's Role in Wealth Creation and Poverty Reduction* (pp. 67–84). Hershey, PA: IGI Global. doi:10.4018/978-1-5225-3117-3.ch004

Yablonsky, S. (2018). Innovation Platforms: Data and Analytics Platforms. In *Multi-Sided Platforms (MSPs) and Sharing Strategies in the Digital Economy: Emerging Research and Opportunities* (pp. 72–95). Hershey, PA: IGI Global. doi:10.4018/978-1-5225-5457-8.ch003

Yaşar, B. (2021). The Impact of COVID-19 on Volatility of Tourism Stocks: Evidence From BIST Tourism Index. In M. Demir, A. Dalgıç, & F. Ergen (Eds.), *Handbook of Research on the Impacts and Implications of COVID-19 on the Tourism Industry* (pp. 23–44). IGI Global. https://doi.org/10.4018/978-1-7998-8231-2.ch002

Yusoff, A., Ahmad, N. H., & Halim, H. A. (2017). Agropreneurship among Gen Y in Malaysia: The Role of Academic Institutions. In N. Ahmad, T. Ramayah, H. Halim, & S. Rahman (Eds.), *Handbook of Research on Small and Medium Enterprises in Developing Countries* (pp. 23–47). Hershey, PA: IGI Global. doi:10.4018/978-1-5225-2165-5.ch002

Zacher, D., & Pechlaner, H. (2021). Resilience as an Opportunity Approach: Challenges and Perspectives for Private Sector Participation on a Community Level. In A. Zehrer, G. Glowka, K. Schwaiger, & V. Ranacher-Lackner (Eds.), *Resiliency Models and Addressing Future Risks for Family Firms in the Tourism Industry* (pp. 75–102). IGI Global. https://doi.org/10.4018/978-1-7998-7352-5.ch004

Zanin, F., Comuzzi, E., & Costantini, A. (2018). The Effect of Business Strategy and Stock Market Listing on the Use of Risk Assessment Tools. In *Management Control Systems in Complex Settings: Emerging Research and Opportunities* (pp. 145–168). Hershey, PA: IGI Global. doi:10.4018/978-1-5225-3987-2.ch007

Zgheib, P. W. (2017). Corporate Innovation and Intrapreneurship in the Middle East. In P. Zgheib (Ed.), *Entrepreneurship and Business Innovation in the Middle East* (pp. 37–56). Hershey, PA: IGI Global. doi:10.4018/978-1-5225-2066-5.ch003

About the Contributors

Harish Chandra Chandan is independent Researcher in Business. He was Professor of Business at Argosy University, Atlanta. He was interim chair of the business program in 2011. He received President's award for excellence in teaching in 2007, 2008 and 2009. His teaching philosophy is grounded in the learner needs and life-long learning. His research interests include research methods, leadership, marketing, and organizational behavior. He has published 20 peer-reviewed articles in business journals and 25 chapters in business reference books. Dr. Chandan has edited 7 books published by IGI Global. Dr. Chandan is currently working on 4 more books with IGI Global. Dr. Chandan has presented conference papers at Academy of Management, International Academy of Business and Management, Southeast Association of Information Systems, and Academy of International Business. Prior to joining Argosy, Dr. Chandan managed optical fiber and cable product qualification laboratories for Lucent Technologies, Bell Laboratories. During his career with Lucent, he had 40 technical publications in the area of Optical Fiber Reliability, a chapter in a book and five U.S. patents.

* * *

Petros Chamakiotis (Ph.D., University of Bath) is an Associate Professor of Management at ESCP Business School in Madrid, Spain, where he is also the Scientific Director of the MSc in Digital Project Management & Consulting. Petros is also affiliated with the Digital Futures at Work (Digit) Research Centrer in the UK and he is a member of the ESCP Chair on "Reinventing Work" in collaboration with BNP Paribas. His research and teaching interests focus on the impact of Information Systems (IS) on work and society. Within this context, his research has focused on IS–mediated envirornments, such as hybrid work, digital platforms, online communities, and virtual teams. He publishes mainly in IS and management journals and he currently serves as an Associate Editor (AE) for the "Information Systems Journal" and for the "International Journal of Information Management".

Melina Seedoyal Doargajudhur (The Business School, Edinburgh Napier University, Craiglockhart Campus, Edinburgh) Melina Seedoyal Doargajudhur is the Programme Lead for the MBA Strategic Project Management Course in the Business School at Edinburgh Napier University. Prior to joining Edinburgh Napier University Business School, Melina was a senior lecturer at the Curtin Mauritius School of Information Technology, an establishment she joined in 2004. She holds a PhD in Information Systems, a Masters in International Business from Curtin University, Australia, and a Bachelor of Science in Information Systems from the University of South Africa. Her current research interests are Bring Your Own Device (BYOD), Mobile/Digital Technology, Information Systems, Structural Equation Modelling, Positive and negative Work Outcomes, Work Behaviour, Employees' well-being and burnout, Job Demands and Resources, Employees' workload and flexibility.

Ridhima Goel is a distinguished research scholar at the Institute of Management Studies and Research, Maharshi Dayanand University, Rohtak. With a dual background in an MBA and an MA in English from the same institution, she brings a unique interdisciplinary perspective to her research. She has been acknowledged with the Second Best Paper Award in a national seminar at Central University Himachal and has presented her work at national and international conferences. Her academic prowess is exemplified by her University Research Fellowship and she is imparting knowledge as a teacher, instructing postgraduate students in open elective subjects like Fundamentals of Management and Fundamentals of Marketing.

Zuberia Hosanoo (The Business School, Edinburgh Napier University, Craiglockhart Campus, Edinburgh) Dr Zuberia Hosanoo is the Programme Lead for 2 Programmes and Lecturer in Strategy at the Edinburgh Napier University. She has previously been a Senior Lecturer and trained psychometrist at Curtin Mauritius. She holds a PhD in Organisations and Management, a Masters in International Business, and a Bachelor of Commerce in Industrial and Organisational Psychology. During her career, she has taught modules at both undergraduate and postgraduate levels; undertaken curriculum development; and trained on Corporate Training Programs. Zuberia is also an active member of the Association of the HR Professionals – Mauritius. Her research interests are Alliances and IORs, Relational Strategy, BYOD, organisational psychology, gender studies and strategy.

Geshwaree Huzooree (The Business School, Edinburgh Napier University, Craiglockhart Campus, Edinburgh) Dr. Geshwaree Huzooree is a Lecturer in Strategy and Project Management for the MBA Programme at the Edinburgh Napier University. Prior to joining Edinburgh Napier University, she was a Senior IT lecturer at Curtin Mauritius for eight years. She also reckons seven years' IT industry experience in software application development and project management. She holds a PhD in Computer Science (2020), an MSc Software Engineering Projects and Management (2014) and a BSc Computer Science and Engineering (2007) from the University of Mauritius. Dr. Geshwaree has a close collaboration with various universities and institutions world-wide. Her main research interests include project management, health Informatics, social informatics, and mobile technologies. She is committed to carry out innovative applied research that tackles national challenges for a better future. She received the Best Oral Presenter Award in the National Research Week in the Higher Education Sector in the field of Science, Technology and Engineering on her innovative work on the "Recent Advances on Sensor Technologies for Monitoring Diabetes Patients" in 2018.

Zidan Kachhi is a Consultant Counselling Psychologist. He is currently associated with the Department of Psychology at PES University, RR Campus, Bengaluru, India as an Assistant Professor. His research interests include gender and sexuality studies, media representation of marginalized communities, biases in science, behavioral finance, and sustainability.

Christian Kaunert is Professor of International Security at Dublin City University, Ireland. He is also Professor of Policing and Security, as well as Director of the International Centre for Policing and Security at the University of South Wales. In addition, he is Jean Monnet Chair, Director of the Jean Monnet Centre of Excellence and Director of the Jean Monnet Network on EU Counter-Terrorism ().

Kritika is a post graduate (Master of Technology(M.Tech)) in computer science and engineering and holds the position of an independent researcher, author and reviewer. She is the recipient of Young Researcher Award 2023, Gold Medallist in International Olympiad of Mathematics and holds accolades from Government of India for obtaining distinction during high school and senior school. The author is serving as Lifetime Member of International Association of Engineers(IAENG), Member of Women In Cybersecurity(WiCys) India Affiliate and and professional member of InSc Institute of Scholars. The author has obtained certifications in cyber security and is a top scorer in examinations like NTSE(India). The area of research includes cyber security, digital forensics, governance, and code smells.

Nitish Kumar Minz is an exceptional Bachelor of Commerce student at K.R. Mangalam University. With an unwavering commitment to excellence and a remarkable work ethic, Mr. Minz has consistently demonstrated his dedication to academic and professional pursuits. His academic achievements extend beyond the classroom, as he has made significant contributions to several research papers and book chapters that are currently in the final stages of publication. His insightful contributions and research expertise have earned him recognition within his field of study. Nitish Kumar Minz, as the President of the Student Council at K.R. Mangalam University, holds a prestigious position. In this leadership role, he has demonstrated exceptional skills in managing various projects and strategic initiatives. His internship at KEIC (KRMU Entrepreneurship & Incubation Center) has allowed him to further refine his project management and strategic planning abilities, providing valuable hands-on experience in the field. Furthermore, he has been selected as NEP SAARTHI- Student Ambassador for Academic Reforms in Transforming Higher Education at K.R. Mangalam University. This esteemed role demonstrates his passion for educational reforms and his commitment to improving the quality of higher education. As a Student Ambassador, Mr. Minz actively engages with fellow students, faculty, and staff, promoting a culture of academic excellence and facilitating positive changes within the university. Recognizing his outstanding accomplishments and exceptional contributions, he has recently been honored with the prestigious Chancellor's Excellence Award. This esteemed award serves as a testament to his dedication, hard work, and exemplary performance throughout his academic journey. Mr. Nitish Kumar Minz's academic prowess, research contributions, leadership roles, and recognition through awards highlight his exceptional abilities and potential for future success. He continues to inspire and make a significant impact within the academic community, demonstrating his commitment to excellence in all his endeavors.

Soujata Rughoobur-Seetah holds a BSc (Hons) in Management (Minor: Information Systems), a MBA in Financial Management and a PhD in Management with specialization in Human Resource Management. Soujata as a Senior Lecturer is actively engaged in teaching the BCom and Masters programme at Curtin Mauritius in the field of Management and Human Resource Management. Soujata has been a part time lecturer in her respective field at the University of Mauritius and University of Technology, Mauritius since 2014 to 2016. Her research interests are Employee turnover, Quality of work life, Organisational support, Organisational justice, Higher Education, Tourism and Gender.

Bhupinder Singh is working as Professor at Sharda University, India. Also, Honorary Professor in University of South Wales UK and Santo Tomas University Tunja, Colombia. His areas of publications as Smart Healthcare, Medicines, fuzzy logics, artificial intelligence, robotics, machine learning, deep learning, federated learning, IoT, PV Glasses, metaverse and many more. He has 3 books, 139 paper publications, 163 paper presentations in international/national conferences and seminars, participated in more than 40 workshops/FDP's/QIP's, 25 courses from international universities of repute, organized more than 59 events with international and national academicians and industry people's, editor-in-chief and co-editor in journals, developed new courses. He has given talks at international universities, resource person in international conferences such as in Nanyang Technological University Singapore, Tashkent State University of Law Uzbekistan; KIMEP University Kazakhstan, All'ah meh Tabatabi University Iran, the Iranian Association of International Criminal law, Iran and Hague Center for International Law and Investment, The Netherlands, Northumbria University Newcastle UK, Taylor's University Malaysia, AFM Krakow University Poland, European Institute for Research and Development Georgia, Business and Technology University Georgia, Texas A & M University US name a few. His leadership, teaching, research and industry experience is of 16 years and 3 Months. His research interests are health law, criminal law, research methodology and emerging multidisciplinary areas as Blockchain Technology, IoT, Machine Learning, Artificial Intelligence, Genome-editing, Photovoltaic PV Glass, SDG's and many more.

Jagdeep Singla is presently employed as an Associate Professor at Institute of Management Studies and Research, Maharshi Dayanand University, Rohtak, India. He has more than 28 years of teaching and industry experience subsequent to the completion of his post-graduation. During his academic career so far, he has authored several books in different areas such as Operations Management, Operations and Supply Chain Management, Brand Management, etc. which are formulated as a part of the course curriculum of M.D. University. He has plenty of research papers/ articles to his credit, published in national and international journals of repute. He has supervised 11 Ph.D. His areas of Specialization are Production and Operations Management, Supply Chain Management, Marketing Management, Human Resource management, and Brand Management.

Nisha Thundiyil is a teaching professional with 22 years of teaching experice including 1 year service in leadership position.

Likitha Venkatesh is an undergraduate psychology student at PES University, RR campus, Bengaluru. Her research interest include specifically on topics related to human behavior, mental health, and psychological abnormalities.

Frederico Vicente Ribeiro de Castro is a Project Manager at the Council of the European Union and founder of Éstudo, a digital communication company in Brussels, Belgium. His work has focused on the public sector, collaborating with EU member states to improve their communication skills and digitalisation. He has been leading multidisciplinary teams in virtual settings for several years and his previous publications have been related to life sciences, associated to his academic background.

Komal Vig is working as the Dean of Sharda School of Law, Sharda University, Greater Noida. She is LL.D. in Criminal Law and MBA (HR). Prof. (Dr.) Komal Vig has served in the field of academics in different capacities for more than 20 years. She has published more than 95 research papers in International and National Journals (including 11 Scopus), 20+ book chapters, 30+ conference paper publications. She supervised 6 scholars who have successfully completed PhD under her guidance and many others are pursuing. She has presented more than 65 research papers in national and international seminars/conferences. Her area of interest is Criminal law, Family law, and laws related to Women's rights. Prof. (Dr.) Komal Vig is a member of the inspection committees of various governing and accrediting bodies. Prof. (Dr.) Komal Vig has chaired numerous sessions in international and national conferences/seminars. She is active member of Board of Studies of various universities, received Life membership of the Indian law Institute (Deemed University) New Delhi, "Women Empowerment Award- 2023" from Council of Research and Sustainable Development, India.

Yan Zou is a Strategy Business Analyst at Nexi Group in Frankfurt, Germany. He has completed the Master in Management program at ESCP Business School with two specializations (Digital Project Management; Artificial Intelligence and Big Data for Business Innovation). Yan's research interests focus on strategic decision making and digitalization. He is actively praticing in the business world and currently working closely with the topic of optimizing organizations' digitalization trajectory.

Index

A

AI 26, 89, 95, 98, 177, 195

B

best practices 35, 58, 135, 138, 158, 166, 171, 194, 208, 210
Bring Your Own Device (BYOD) 117-118, 140, 149
Burnout 1, 8-9, 12-13, 15, 18-19, 22, 24, 35-38, 40-41, 43-44, 46-47, 119, 125, 132, 136, 138, 140, 150, 163, 194, 236
BYOD trend 119, 121, 140

C

cloud-based technologies 152, 154
contemporary workplace 153, 156
corporate sector 99-100, 108, 114
COVID-19 33, 36, 39, 41, 44-46, 57-58, 60, 67, 82, 98, 117-119, 121-123, 125, 133-134, 140, 155, 157, 159, 172, 179, 181-182, 187, 199, 208-211, 214, 217-218, 220, 229-230, 235, 239
COVID-induced lockdowns 212, 218
cyber threats 123, 151, 158, 167, 169, 176-177
Cybersecurity 123, 138, 151, 158-159, 165-169, 171, 176

D

Data Protection 24, 90, 119, 139, 158
Decision-Making 4, 20, 23, 81-86, 90-95, 123
digital self-replica 90, 92-93
direct supervision 107, 130, 134
dynamic interplay 182, 195

E

emotional impact 81, 95
Employee Engagement 22-23, 76, 138, 173, 194-195, 208-211, 213, 218, 223, 238, 240
employee preferences 181, 188, 199
Employee Psychological Health 32, 39-41, 43-45, 47-48
employee well-being 2-3, 8, 11, 15, 18, 21, 24-25, 59, 124, 136-137, 139, 153, 187, 194-195, 197, 200, 213, 223, 235
employer-employee relationship 212, 218
Endpoint Security Solutions 151, 165, 176
enhanced confidence 81, 95

F

F2F 84, 92-94, 98
face-to-face interaction 62, 105-106, 181
face-to-face supervision 65, 232, 234
Factor Analysis 42
Flexibility 1, 6-7, 9-10, 12-13, 15, 25, 34, 42, 44, 46, 59-61, 63, 66-67, 69, 72, 74-76, 111, 119, 122, 128-129, 132-134, 138, 152-153, 156-157, 169, 173, 177-179, 181, 186-189, 193, 199, 211, 213, 220-221, 223, 229, 235-236, 238-240
Flexible Work Culture 155, 179-181, 195, 199-200

G

general well-being 2, 10, 12-14, 26

H

Huddles 109, 113
human resource 62, 216, 218, 231, 240
hybrid model 75, 213
Hybrid Work 61, 99-102, 156, 173, 178-181, 186-187, 189, 191-195, 197-199, 229-232, 238-240

I

ICT 81-82, 84, 90-91, 93-95, 98, 101, 159, 182-183
information integration 81, 83, 90, 95
In-Office Work 75, 186-187, 229-231, 240
International Labor 101-102
Isolation 2-3, 10, 12-13, 26, 35, 58, 63, 70, 76, 99-100, 108-112, 131, 200, 232, 234

J

JD-R Framework 124, 149
JD-R theory 124, 140
job autonomy 36, 44, 124-125, 127, 132, 140, 150
job contentment 68-70
job design 36-37, 41-43, 45-47, 124
job satisfaction 3, 9, 18, 21, 36-37, 39-41, 43, 45-47, 57-59, 63-65, 67, 69, 71-72, 74-76, 112, 120, 122, 125, 129-130, 133, 136, 140, 150, 155, 157, 187, 193, 195, 208, 211, 235-237

K

knowledge gap 81-82, 85, 93
knowledge workers 117, 120, 125, 132, 135, 138-140, 150, 182, 199

M

Manager-Employee Relationships 208-210, 217-221, 223
Mental health 2-4, 9-10, 12-15, 18, 21, 26, 45, 73, 76, 108, 110, 133, 136-137, 139-140, 194-195, 235
MST 84, 93, 98
multifaceted nature 211-214
multi-factor authentication 159, 165, 170

N

New Normal 46, 63, 99-100, 114, 179, 230
non-monetary reward 213, 218
non-working hours 9, 12

O

occasional telecommuting 61
office space 59, 67, 185, 231
online platforms 61, 73, 101
On-site presence 232, 239
organizational commitment 64, 119, 125, 129, 138, 140
organizational resilience 188, 199-200
overarching RQ 85, 93

P

Pandemic 33-34, 36, 38-41, 44-46, 58, 60-61, 63, 67, 82, 98, 117-123, 133-134, 138, 140, 152, 155-157, 159, 172, 179, 181-182, 187, 199, 208-210, 212, 214, 217-218, 220, 229-230, 235, 239
Performance Indicators 22
personal life 1-3, 5-6, 8, 12, 15, 38-39, 58, 69, 76, 111, 131-133, 135, 150, 187, 194-195, 209, 235
personal needs 7, 232, 235
potential biases 211-213
psychological health 32-34, 36-37, 39-48

R

rational decision 82, 90, 93-94
Rationality 82-84, 86, 91-95
Remote Management 211, 215
Remote Work 1-3, 6-8, 10-13, 19-20, 22, 24-26, 33-34, 40, 47, 57-66, 68-69, 72-76, 99-109, 113-114, 117-118, 122, 127, 129-130, 134-136, 138-140,

151-158, 161, 163, 165-167, 170, 172-174, 179-186, 191, 194-195, 208-215, 217-218, 220-221, 223, 230-232, 234, 237-238, 240
Remote Workers 2-4, 6-7, 10-12, 19, 22-24, 34, 58-59, 63, 70, 73-75, 104-107, 109-110, 112-113, 122-123, 127-134, 137, 139, 166-170, 173, 211-213, 215, 221, 234, 237, 239-240
Remote Working 1, 5-6, 10-11, 16, 19, 22, 26, 33-34, 40-41, 43, 45-47, 58-60, 100, 102, 117-118, 121-122, 125-136, 139-140, 149-154, 157, 159, 163, 165, 167, 169, 172, 177, 179, 185, 237, 239

S

sense of belonging 193, 219, 239-240
sensitive data 158-159, 165-166, 168-169, 171
Sleep Patterns 2-3
social animals 100
Social engineering 158-159, 168, 172
social exchanges 44, 46-47
Social Interactions 62, 131, 133, 221, 233-234, 238
social isolation 2-3, 10, 12, 63, 232, 234
strategic alignment 20
Stress 1, 3, 5, 8-10, 12-15, 34, 38-39, 45, 47, 64, 66, 103, 106, 108, 110, 112, 119-120, 122-125, 129-133, 136, 138-140, 149-150, 157, 163, 194, 200, 218, 236, 239

T

team performance 64, 208, 210, 213, 220
technological advancements 58, 95, 100, 136, 153-154, 180-182, 195, 199
Telecommunicating 84, 158
Telework 33, 60, 63-65, 99-102, 121, 149, 211-212, 214, 216, 229-236, 238-239
Trends 5-6, 60, 108, 118, 157, 167, 182-183, 185, 195

U

unintended consequences 211, 215, 217

V

video conference 84, 90
Virtual Collaboration 81, 210, 217
virtual environments 216
virtual settings 212, 216
Virtual Teams 81-83, 98, 214, 217

W

Web 3.0 151-152, 177
wellness initiatives 7, 10
Work From Home 72-73, 102, 104, 119, 122, 133, 152, 154, 163, 180, 232, 239-240
Work Outcomes 35, 117, 123-126, 135, 138-139, 150
Working remotely 1-2, 6-7, 11, 14, 36, 40, 44, 60-61, 63-70, 73-75, 101-102, 117-118, 121, 125, 128, 130-131, 137, 154-155, 163, 166, 170-171, 173-174, 181
Work-Life Balance 1-3, 5-15, 18-19, 22, 24-26, 32-33, 37-39, 63, 66, 69, 105, 122-123, 129, 133-134, 136-137, 139, 152, 155, 163, 173, 179, 181, 187, 194-195, 221, 235
work-life conflict 120, 123-125, 131, 133, 136, 138, 140, 150, 236
work-life integration 119, 140, 174, 185, 199

www.ingramcontent.com/pod-product-compliance
Lightning Source LLC
LaVergne TN
LVHW080639011026
849160LV00014B/1269